Data Mining for Global Trends in Mountain Biodiversity

Data Mining for Global Trends in Mountain Biodiversity

EDITED BY
Eva M. Spehn and Christian Körner

CRC Press
Taylor & Francis Group
Boca Raton London New York

CRC Press is an imprint of the
Taylor & Francis Group, an **informa** business

CRC Press
Taylor & Francis Group
6000 Broken Sound Parkway NW, Suite 300
Boca Raton, FL 33487-2742

First issued in paperback 2017

ISBN 13: 978-1-138-11263-6 (pbk)
ISBN 13: 978-1-4200-8369-9 (hbk)

Library of Congress Cataloging-in-Publication Data

Data mining for global trends in mountain biodiversity / editors, Christian Korner, Eva M. Spehn.
p. cm.
Includes bibliographical references and index.
ISBN 978-1-4200-8369-9 (hardcover : alk. paper)
1. Mountain ecology--Research. 2. Biodiversity--Research. 3. Mountain ecology--Databases. 4. Biodiversity--Databases. 5. Data mining. I. Körner, Christian, 1949- II. Spehn, E. M. (Eva M.) III. Title.

QH541.5.M65D38 2010
577.5'3--dc22 2009030136

Visit the Taylor & Francis Web site at
http://www.taylorandfrancis.com

and the CRC Press Web site at
http://www.crcpress.com

Contents

Preface vii

Editors ix

Contributors xi

Chapter 1 Exploring and Explaining Mountain Biodiversity: The Role and Power of Geophysical Information Systems 1

Christian Körner and Jens Paulsen

Chapter 2 Primary Biodiversity Data—The Foundation for Understanding Global Mountain Biodiversity 11

Larry Speers

Chapter 3 Using Primary Biodiversity Data in Mountain Species Numbers Assessments 17

Jorge Soberón M.

Chapter 4 The Global Need for, and Appreciation of, High-Quality Metadata in Biodiversity Database Work 25

Falk Huettmann

Chapter 5 A Possible Correlation between the Altitudinal and Latitudinal Ranges of Species in the High Elevation Flora of the Andes 29

Mary T. Kalin Arroyo, Leah S. Dudley, Patricio Pliscoff, Lohengrin A. Cavieres, Francisco A. Squeo, Clodomiro Marticorena, and Ricardo Rozzi

Chapter 6 Exploring Patterns of Plant Diversity in China's Mountains 39

Jingyun Fang, Xiangping Wang, Zhiyao Tang, Zehao Shen, and Chengyang Zheng

Chapter 7 Elevational Pattern of Seed Plant Species Richness in the Hengduan Mountains, Southwest China: Area and Climate 49

Da-Cai Zhang and Hang Sun

Chapter 8 Elevational Gradients of Species Richness Derived from Local Field Surveys versus "Mining" of Archive Data 57

Michael Kessler, Thorsten Krömer, Jürgen Kluge, Dirk N. Karger, Amparo Acebey, Andreas Hemp, Sebastian K. Herzog, and Marcus Lehnert

Chapter 9 Species Richness of Breeding Birds along the Altitudinal Gradient—An Analysis of Atlas Databases from Switzerland and Catalonia (NE Spain) 65

Niklaus Zbinden, Marc Kéry, Verena Keller, Lluís Brotons, Sergi Herrando, and Hans Schmid

Chapter 10 Diverse Elevational Diversity Gradients in Great Smoky Mountains National Park, U.S.A. 75

Nathan J. Sanders, Robert R. Dunn, Matthew C. Fitzpatrick, Christopher E. Carlton, Michael R. Pogue, Charles R. Parker, and Theodore R. Simons

Chapter 11 Integrating Data across Biodiversity Levels: The Project IntraBioDiv 89

Andreas Tribsch, Thorsten Englisch, Felix Gugerli, Rolf Holderegger, Harald Niklfeld, Katharina Steinmann, Conny Thiel-Egenter, Niklaus E. Zimmermann, Pierre Taberlet, and IntraBioDiv Consortium

Chapter 12 A Plant Functional Traits Database for the Alps—Application to the Understanding of Functional Effects of Changed Grassland Management 107

Sandra Lavorel, Sophie Gachet, Amandine Sahl, Marie-Pascale Colace, Stéphanie Gaucherand, Mélanie Burylo, and Richard Bonet

Chapter 13 Using Species Occurrence Databases to Determine Niche Dynamics of Montane and Lowland Species since the Last Glacial Maximum 125

Robert Guralnick and Peter B. Pearman

Chapter 14 A Georeferenced Biodiversity Databank for Evaluating the Impact of Climate Change in Southern Italy Mountains 137

Roberto Pizzolotto, Maria Sapia, Francesco Rotondaro, Stefano Scalercio, and Pietro Brandmayr

Chapter 15 Using Georeferenced Databases to Assess the Effect of Climate Change on Alpine Plant Species and Diversity 149

Christophe F. Randin, Robin Engler, Peter B. Pearman, Pascal Vittoz, and Antoine Guisan

Chapter 16 The "Mountain Laboratory" of Nature—A Largely Unexplored Mine of Information: Synthesis of the Book 165

Eva M. Spehn and Christian Körner

Chapter 17 Creative Use of Mountain Biodiversity Databases: The Kazbegi Research Agenda of GMBA-DIVERSITAS 171

Christian Körner, Michael Donoghue, Thomas Fabbro, Christoph Häuser, David Nogués-Bravo, Mary T. Kalin Arroyo, Jorge Soberón M., Larry Speers, Eva M. Spehn, Hang Sun, Andreas Tribsch, Piotr Tykarski, and Niklaus Zbinden

Index 179

Preface

Biogeographers collect and map biological organisms; experimental ecologists explore environmental preferences and responses of species, either in the field or in controlled environments; and evolutionary ecologists construct phylogenetic trees and interpret patterns out of current and past occurrence of taxa. Space and time are major constraints for all these activities. Biological archives cover areas far bigger than any researcher can dream of surveying, they cover periods far longer than a researcher's active period or a given funding period lasts, and the variety of environmental conditions covered by such archive data exceed anything one can reasonably think of simulating in growth chambers. Most importantly, the plants and animals collected or observed for archive data grew in a natural matrix of companion species and soil conditions; hence, they can be expected to be in a sort of balance with their environmental requirements—more than we can ever hope to achieve in a greenhouse.

Now, as electronic archiving of biodiversity data has made great advances, and climate and other geophysical data are digitized, we enter a new era of biogeography, functional ecology, and evolutionary ecology. Nature itself becomes an accessible test field for ecological and evolutionary hypotheses. Real world data of global coverage of biological organisms become available for theory development. Biological collections, often collected for other purposes or for their own sake, suddenly become an arena for empirical research that the founders and curators of these archives probably never dreamed of.

Mountains are one of the major "experimental fields of nature" because of the steep environmental gradients they cover and their spatial fragmentation at otherwise global occurrence of habitat types. This volume aims at highlighting the scientific power of nine biological databases for furthering ecological and evolutionary theory related to mountain biota. The Global Mountain Biodiversity Assessment (GMBA), a cross-cutting research network of DIVERSITAS, assembled experts from around the world at two workshops on "Altitude referenced biological databases—A tool for understanding mountain biodiversity," one held in Kazbegi, Central Caucasus, Republic of Georgia, from July 25 to July 30, 2006. The second workshop took place at the Global Biodiversity Information Facility (GBIF) Secretariat in Copenhagen from September 26 to September 28, 2007.

We would like to thank the Institute of Botany, Ilia Chavchavadze State University in Tbilisi, Republic of Georgia, for hosting our workshop at its field station in Kazbegi; especially the organizers, Professor Otar Abdalaze and Professor Gia Nakhutsrishvili; and the GBIF Secretariat, especially Éamonn O'Tuama, for hosting our second workshop in Copenhagen. Many of the contributors of this book took part in one of the workshops, some at their own expense, and we would like to thank them for their valuable contributions and discussions that lead to this book.

This book is the third publication of the GMBA series with CRC Press (Vol I: Körner, C., and E.M. Spehn, Eds. 2002. *Mountain Biodiversity: A Global Assessment*. Parthenon Publishers; Vol II: Spehn, E.M., M. Liberman, and C. Körner, Eds. 2005. *Land Use Change and Mountain Biodiversity*. CRC Press). The GMBA office has been supported by the Swiss National Science Foundation (SNSF) since 2004, and SNSF and DIVERSITAS both cofinanced the workshops. We wish to thank Katrin Rudmann-Maurer for her help in the process of editing this book, especially for her text and graphic layout.

With this book, we hope for a wave of enthusiasm and activities these new options for mountain research are now offering. The thoughts and examples presented here will hopefully stimulate many to creatively use archive data to answer old questions with new tools, and advance our knowledge and understanding of mountain biodiversity worldwide.

Eva M. Spehn and Christian Körner

Editors

Christian Körner is a professor of plant ecology at the University of Basel, Switzerland. He received his academic degrees from the University of Innsbruck, Austria. Beginning with his PhD (1977), he had an interest in alpine ecology, which culminated in his textbook *Alpine Plant Life* (Berlin: Springer, 2003), now a standard textbook in functional ecology. In this, and his other ca. 80 publications on alpine ecology, he aimed at large scale multi-species comparisons, which also led to his 2000 engagement as chair of DIVERSITAS' cross-cutting network, Global Mountain Biodiversity Assessment. The volume presented here is the third to have emerged from this international networking activity.

Dr. Körner was also responsible for the Mountain Chapter of the Millennium Ecosystem Assessment, which is considered an IPCC-type report of the world's major ecosystems. His group is very active in exploring the consequences of elevated CO_2 on natural ecosystems, with nearly 100 publications in major journals—including the first and only in situ test in an alpine ecosystem. His novel theories on the causes of high elevation treeline led to a recent revival in the international interest in this field. Körner is coeditor-in-chief of *Oecologia* (Springer) and a member of various academies of sciences. In 2007, he received the Marsh Award from the British Ecological Society for outstanding research in ecology, and he also receives among the highest citation rates in the field of functional ecology. This book project is an attempt to bridge from archive-oriented scientific works to a novel way of testing and furthering ecological and evolutionary theory—thus attributing new value and respect to centuries of collectors' efforts. In Körner's view, mountains represent an ideal testing ground for the advance of this field of science, given that their steep environmental gradients represent a sort of "globally replicated experiment by nature," and, at the same time, are hot spots of biodiversity.

Eva M. Spehn coordinates the Global Mountain Biodiversity Assessment (GMBA) of DIVERSITAS, hosted at the University of Basel, Switzerland. She completed her PhD in functional ecology, studying the effects of plant biodiversity on ecosystem functions in a European-wide, replicated experiment (BIODEPTH) at the University of Basel. She is involved in several research projects on land-use and climate change on mountain biodiversity in the Swiss Alps, the Central Caucasus, and on the Tibetan Plateau. She has published 15 peer-reviewed articles, 10 book chapters, and coedited the previous two GMBA synthesis books with CRC Press. She was member of the Swiss delegation for SBSTTA 8 & 9 (for Mountain Biodiversity) of the Convention on Biological Diversity, lead author of the Millennium Ecosystem Assessment Mountain chapter, and has organized over a dozen thematic workshops and an international conference on mountain biodiversity. This book is the outcome of two GMBA workshops and will be soon accompanied by a thematic mountain online portal as part of GBIF (Global Biodiversity Information Facility), which allows extraction and mining of mountain biodiversity data.

Contributors

Amparo Acebey
Institute for Crop and Animal Production in the Tropics
Georg-August University of Göttingen
Göttingen, Germany

Mary T. Kalin Arroyo
Instituto de Ecología y Biodiversidad
Santiago, Chile

and

Departamento de Ciencias Ecológicas
Facultad de Ciencias
Universidad de Chile
Santiago, Chile

Richard Bonet
Parc National des Ecrins
Gap, France

Pietro Brandmayr
Dipartimento di Ecologia
Università della Calabria
Rende, Italy

Lluís Brotons
Área de Biodiversidad
Grupo de Ecología del Paisaje
Centre Tecnològic Forestal de Catalunya
Pujada del Seminari s/n
Solsona, Spain

Mélanie Burylo
CEMAGREF Ecosystèmes Montagnards
Grenoble, France

Christopher E. Carlton
Department of Entomology
Louisiana State University
Baton Rouge, Louisiana

Lohengrin A. Cavieres
Instituto de Ecología y Biodiversidad
Santiago, Chile

and

Departamento de Botánica
Facultad de Ciencias Naturales y Oceanográficas
Universidad de Concepción
Concepción, Chile

Marie-Pascale Colace
Laboratoire d'Ecologie Alpine
Station Alpine Joseph Fourier
Université Joseph Fourier
Grenoble Cedex, France

Michael Donoghue
Department of Ecology and Evolutionary Biology
Yale University
New Haven, Connecticut

Leah S. Dudley
Instituto de Ecología y Biodiversidad
Santiago, Chile

Robert R. Dunn
Department of Zoology
North Carolina State University
Raleigh, North Carolina

Robin Engler
Department of Ecology and Evolution
Laboratory for Conservation Biology
Lausanne, Switzerland

Thorsten Englisch
Department of Biogeography and Botanical Garden
University of Vienna
Vienna, Austria

Thomas Fabbro
Unit of Evolutionary Biology
University of Basel
Basel, Switzerland

Jingyun Fang
PKU-PSD Project Team
Department of Ecology
College of Environmental Sciences
Peking University
Beijing, China

Matthew C. Fitzpatrick
Department of Ecology and Evolutionary Biology
University of Tennessee
Knoxville, Tennessee

Sophie Gachet
Muséum National d'Histoire Naturelle
Paris, France

Stéphanie Gaucherand
CEMAGREF Ecosystèmes Montagnards
Grenoble, France

Felix Gugerli
Swiss Federal Research Institute WSL
Birmensdorf, Switzerland

Antoine Guisan
Department of Ecology and Evolution
Laboratory for Conservation Biology
Lausanne, Switzerland

and

Land Use Dynamics
Swiss Federal Research Institute WSL
Birmensdorf, Switzerland

Robert Guralnick
Department of Ecology and Evolutionary Biology
CU Museum of Natural History
University of Colorado at Boulder
Boulder, Colorado

Christoph Häuser
Staatliches Museum für Naturkunde
Stuttgart, Germany

Andreas Hemp
Ecological Botanical Garden
University of Bayreuth
Bayreuth, Germany

Sergi Herrando
Technical Office
Catalan Ornithological Institute
Barcelona, Spain

Sebastian K. Herzog
Asociación Armonía—BirdLife International
Santa Cruz de la Sierra, Bolivia

Rolf Holderegger
Swiss Federal Research Institute WSL
Birmensdorf, Switzerland

Falk Huettmann
EWHALE Lab
Biology and Wildlife Department
Institute of Arctic Biology
University of Alaska
Fairbanks, Alaska

Dirk N. Karger
Albrecht-von-Haller-Institut für Pflanzenwissenschaften
Abteilung Systematische Botanik
Göttingen, Germany

Verena Keller
Swiss Ornithological Institute
Sempach, Switzerland

Marc Kéry
Swiss Ornithological Institute
Sempach, Switzerland

Michael Kessler
Albrecht-von-Haller-Institut für Pflanzenwissenschaften
Abteilung Systematische Botanik
Göttingen, Germany

Jürgen Kluge
Albrecht-von-Haller-Institut für Pflanzenwissenschaften
Abteilung Systematische Botanik
Göttingen, Germany

Christian Körner
Institute of Botany
University of Basel
Basel, Switzerland

Thorsten Krömer
Centro de Investigaciones Tropicales
Universidad Veracruzana
Veracruz, México

Sandra Lavorel
Laboratoire d'Ecologie Alpine
Station Alpine Joseph Fourier
Université Joseph Fourier
Grenoble Cedex, France

Marcus Lehnert
Albrecht-von-Haller-Institut für Pflanzenwissenschaften
Abteilung Systematische Botanik
Göttingen, Germany

Clodomiro Marticorena
Departamento de Botánica
Facultad de Ciencias Naturales y Oceanográficas
Universidad de Concepción
Concepción, Chile

Harald Niklfeld
Department of Biogeography and Botanical Garden
University of Vienna
Vienna, Austria

David Nogués-Bravo
Center for Macroecology
Copenhagen, Denmark

Charles R. Parker
Great Smokies Field Station
U.S. Geological Survey
Gatlinburg, Tennessee

Jens Paulsen
Institute of Botany
University of Basel
Basel, Switzerland

Peter B. Pearman
Department of Ecology and Evolution
University of Lausanne
Lausanne, Switzerland

Roberto Pizzolotto
Università della Calabria
Dipartimento di Ecologia
Rende, Italy

Patricio Pliscoff
Instituto de Ecología y Biodiversidad
Santiago, Chile

Michael R. Pogue
Systematic Entomology Laboratory
PSI, Agricultural Research Service
U.S. Department of Agriculture
National Museum of Natural History
Washington, D.C.

Christophe F. Randin
Laboratory for Conservation Biology
Department of Ecology and Evolution
University of Lausanne
Lausanne, Switzerland

Francesco Rotondaro
Parco Nazionale del Pollino
Rotonda, Italy

Ricardo Rozzi
Instituto de Ecología y Biodiversidad
Santiago, Chile

and

University of North Texas
Denton, Texas

Amandine Sahl
Muséum National d'Histoire Naturelle
Paris, France

Nathan J. Sanders
Department of Ecology and Evolutionary Biology
University of Tennessee
Knoxville, Tennessee

Maria Sapia
Dipartimento di Ecologia
Università della Calabria
Rende, Italy

Stefano Scalercio
Istituto Sperimentale perl'Olivicoltura
Rende, Italy

Hans Schmid
Swiss Ornithological Institute
Sempach, Switzerland

Zehao Shen
PKU-PSD Project Team
Department of Ecology
College of Environmental Sciences
Peking University
Beijing, China

Theodore R. Simons
USGS NC Cooperative Fish and Wildlife Research Unit
Department of Zoology
North Carolina State University
Raleigh, North Carolina

Jorge Soberón M.
Department of Ecology and Evolutionary Biology
Biodiversity Research Center
University of Kansas
Lawrence, Kansas

Larry Speers
Global Biodiversity Information Facility
Copenhagen, Denmark

Eva M. Spehn
Global Mountain Biodiversity Assessment
Institute of Botany
University of Basel
Basel, Switzerland

Francisco A. Squeo
Instituto de Ecología y Biodiversidad
Santiago, Chile

and

Departamento de Biología
Facultad de Ciencias
Universidad de La Serena
La Serena, Chile

Katharina Steinmann
WSL Swiss Federal Research Institute
Birmensdorf, Switzerland

Hang Sun
Kunming Institute of Botany
Chinese Academy of Sciences
Yunnan, China

Pierre Taberlet
Laboratoire d'Ecologie Alpine
University Joseph Fourier
Grenoble Cedex, France

Zhiyao Tang
PKU-PSD Project Team
Department of Ecology
College of Environmental Sciences
Peking University
Beijing, China

Conny Thiel-Egenter
WSL Swiss Federal Research Institute
Birmensdorf, Switzerland

Andreas Tribsch
Department of Organismic Biology
Ecology and Diversity of Plants
University of Salzburg
Salzburg, Austria

and

Department of Systematic and Evolutionary Botany
Wien, Austria

Piotr Tykarski
Department of Ecology
Warsaw University
Warsaw, Poland

Pascal Vittoz
Department of Ecology and Evolution
Faculty of Geosciences and Environment
University of Lausanne
Lausanne, Switzerland

Xiangping Wang
PKU-PSD Project Team
Department of Ecology
College of Environmental Sciences
Peking University
Beijing, China

Niklaus Zbinden
Swiss Ornithological Institute
Sempach, Switzerland

Da-Cai Zhang
Key Laboratory of Biogeography and Biodiversity
Kunming Institute of Botany
Chinese Academy of Sciences

and

Southwest Forestry College
Yunnan, China

Chengyang Zheng
PKU-PSD Project Team
Department of Ecology
College of Environmental Sciences
Peking University
Beijing, China

Niklaus E. Zimmermann
WSL Swiss Federal Research Institute
Birmensdorf, Switzerland

1 Exploring and Explaining Mountain Biodiversity
The Role and Power of Geophysical Information Systems

Christian Körner and Jens Paulsen

CONTENTS

The Task Ahead ... 1
Some Conventions Are Needed ... 3
Global Mountain Geostatistics for Testing Biodiversity Theory ... 4
Global Area × Seasonality Patterns ... 5
Global Ruggedness Patterns ... 5
Latitudinal and Altitudinal Patterns of "Warmth" ... 6
Conclusions ... 8
Summary ... 10
References ... 10

THE TASK AHEAD

Evolution of life needs space and time. Additional large-scale drivers of diversification are climate history, actual climate, and continental connectivity. Mountains represent a special case for all these drivers: Land area declines very rapidly with altitude—above the treeline, the global land area on average is halved every additional 167 m of altitude (Körner, 2007). Outside the tropics, the duration of the growing season also declines rapidly with altitude (less time available per year). Mountains also exhibit the most pronounced climatic gradients on earth, and glacial–interglacial cycles affected mountains more than most lowlands. Finally, mountains represent islands, archipelagos of high-altitude habitats, isolated from each other by valleys or lowlands, hence their species diversity should decline irrespective of other conditions, as is known from island biogeography (MacArthur and Wilson, 1967).

For a number of mountains the altitudinal decline in plant species diversity above treeline does indeed follow the reduction in land area—a rather surprising (not far from) linear correlation—suggesting that the number of species per unit of land area is constant (Körner, 2000), despite the dramatic changes in life conditions. Whether biogeographic space rules are the exhaustive explanation or the patterns observed are autocorrelations with other altitudinal changes is unknown. Global comparisons of altitudinal biodiversity trends using georeferenced electronic biodiversity data (Global Biodiversity Information Facility [GBIF], http://www.gbif.org) in connection with Geographical Information Systems (GIS) databases on topography and climate offer a global test system for biodiversity theory in general, and for mountains in particular. Mountains may be seen as "experiments by nature," and the results of this highly replicated "experiment" can be explored by the e-mining of existing data pools. Advocating such use of existing databases is what this volume aims at.

Worldwide, alpine biota evolved even more species of flowering plants than would be expected from alpine land area alone. About 4% of all taxa of flowering plants grow on ca. 3% of land area, disregarding Antarctica,

Greenland, and the great deserts such as the Sahara (Körner, 1995). Why these cool, climatically hostile environments show more plant species than the average lowland ecosystems may be related to small individual plant size (less space per individual), or to greater habitat diversity (a function of topography, i.e., geodiversity driven by geology and gravity), or to some other factors specific to mountains. Plant size may indeed be the major explanation, not just because of smaller spatial niche per individual, but also because of mating systems (more selfing) and per generation mutation rate, which has been considered a function of the number of mitoses that occur from zygote to gamete (Scofield and Schultz, 2006).

Sufficient spatial resolution provided, electronic data mining permits testing of all these potential relationships by comparing data sets from various regions in which the significance of certain drivers of evolution differs. For instance, one can test large versus small mountain systems, connected versus isolated mountains, those which experienced strong glaciation and those which did not, mountains which differ in season length (latitude, factor time), and mountains which are surrounded by rich or poor lowland biota (source dependency of biodiversity). Nearly all hypotheses for global patterns of biodiversity and diversification (e.g., Currie et al., 2004; Storch et al., 2006) can be tested using the broad spectrum of mountain biota across the globe.

One can further explore the theory of adaptation to life at high altitudes, e.g., the presence or absence of C4 versus C3 plants, the significance of certain tussock types, of cushion plants, specific pollinators, ratios between numbers of species among taxonomic groups, etc. For instance, the debate on which facet of the climate is responsible for the latitudinal decline in species diversity (temperature or solar energy discussed for bird diversity by Storch et al., 2006) can be substantiated by using biodiversity trends from mountain transects along which it always gets cooler with increasing altitude, but solar radiation (the dose, rather than peak rates) may change with altitude in any direction, depending on regional cloudiness (e.g., decreasing to one-third of lowland values at treeline in New Guinea or doubling with altitude along the tropical Andes E-slope, or no change with altitude as in the Swiss Alps; Körner, 2003, 2007b).

Another arena is anthropogenic influence. Does traditional (sustainable) land use affect regional biodiversity? Are pristine mountain biota richer in species than those that had been managed by humans for millennia? Do certain engineer types of plants (e.g., tussock grasses) control slope stability in steep terrain worldwide? As an example, what would the landscape shown in Figure 1.1 look like if the small tussock grass *Festuca valesiaca* were not present? It controls the edges of erosion canyons over large mountain terrain. Is the larger-scale distribution of this species related to grazing intensity? A single species may determine whether the rest of the taxa have a space to live on. Combining land use information with biodiversity data may offer answers to these questions.

What makes mountains superior to any other test system are the steep, physics driven, climatic gradients nested in otherwise different flora and fauna and spatial settings within continents. Furthermore, nature's "mountain experiments" are highly replicated across the globe. We will present some important conventions and definitions and will introduce the task by discussing a few examples of powerful geostatistical data. These criteria and bioclimatological reference lines will become part of the search routines of the Global Mountain Biodiversity Assessment (GMBA)–GBIF electronic mountain biodiversity portal.

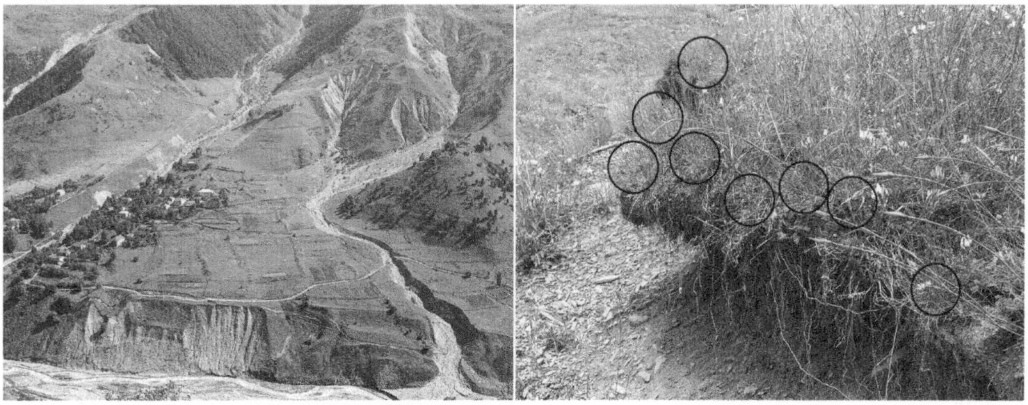

FIGURE 1.1 *Festuca valesiaca* (circled) engineers the resistance to erosion at the edge of large erosion canyons in the Central Caucasus (Republic of Georgia). Left, a village threatened by accelerating slope erosion, and the erosion edge (right) almost 100% covered by *F. valesiaca*.

SOME CONVENTIONS ARE NEEDED

What is a "mountain," and how can altitudinal life zones be defined in a way that they can be applied and compared globally? Meters of altitude are inappropriate, because of the depression of isotherms with latitude and the existence of vast high-altitude plateaus. Temperatures during periods of active life may be similar during the ten weeks arctic–alpine summer at 600 m altitude at 70° N, or the year-round season at 4,000 m near the equator. If defined by ruggedness alone, any steep lowland hill could be a mountain just like Mt. Everest. If ruggedness is ignored, the short-grass prairie and the large central Asian high plateaus would become mountains. It is obvious that this problem cannot be solved on scientific grounds, but rather is a matter of agreement and convention for the sake of communication (Kapos et al., 2000; Körner and Ohsawa, 2005; Körner, 2007a, 2007b), but also for the targeted use of georeferenced biodiversity databases.

A climatic reference line is needed against which certain mountain life zones across the globe can be defined. Of all bioclimatic boundaries, the high altitude limit of tree growth, the alpine treeline, is the best reference for a global mountain concept. Climatic treelines (not to be confused with other, mostly anthropogenic reasons for forest boundaries at high altitude) can be defined as a line connecting the uppermost patches of trees of at least 3 m height. This convention makes a distinction between shrub-shaped and upright trees (which clearly emerge from the sheltering ground cover), acknowledges that isolated tree individuals may be found at higher places at peculiar microhabitats (outposts), and, since the line is partly virtual, accounts for the fact that forests gradually open and become fragmented as they approach the treeline. Adopting this convention, the climatic treelines of the humid parts of the world occur at the 6.6 + 0.8°C isotherm during the growing season (Körner and Paulsen, 2004, and newer unpublished data). The same isotherm may be applied for a threshold for arid or semiarid regions, despite the absence of trees. Under such conditions, this thermal boundary may still serve as a separator for alpine versus mountain terrain that could support trees if there were sufficient moisture. Quite often such terrain does in fact support tree growth along deep gullies. In order to meet nonhumid conditions, moisture criteria need to be applied, with a precipitation of < 200 mm at the thermal tree limit considered prohibitive of any tree growth.

"Growing season" is defined here as the period during which plants are photosynthetically fully active and have green foliage, should they be deciduous. Growth subsumes both the meristematic activity (production of new tissue) and the accumulation of storage reserves ("stored growth"). Outside the tropics, the growing season often coincides with the snow-free period (which may be a few days longer than the actual growing season). In order to support tree growth, the duration of the snow-free period must not be < 90 days, as may happen in heavily snow-packed regions. With these definitions, the growing season at the climatic treeline starts and ends when the weekly mean air temperature passes through 0°C. This air temperature coincides with a ca. 3°C root zone temperature threshold in soils shaded by dense vegetation.

With these conventions, standardized (comparable) life conditions can be defined across the world's mountain biota. Ecosystems above the climatic treeline altitude (or the corresponding isotherm in arid regions) are termed "alpine"; those immediately below are termed "montane." The upper limit of the alpine belt is defined by the snow line, the line at which snow can persist year round (this coincides with a year-round monthly mean air temperature at or below 0°C). The uppermost part of the alpine zone often is termed nival (or made a separate belt), referring to scattered islands of life among rocks and scree fields. As a rule of thumb, the alpine belt ranges over ca. 1,000 m of altitude. So when the treeline is at 600 m (some subpolar or boreal mountains) the alpine/nival belt ends at ca. 1,600 m, when the treeline is at 4,000 m, the alpine/nival belt ends at ca. 5,000 m.

Unfortunately, there is no such general definition of the lower limit of the montane belt because it is largely defined through specific forest belts by different schools of vegetation ecologists. Naturally, the montane belt is forested (except for mires, scree, or rock). A good climatologically coherent approximation may be a width (elevational amplitude, starting from treeline) of 300 m in subpolar and boreal regions (near 70° N) and 2,500 m near the equator. The upper montane forest, i.e., the type of forest that will eventually form the treeline, often spans a range of about 600 m of altitude downslope from the treeline. A climatic approximation for the lower limit of this upper montane forest is the 10°C isotherm for the warmest six months of the year. The lowermost limit of montane forests does not match climatological criteria. One may consider any terrain with a certain degree of ruggedness down to sea level, e.g., 300 m above sea level, as "montane" (Kapos et al., 2000). It is quite obvious that this is not a question of science but of taste.

The forest–alpine transition is called treeline ecotone, a belt between timberline (limit of closed, tall forest) and the tree species line (the uppermost crippled tree individuals), with the treeline in between (Körner, 2007a).

This is a belt of particular biological richness because of the merging of two types of biota. It may span a range of up to 200 m of altitude, but mostly covers no more than 50 m of altitude around the treeline position. Hence, assuming a 0.6 K altitudinal laps rate of temperature per 100 m of altitude, the forest–alpine transition occurs within ca. ± 0.2 K.

GLOBAL MOUNTAIN GEOSTATISTICS FOR TESTING BIODIVERSITY THEORY

Using a digital world topography database of 20 m areal resolution and a world climate database (e.g., Worldclim) adjusted from a 30 sec resolution to a practical 2.5 min. resolution, it is possible to identify land area that meets the aforementioned thermal boundaries for the alpine life zone, or for forests in general, or for any other type of vegetation for which the climatic envelope had been defined. For the current purpose, a model was adopted that accounts for moisture limitation (in most cases irrelevant at montane and alpine altitudes) and selects pixels by thermal criteria within the nonarid fraction of all terrestrial pixels. As an additional criterion, ruggedness can be defined as the maximum altitudinal difference between a target pixel and its surrounding pixels. Terrain is considered predominantly mountainous if the maximum altitude difference between

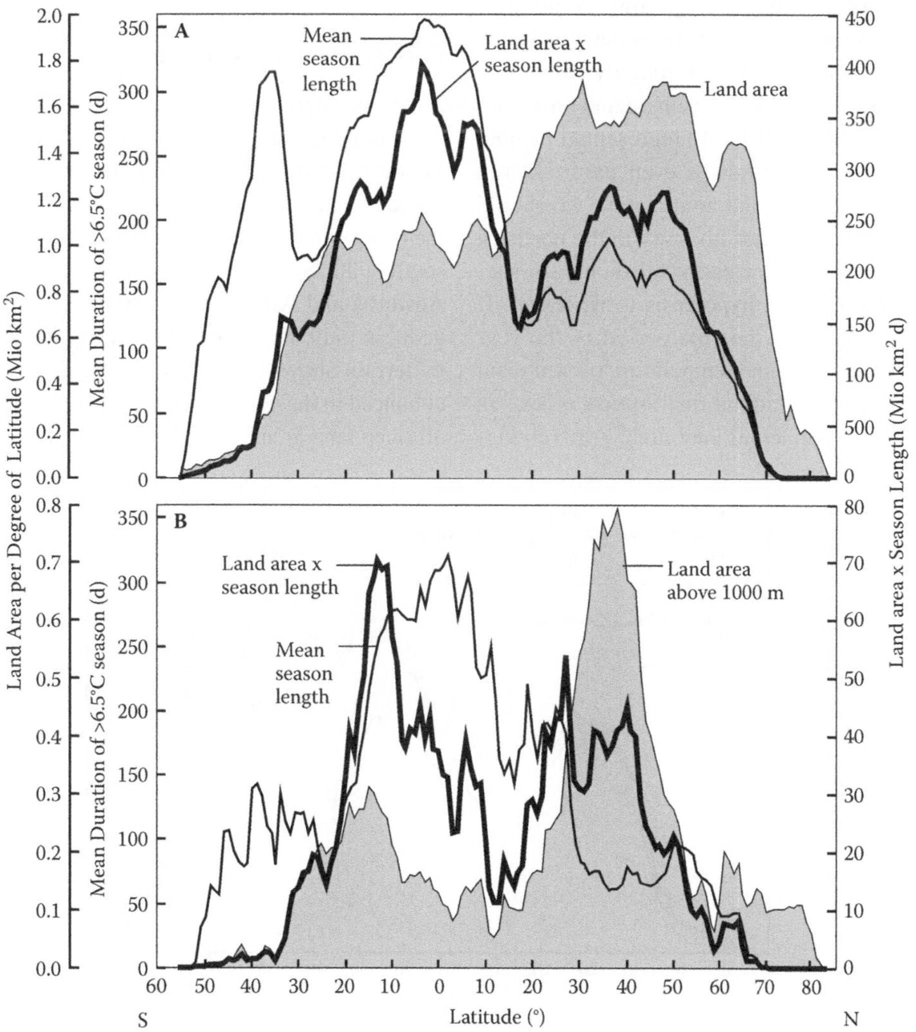

FIGURE 1.2 (A) Latitudinal patterns of land area distribution, mean season length per latitudinal belt, and the combination of land area times season length per 1° latitude belt. (B) As before, but for the global land area higher than 1,000 m disregarding any flat lands (see Figure 1.2), i.e., mountainous terrain only defined by altitude and a minimum slopiness. Note the different scales between the two diagrams and the differences in hemispherical asymmetry in the various latitudinal transects.

neighboring 30" pixels within a central cluster of nine 30" pixels in each 2'30" grid is > 250 m at the 30" resolution, moderately slopy at 50 to 250 m and close to flat at < 50 m. It is important to combine thermal and moisture criteria with ruggedness to separate alpine from tundra, and montane from boreal lowland, because temperatures are similar between those pairs of biota.

Global Area × Seasonality Patterns

A first important question is where, at which latitudes, the global land area and mountain areas are situated. For the analysis presented here, 100% of terrestrial area corresponds to 135 million km^2, disregarding Antarctica. Figure 1.2 illustrates the well-known asymmetry of hemispherical land area distribution, with roughly two-thirds of all land area in the northern hemisphere and only one-third in the southern hemisphere. As represented by land above 1,000 m, mountain terrain is even more skewed to the north. Hence, should land area matter, we should find two-thirds of global species diversity in the northern hemisphere and ca. three-quarters of all mountain species diversity in the north, a hypothesis to be tested. If we further account for time, i.e., the period of the year during which both moisture and temperature permit plant growth, we arrive at a latitudinal distribution of season length irrespective of the actual land area. A pixel-wise account of season length *x* available land area per one degree latitude belt yields a global pattern for "opportunities for life" in a space *x* time matrix. Remarkably, this line is more hemispherically symmetric compared to the land area distribution.

This simple space x time analysis, which does not yet account for specific mountain life zones, already offers a suite of possibilities of testing theories for global biodiversity patterns. In essence, the data presented in Figure 1.2 suggest a clear north hemisphere center of mountain biodiversity, if land area is the dominant determinant, and a more balanced distribution, or a south hemisphere peak, if time (season length) is the dominant driver.

Global Ruggedness Patterns

A second step is more specifically mountain oriented, accounting for ruggedness of terrain. As can be seen from Figure 1.3, the fractions of flat, medium rugged, and very rugged terrain do not show strong global variation, with the exception that very rugged terrain is underrepresented south of 20° S, owing to the old Gondwana land masses in Australia and Africa. If mountains were defined by ruggedness only, the hemispherical asymmetry as captured by terrain above 1,000 m in Figure 1.2 becomes strongly enhanced in the northern hemisphere. In fact, four-fifths of all steep land area is found in the northern hemisphere.

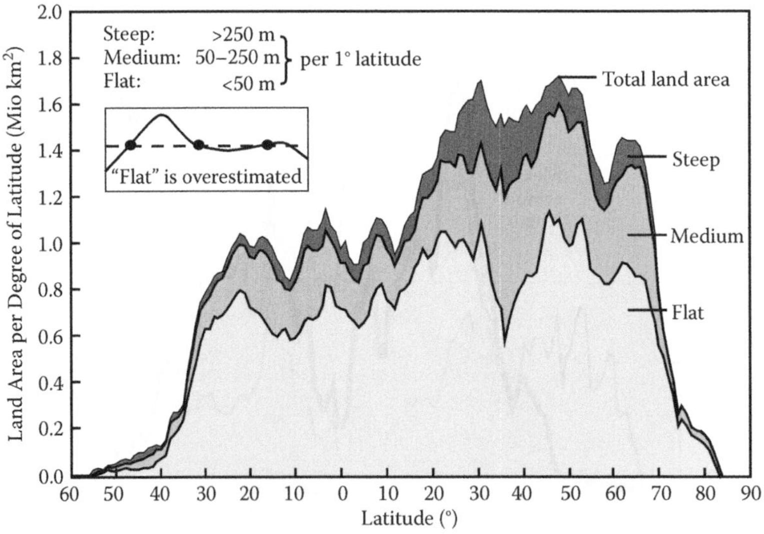

FIGURE 1.3 Latitudinal patterns of ruggedness of land area. Note the flat boreal-subarctic zone and the lack of a significant amount of rugged terrain south of ca. 20° S. Rugged terrain represents "mountains," irrespective of actual climate. "Medium" may either include medium rugged terrain, or a mix of very rugged and flat terrain. The statistics applied through a constant grid, by necessity, will less likely hit rugged compared to flat terrain (inset diagram), hence, "flat" is likely to be over-represented. This is a systematic bias that will not affect the general global picture.

A statistics of land area for various climatic zones accounting for ruggedness illustrates that most of the 81% of the not permanently arid terrestrial area falls in the "flat" category, namely 68 Mio km^2, 30 Mio km^2 are medium, and only 11 Mio km^2 very rugged terrain (Figure 1.4). Tundra and alpine ecosystems make up ca 8 Mio km^2 in the flat category (largely tundra) and 4 Mio km^2 in the rugged category (largely alpine).

Treeline and snow line modeled by these climatic criteria fit the real world pattern very well (Figure 1.5). The equatorial treelines are at 4,000 m, and the highest treelines are found in the Himalayan region (Tibet, 4,700 m) at ca. 32° N. The similarly or even higher treelines in the Andes (Bolivia, ca. 4,800 m) cover too small areas to be depicted. The snow line envelopes the alpine (including nival) belt by a parallel isotherm. The treeline and nival zone lines in Figure 1.5 are enveloping the alpine belt globally, thus permitting to compare taxonomic diversity globally, across all latitudes. High resolution georeferenced biodiversity data can be specifically constrained by this climatic envelope. Note the narrowing of the alpine belt from nearly 2,000 m at 30° N to 500 m at 60° N, offering a test case for climate driven space constraints of biodiversity in mountains.

Latitudinal and Altitudinal Patterns of "Warmth"

If one defines thermal envelopes by annual day degrees (sums of daily mean temperatures) above a certain threshold (e.g., a growing season mean of above 6.5°C) one can select pixels of a certain altitude across latitudes, and explore the mean and maximum day-degrees. Figure 1.6 shows the results for sea level, 2,000 m and 3,000 m. At each of these common altitudes, tropical pixels are obviously much warmer than polar pixels. An organism that has a certain thermal requirement would thus find similar thermal sums across the globe, but at rather different altitudes during the growing season. However, in the tropics, the same sum is composed of much lower temperatures, because of the twelve-month season. Would longer but cooler conditions end up with similar opportunities for life than a short but warmer situation, similar to the nearly constant day degrees at 3,000 m altitude between 35° N and 30° S? There is a short peak at 20° S, largely due to the (positive) thermal anomaly in the Andes at this latitude and a clear depression at the equator caused by greater fractions of cloudy weather and the lack of large mountain areas at the equator. In fact, mean day degrees at 3,000 m altitude double from 0° to 20° S. Is the biological richness only half as large at respective equatorial

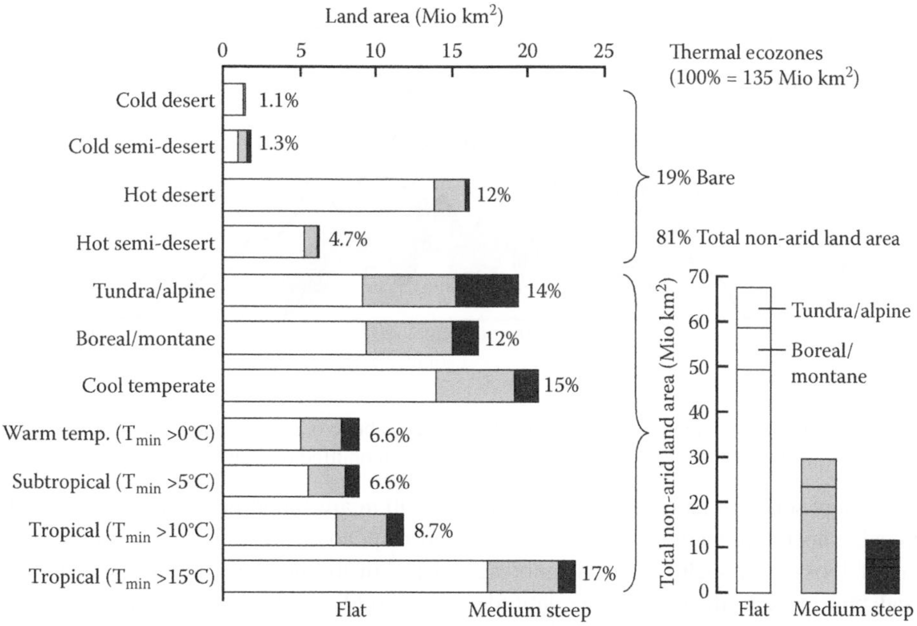

FIGURE 1.4 Ruggedness of terrain in the thermal ecozones of the globe without Antarctica. Of the remaining 135 million km^2, 19% are desert (cold or arid), and 81% carry some vegetation. Note the small fraction of land area which is on average "rugged" as defined by a > 250 m mean difference in altitude between adjacent 30" pixels of land area.

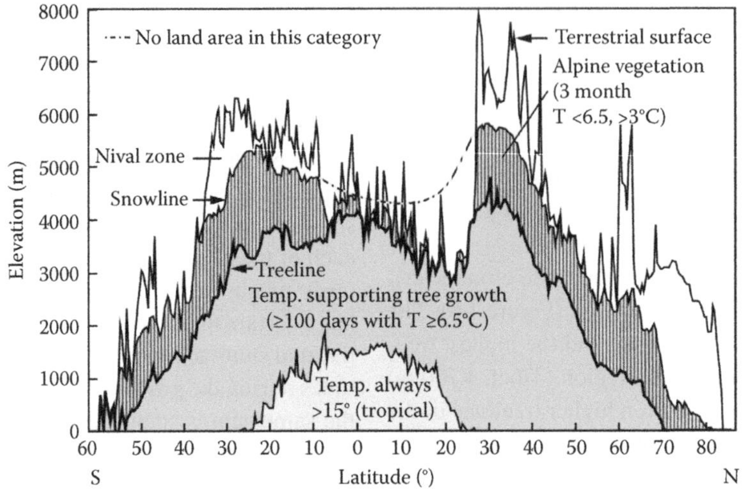

FIGURE 1.5 The latitudinal position of the climatic treeline and the snowline envelope the alpine belt. Note the latitudinal narrowing of the alpine belt in the northern hemisphere. The dotted line indicates latitudes which lack land area in the respective categories large enough to be shown at this resolution. The area below treeline illustrates the potentially forested land as defined by temperature alone. For comparison, the tropical zone is illustrated at the bottom.

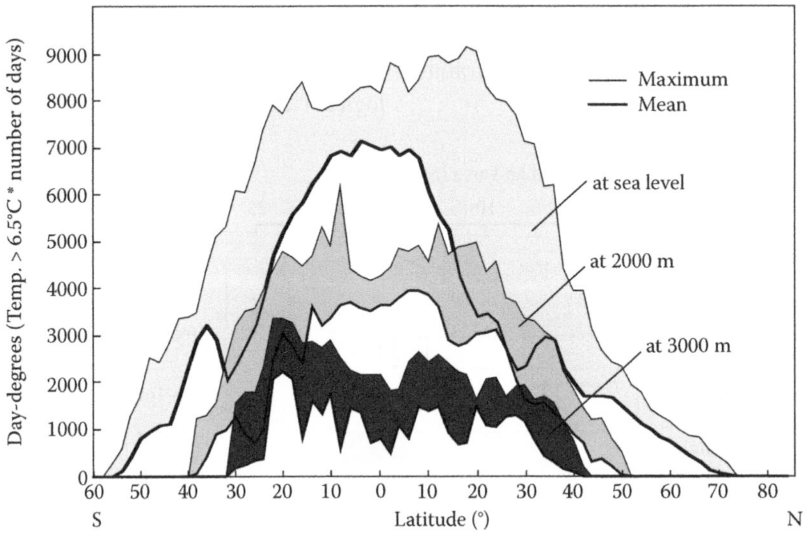

FIGURE 1.6 Latitudinal trends in annual thermal sums (day degrees above 6.5°C) for land area at sea level, at 2,000 m, and 3,000 m of altitude. Mean and maxima are shown (the minima have been omitted for clarity).

compared to these more southern latitudes (as one could expect if integrated temperature matters)?

Another way of looking at such trends is ignoring hemispheres and plotting day degrees across latitude for 1,000 m steps in altitude (Figure 1.7). The lines in Figure 1.7 illustrate broad plateaus, roughly up to 30° latitude, followed by a nearly linear polewards decline. The warmest places on the globe above 4,000 and above 5,000 m are found between 25° and 35° latitude, which corresponds to Bolivia in the southern hemisphere and Tibet in the northern hemisphere, both regions where the world's highest situated treelines are found (4,700 to 4,800 m).

These few examples show how the world's terrestrial area and the world's mountain area can be subdivided into coherent categories of thermal regimes (temperature

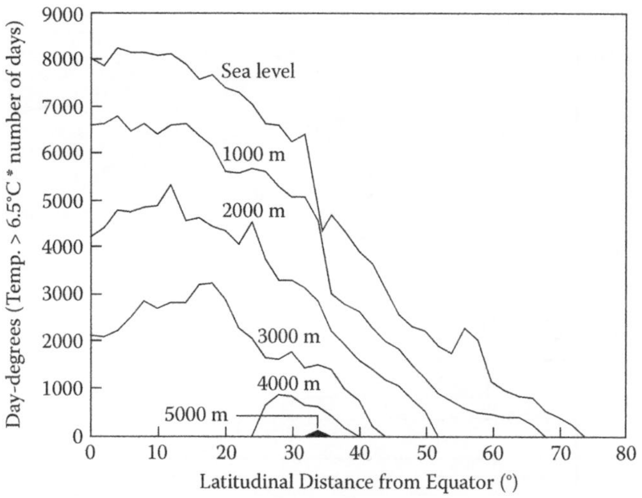

FIGURE 1.7 As in Figure 1.6, but for the southern and northern hemispheres combined and for a broader spectrum of altitudes.

envelopes), seasonalities (time), ruggedness ("mountainess"), and areal extent (space) to test hypotheses on patterns and causes for biodiversity. In addition to horizontal land area data, land area losses attributed to altitude can be accounted for in the vertical. These losses are evident, as was previously pointed out. As an example, Figure 1.8 illustrates the land area in the European Alps in slices of 100 m altitude.

CONCLUSIONS

The geostatistical tools available today permit answering some of the big questions related to the mechanisms of the evolution of biological diversity in a spatial context, provided georeferenced biodiversity data are available at the appropriate scale. These geostatistical tools are much further advanced, as are the organismic inventory data. All sorts of climatic factors related to temperature, moisture, and their combination can be attributed to scales down to a 30" resolution or even below.

Because biodiversity data are commonly not available globally, but regionally, mountains offer the unique chance to test climate–biodiversity relationships over small areas, which cover climatic gradients otherwise only found over thousands of kilometers of latitudinal distance. Thanks to treeline ecology, a robust bioclimatic reference line does exist, which permits applying a common protocol, i.e., conventions about climatic zones. In addition, latitudinal and altitudinal variation in seasonality offers variation in "time for life," and topographic ruggedness indices permit constraining land area to mountain areas. The scattered global occurrence of mountains across the full matrix of conditions (Figure 1.9) offers possibilities for highly replicated "experiments" for testing species or functional type diversity in various groups of organismic taxa against space and time gradients and bioclimatic effects.

Altitude referenced biological data are key to any such undertaking. Because taxa can be associated (electronically) with morphological, physiological, and phylogenetic characteristics, a wide spectrum of hypotheses related to environment–evolution linkages can be explored (see Chapter 17). Ultimately such data-mining–based research can help to identify and partly explain common adaptive traits, typical for mountain biota. It is proposed that space (land area per climate category) and time (duration of the growing season) provide the major explanation of mountain biodiversity, and perhaps biodiversity in general. Disturbance regimes and habitat diversity (geodiversity) are more regional (azonal) drivers. It may then appear that the physiological (adaptive) constraints operate at the (altitudinal, i.e., climatic) boundaries of species ranges, but their significance is generally overvalued for the interior of a species' distributional area (elevational niche).

The list of interesting scientific questions to be answered by selective use of database information and innovative linking of various kinds of databases is endless. For mountain biodiversity, major directions of data-mining research had been summarized in a recent expert meeting organized by the Global Mountain Biodiversity Assessment (GMBA) of DIVERSITAS (Chapter 17). In

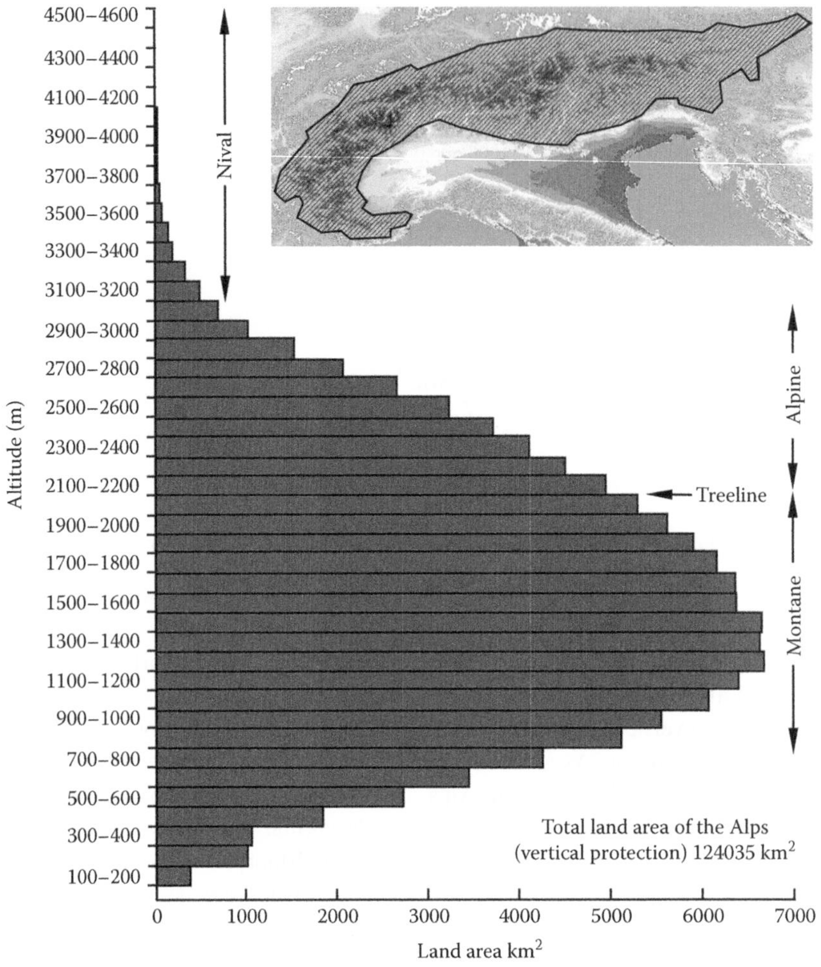

FIGURE 1.8 The "loss" of land area with elevation above a certain threshold elevation (here, 1200 m), as exemplified for the European Alps (inset polygon map).

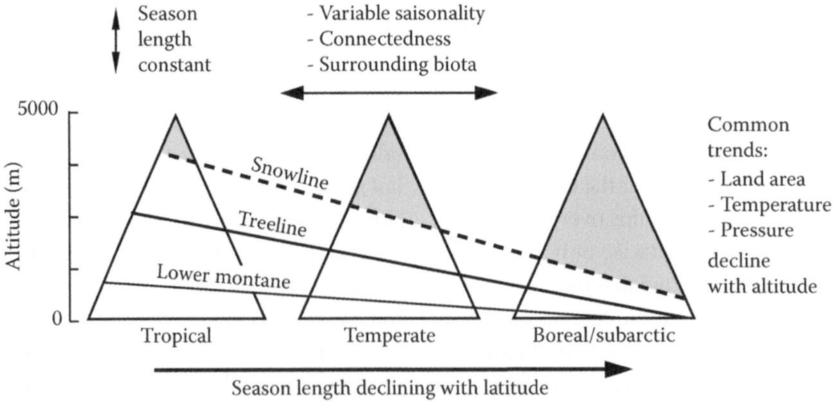

FIGURE 1.9 The matrix of environmental conditions and gradients across the world's mountains.

cooperation with GBIF, GMBA is strongly advocating the use of these new tools for biodiversity research in an ecological and evolutionary context that goes far beyond counting taxa.

SUMMARY

Georeferenced biodiversity databases can be linked with geostatistical information systems to answer basic questions of the causes and trends of biodiversity. This article presents some ideas on the application of these novel tools with a mountain biodiversity focus. It is supposed that mountains represent a globally, highly replicated "experiment by nature" that offers testing a broad spectrum of hypotheses related to global biodiversity patterns. As "islands in the sky," mountains permit exploring both space and time (season length) constraints to species richness and diversification within a broad matrix of climatic conditions. A tool kit of globally applicable climatic reference lines is presented that permits meaningful, large-scale comparisons. Examples are provided for the geostatistical analysis of climatic zones, seasonality, and shrinking of land area with altitude and topographic ruggedness, all in support of such global biodiversity comparisons.

REFERENCES

Currie, D.J., G.G. Mittelbach, H.V. Cornell, R. Field, J. F. Guegan, B.A. Hawkins, D.M. Kaufman, J.T. Kerr, T. Oberdorff, E. O'Brien, and J.R.G. Turner. 2004. Predictions and tests of climate-based hypotheses of broad-scale variation in taxonomic richness. *Ecol. Lett.* 7:1121–34.

Kapos, V., J. Rhind, M. Edwards, M.F. Price, and C. Ravilious. 2000. Developing a map of the world's mountain forests. In *Forests in Sustainable Mountain Development* (IUFRO Research Series 5). Price, M.F., and N. Butt, eds. pp. 4–9. Wallingford Oxon: CABI-Publishing.

Körner, Ch. 1995. Alpine plant diversity: A global survey and functional interpretations. In *Arctic & Alpine Biodiversity: Patterns, Causes & Ecosystem Consequences.* Chapin III, F.S., and Ch. Körner, eds. *Ecol. Studies.* 113:45–62, Berlin: Springer.

Körner, Ch. 2000. Why are there global gradients in species richness? Mountains might hold the answer. *Trends Ecol. Evol.* 15:513–14.

Körner, Ch. 2003. *Alpine Plant Life.* 2nd ed. Berlin: Springer.

Körner, Ch. 2007a. Climatic treelines: Conventions, global patterns, causes. *Erdkunde.* 61:316–324.

Körner, Ch. 2007b. The use of "altitude" in ecological research. *Trends Ecol. Evol.* 22:569–74.

Körner, Ch., and m. Ohsawa. 2005. Mountain systems. In *Ecosystems and Human Well-Being: Current State and Trends.* Vol. 1. pp. 681–716. Hassan, R., R. Scholes, and N. Ash, eds. Washington, D.C.: Island Press.

Körner, Ch., and J. Paulsen. 2004. A world-wide study of high altitude treeline temperatures. *J. Biogeogr.* 31:713–32.

MacArthur, R.H., and E.O Wilson. 1967. *The Theory of Island Biogeography.* Princeton: Princeton University.

Scofield, D.G., and S.T. Schultz. 2006. Mitosis, stature and evolution of plant mating systems: Low-Phi and high-Phi plants. *Proc. R. Soc. Lond. Ser. B-Biol. Sci.* 273:275–82.

Storch, D., R.G. Davies, S. Zajícek, C.D.L. Orme, V.A. Olson, G.H. Thomas, T.-S. Ding, P.C. Rasmussen, R.S. Ridgely, P.M. Bennett, T.M. Blackburn, I.P.F. Owens, and K.J. Gaston. 2006. Energy, range dynamics and global species richness patterns: Reconciling mid-domain effects and environmental determinants of avian diversity. *Ecol. Lett.* 9:1308–20.

2 Primary Biodiversity Data—The Foundation for Understanding Global Mountain Biodiversity

Larry Speers

CONTENTS

Introduction 11
The Advantage of Sharing Biodiversity Data 12
Data Quality and Limitations 13
Summary 14
References 14

INTRODUCTION

Mankind's knowledge of the distribution and abundance both current and historical of any particular species is exclusively based on observing individuals or groups of individuals, or collecting individual specimens of that species at different points in time and at various locations. These individual observations and the information associated with the preserved specimens deposited in our natural history collections form the primary data upon which our understanding of the temporal and spatial distribution of species-level biodiversity is based. This massive number of individual data points forms the base of the species-level biodiversity Data–Information–Knowledge–Wisdom (DIKW) hierarchy (Rowley, 2007) (Figure 2.1), from which other levels of understanding can be developed. The data, information, and knowledge documenting mountain biodiversity represent a subset of the broader biodiversity DIKW hierarchy.

It has been estimated that we share the planet with 5 million, if not even 50 million, other species (May 1998). Although the distributions of only a tiny fraction of these species have been described in any detail, we must appreciate that the number of individual data points that have been accumulated by mankind to provide even this limited amount of information is truly enormous. Unfortunately, very little of the original primary data that is the actual basis of our knowledge of individual species distributions is available to the global scientific community for re-evaluation and reassessment (one of the foundation of scientific concepts, and also allowing for synergy and applications not even comprehended). In general, it is only an outline of these data sets that is presented in the scientific literature as schematic range maps or as summarized and text-based descriptions of species ranges, population size, temporal distributions, and so on. This species-level information (referred to as level two in the DIKW) which is developed from the primary observations and specimen-based data sets is an attribute of the species rather than any particular individual. Documenting changes in species ranges or changes in population size expands our knowledge (level three in the DIKW hierarchy) of particular species or species communities. However, the real challenge is to develop this knowledge into wisdom (level four of the DIKW hierarchy) so we actually can understand the factors causing change, and then implement practices and policies to manage these changes using the best available information and science. However, it is important to emphasize that primary specimen and observational data are the basis of the complete DIKW hierarchy.

The most significant source of historical primary biodiversity data are the estimated 2 to 3 billion specimens (Krishtalka and Humphrey, 2000) that are currently preserved in the world's natural history collections. Each

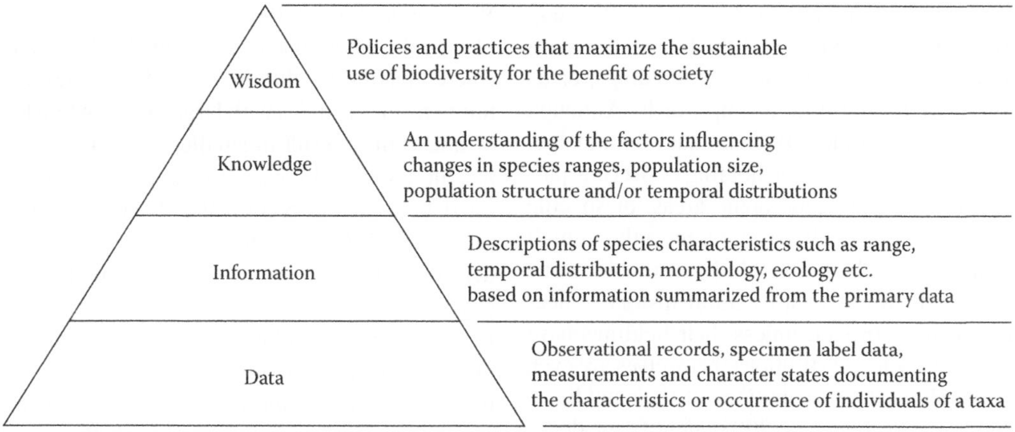

FIGURE 2.1 The species-level biodiversity Data–Information–Knowledge–Wisdom (DIKW) hierarchy (Modified from Rowley, 2007).

of these specimens provides us with a single data point documenting the occurrence of an individual of a particular species at a particular geographic locality and at a particular point in time. In contrast to most observational data sets that are not created with the intent of long-term preservation, all of the specimens in natural history collections have been deposited with the expectation that these specimens would contribute to future scientific studies, and in principle they are all available for study by the broader scientific community. Numerous authors (Funk et al., 2005; Huber, 1998; Lane, 1996; Winkler, 2004; Graham et al., 2004) have discussed the importance of natural history collections. In comparison with observational data, the critical advantage of having a voucher specimen is that the taxonomic identification of any particular specimen can be thoroughly re-evaluated through time. Although taxonomic names are basically just "tags" we use to communicate biodiversity information, these tags are based on particular species concepts. Each species concept is a hypothesis, and as new information is accumulated, the circumscription of these concepts may change (Thompson, 1996). Unlike an observational record where it is generally impossible to correct a misidentification or apply a different species concept after a particular observation is recorded, having a physical voucher allows us to reevaluate identifications or update species concepts at any time.

THE ADVANTAGE OF SHARING BIODIVERSITY DATA

If all of these primary observational and specimen-based data points could be combined into a single data set, they would provide us with our most comprehensive understanding of the distributional ranges, both present and historical, of the world's flora and fauna. By overlaying this massive data set on regional digital elevation models and mountain polygons, for instance, and then analyzing species distributions through time, we could easily document global trends in mountain biodiversity.

Unfortunately, since very few collections have been able to make the primary data documenting their holdings available in a searchable digital format, straightforward, easy access to this massive information resource is currently not possible. Lack of digital accessibility to this critical information may come as a surprise since it is clear that many countries, particularly ones with large gross domestic products and significant research budgets, could easily support the development of such digital access, and this access would prove to be very beneficial for their citizens and the world as a whole. In fact, without access to such a searchable digital catalog of the holdings of any particular institution, it is very difficult to assess and predict the taxonomic, temporal, and geographic scope of that institute's holdings. This is due to the fact that the holdings of few natural history collections have been accumulated based on any long-term strategic planning process (Lane, 1996). Additions to any natural history collection holdings at any particular time have been dependent on the taxonomic and geographic interests of particular staff members, and the availability of support for the staff's collecting and curation activities. They are a reflection of *ad hoc* science, which cannot be judged as efficient and hypothesis-driven. Thus, the taxonomic, geographic, and temporal distribution of the holdings of any collection reflects changes in staff and staff interests, and funding support for collecting activities throughout the history of that collection. As a result of this erratic way in which any particular collection has developed, it is impossible, without having someone physically study the collection, to assess and predict the taxonomic, temporal, and geographic distribution of its holdings, or its scientific merit even, and how to improve it. Although it is usually possible to identify individual natural history collections with significant holdings covering particular taxonomic groups, geographic areas, or specific time periods, these are likely to represent only a small proportion of the records that would be available if one could access the holdings of all natural history collections simultaneously. In fact, unless their holdings have been databased, very few institutions themselves are even knowledgeable about the distributions of their holdings. Peterson and Navarro-Sigüenza (2003) identify the documentation of a collection's holdings as one of the most significant benefits of databasing a collection. However, many other uses can be named, but many of them are just developing with the development in other disciplines (Graham et al., 2004) and are not even foreseeable.

DATA QUALITY AND LIMITATIONS

However, users of collections based on primary data need to be aware of the origin of this data and its limitations. Unlike observational data sets which may include zero values indicating that a particular taxon was not observed, specimen-based data only document the presence of a taxon. Chapman (2005a, 2005b) reviews issues related to data quality and discusses various approaches for cleaning primary species-occurrence data. Accurate georeferencing of the locality information associated with specimens is particularly critical for mountain biodiversity studies. (See Chapman and Wieczorek, 2006, for a discussion of georeferencing specimen-based locality information.) Rowe (2005) has shown that imprecision in georeferencing specimen data can have a particularly significantly impact on using this data for mountain biodiversity studies.

The reason for this is that in mountainous areas relatively small errors in geospatial accuracy can result in significant altitudinal differences when plotted on topographic maps or digital elevation models. Since having access to accurate altitudinal information is so important for Global Mountain Biodiversity Assessment, this aspect of data quality of archived primary biodiversity data is of critical importance for the GMBA Kazbegi Research Agenda (Körner et al 2007).

To address this difficulty, it is recommended that when georeferencing primary data each point be associated with a measure of "uncertainty" utilizing the "point–radius" method for georeferencing as outlined in Wieczorek et al. (2004). This method defines the radius of a circle, where one can be completely certain that the true collecting locality is found. Higher levels of uncertainty result in circles with larger areas, while lower levels of uncertainty result in smaller areas. Once each record has an associated measure of uncertainty, one can filter the data to exclude records that do not meet the precision requirements for any particular study. This allows one to determine the "fitness of use" of individual records for any particular study (Chapman and Wieczorek, 2006). Due to the potential large impact on altitudinal readings of small errors in georeferencing, these approaches are particularly critical for utilizing archived primary biodiversity data in support of the GMBA Kazbegi Research Agenda (Chapter 17).

Unlike specimen-based data, many observational data sets, particularly those designed to monitor changes in population size or distribution, can document the absence of taxa (confirmed absence). However, for these studies to be of long-term value, it is of the uttermost importance to document survey effort in space, time, and the methodology used. Regrettably, the primary data that forms the basis of these observational studies is seldom archived or made available to the broader community. This is particularly unfortunate, since most such observational studies are limited to relatively small geographic areas, short time periods, and particular taxonomic groups. The lack of access to these primary data sets makes it impossible to reevaluate the conclusions of these studies or to combine their primary data with other data to provide a broader temporal, geographic, and taxonomic understanding. Failure to preserve and share these valuable primary data sets is to a large degree a legacy of how data was managed using paper-based processes prior to the development of our current electronic capacity. This represents a cultural and institutional topic, too. Before the advent of powerful, inexpensive computers, it was impractical to publish, archive, and share large primary data sets. With the introduction of computer technology, this is no longer the case. Unfortunately, very few funding agencies have moved toward digital preservation, or would even recognize the importance of the long-term preservation and sharing of primary scientific data. In general, the criteria for career development for scientific professionals is still based on the number of peer-reviewed hardcopy publications and not sharing the data that these publications are based on. Very few biodiversity scientists have any background training in digital information management and as a result are not well prepared to educate their students with these skill sets.

In 1999, the Organisation for Economic Co-operation and Development (OECD) Megascience Forum Working Group on Biological Informatics recommended the formation of a Global Biodiversity Information Facility (GBIF) to "co-ordinate . . . the standardisation, digitisation and global dissemination (within an appropriate property rights framework) of the world's biodiversity data" (http://www.gbif.org/GBIF_org/facility/OECD_Endorsement). This recommendation recognized that the most significant barrier in the study of species-level biodiversity was actually not the lack of primary biodiversity data but the difficulty the scientific community had in accessing this data. Since its establishment in 2001, GBIF (www.gbif.org) has endeavored to facilitate access to the primary biodiversity data that is the basis of the biodiversity DIKW hierarchy. Currently (early 2008), more than 100 million primary biodiversity data points have been indexed through the GBIF portal. Obviously, this is still only a tiny fraction of the potential data that could be made accessible with further effort.

However, lack of precise georeferencing and data inconsistency, including their accurate description and documentation, have been identified as issues in utilizing the historical data currently available through GBIF. It is important that these problems be addressed when planning, funding, and budgeting for new collecting and observational studies. If society is to obtain maximum benefit from our investment of public funds in natural history collections and in observational biodiversity science, every effort should be made so that future scientists do not have to deal with these types of problems in data sets that are currently being collected or set up. Therefore, there is an urgent need for funding agencies, educators, professional scientific societies, and science policy administrators to reevaluate how the primary data sets that are a product of this research are documented, archived, and shared (Arzberger et al., 2004). The benefits of good data management and data documentation in scientific research include increased research efficiency, long-term data preservation, increased data quality, and complete sharing and increased collaboration in order to return the public investment. The next generation of biodiversity scientists needs better training in digital data management, and the sharing of primary biodiversity data needs to be fully encouraged. Data management initiatives like those supported by the International Polar Year (IPY; Huettmann, 2007) provide us with a good model and deserve more attention and long-term support.

If we can liberate the wealth of primary biodiversity data in the world's museums and georeference this data with appropriate measures of uncertainty, it will then be possible to overlay this information with other geophysical information, particularly climate data, to utilize the data in ways that original collectors never dreamed of. Beyond uses for basic science questions, information distilled from electronic archives can address a suite of issues, such as identification of indicator taxa for environmental conditions, conservational studies, aspects of landscape engineering and impact of assessments of environmental hazards, distribution of medicinal and food plants, invasive organisms, and all aspects of global change research. Realizing this vision will only be possible if the data is digital, easily accessible, and georeferenced using standardized approaches. The vision of GBIF is to make large documented primary biodiversity databases easily accessible. Sharing this data using standardized data sharing protocols will make it possible for specialized thematic interest groups, such as GMBA, to develop gateways, such as a GMBA Mountain Biodiversity Portal that can filter all data that does not relate to mountain biodiversity, and provide the data that is of particular interest for mountain biodiversity research, and to link and display this data in ways that support the Kazbegi Research Agenda of GMBA–DIVERSITAS (Chapter 17).

SUMMARY

Mankind's knowledge and understanding of the distribution and abundance of the world's biota is based on the cumulative observations of single individuals or groups of individual taxa at different points in time and at various geographic locations. These primary data points are the basis of a biodiversity *Data–Information–Knowledge–Wisdom* (DIKW) hierarchy. The most significant source of historical primary biodiversity data points are the estimated 2 to 3 billion specimens preserved in the world's natural history collections. Georeferencing using standardized approaches and sharing this wealth of primary archived data will open up new approaches for studying global mountain biodiversity and directly contribute to the GMBA Kazbegi Research Agenda (Chapter 17).

REFERENCES

Arzberger, P., P. Schroeder, A. Beaulieu, G. Bowker, K. Casey, L. Laaksonen, D. Moorman, P. Uhlir, and P. Wouters. 2004. Promoting access to public research data for scientific, economic, and social development. *Data Science Journal* 3:135–52.

Chapman, A.D. 2005a. Principles of data quality. Version 1.0. Report for the Global Biodiversity Information Facility, Copenhagen. Available at http://www.gbif.org/prog/digit/data_quality

Chapman, A.D. 2005b. Principles and methods of data cleaning–primary species and species-occurrence data. Version 1.0. Report for the Global Biodiversity Information Facility, Copenhagen. Available at http://www.gbif.org/prog/digit/data_quality

Chapman, A.D. and J. Wieczorek, eds. 2006. *Guide to Best Practices for Georeferencing*. Copenhagen: Global Biodiversity Information Facility. Available at http://www.gbif.org/prog/digit/Georeferencing

Graham, C.H., S. Ferrier, F. Huettmann, C. Moritz, and A.T. Peterson. 2004. New developments in museum-based informatics and applications in biodiversity analysis. *Trends in Ecology & Evolution* 19:497–503.

Funk, V.A., P.C. Hoch, L.A. Prather, and W.L. Wagner. 2005. The importance of vouchers. *Taxon* 54:127–29.

Huber, J.T. 1998. The importance of voucher specimens, with practical guidelines for preserving specimens of the major invertebrate phyla for identification. *Journal of Natural History* 32:367–85.

Huettmann, F. 2007. The digital teaching legacy of the International Polar Year (IPY): Details of a present to the global village for achieving sustainability. In Proceedings 18th International Workshop on Database and Expert Systems Applications (DEXA), Sept. 3–7, 2007, Regensburg, Germany, pp. 673–77. Tjoa, M., and R.R. Wagner, eds. Los Alamitos, Calif.: IEEE Computer Society.

Körner, C., M. Donoghue, T. Fabbro, C. Häuser, D. Nogués-Bravo, M.T.K. Arroyo, J. Soberon, L. Speers, E.M. Spehn, H. Sun, A. Tribsch, P. Tykarski, and N. Zbinden. 2007. Creative use of mountain biodiversity databases: The Kazbegi Research Agenda of GMBA-DIVERSITAS. *Mountain Research and Development* 27(3):276–81

Krishtalka, L., and P.S. Humphrey. 2000. Can natural history museums capture the future? *Bioscience* 50:611–617.

Lane, M.A. 1996. Roles of natural history collections. *Annals of the Missouri Botanical Garden* 83:536–45.

May, R.M. 1998. How many species are there on earth? *Science* 241:1441–49.

Peterson, A.T., and A.G. Navarro-Sigüenza. 2003. Computerizing bird collections and sharing collection data openly: Why bother? *Bonner Zoologische Beiträge* 51:205–12.

Rowe, R.J. 2005. Elevational gradient analyses and the use of historical museum specimens: A cautionary tale. *Journal of Biogeography* 32:1883–97.

Rowley, J. 2007. The wisdom hierarchy: Representations of the DIKW hierarchy. *Journal of Information Science* 33:163–80.

Thompson, F.C. 1996. Names: The keys to biodiversity. In *Biodiversity II: Understanding and Protecting Our Biological Resources*, pp. 199–216. Reaka-Kudla, M.L., D. Wilson, and E.O. Wilson, eds. Washington: Joseph Henry Press.

Wieczorek, J., Q. Guo, and R. Hijmans. 2004. The point-radius method for georeferencing locality descriptions and calculating associated uncertainty. *International Journal of Geographical Information Science* 18:745–67.

Winkler, K. 2004. Natural history museums in a post biodiversity era. *Bioscience* 54:455–459.

3 Using Primary Biodiversity Data in Mountain Species Numbers Assessments

Jorge Soberón M.

CONTENTS

Introduction.....17
Methods.....18
Debugging a Heterogeneous-Origin Database.....18
Results and Discussion.....20
Summary.....22
Acknowledgments.....22
References.....22

INTRODUCTION

Mountainous areas of the world are extremely important, among other things, because of the services their ecosystems provide and because they often constitute areas of high concentration of endemic species (Peterson et al., 1993; Jetz et al., 2004; Körner and Ohsawa, 2004). Despite the fact that the inventory of species on earth is far from finished (May, 1990), recent and misguided policies by many countries are making the collection of new specimens an increasingly difficult enterprise (Grajal, 1999). Although it is imperative that efforts to increase existing collections are maintained and strengthened, and to augment the taxonomic capacity in the developing world, it is paradoxical that access to the billions of specimens (Chalmers, 1996) that have already been collected was, until recently, very limited, and therefore, access to primary biodiversity data was restricted to a handful of experts. For scientists in the developing countries, it was difficult and expensive to have access to data in the specimens, because they are scattered over many institutions, countries, and continents. Besides, most specimens were not computerized, and even if they were, the data remained locked in the computers of the museums or herbaria (Soberón et al., 1996). Although some of the data was partially published in catalogues, monographs, and other specialized work that often included the locality of the specimens used to do a revision or new descriptions, such specialized literature also tended to remain in the libraries of the major institutions of the North and therefore was almost inaccessible to nonexperts or to experts in the poor countries.

Fortunately, ten years ago a couple of pilot initiatives demonstrated the feasibility and usefulness of sharing, online, primary biodiversity data. One was the North America Biodiversity Information Network (NABIN) that, using the libraries protocol Z39.50, began demonstrating the power of sharing specimen's label data of museums in Mexico, Canada, and the United States (Soberón, 1999). The other was the Red Mexicana de Biodiversidad (REMIB) that began joining databases in Mexican museums and herbaria, and later included institutions in the U.S.A., Costa Rica, Peru, and Spain (Soberón et al., 1996). REMIB was based on the use of TPCP/IP sockets. After a few years of existence, these two networks enabled access to a few dozen million primary biodiversity data of almost one hundred collections scattered mostly over North America (including Mexico).

Those two pilot initiatives were soon followed by many others, and nowadays there are significantly more than one hundred million registers of biological species publicly available on the Web. The major portal of entrance is the Global Biodiversity Information Facility, or GBIF (http://www.gbif.org), which allows access to more than 140 million records, about 70% of which are georeferenced (Edwards, 2004). This tremendous amount of primary data potentially represents a significant resource for studies related to mountainous areas, regardless of the definition of "mountainous areas," (Körner and Ohsawa, 2004) because of its sheer volume. Very simple queries can now be performed online and, for certain taxonomic groups, large amounts of data can be retrieved in seconds. A number of caveats have to be observed, however. Most emphatically, the data extracted from Web sites should not be used directly to perform analysis without first observing a number of precautions. The taxonomy of such queries must be checked, the georeferencing has to be performed or checked, and, generally speaking, consultations with experts are advisable before the data are used in applications. In this chapter I will demonstrate, using an example with birds of Mexico, how a specimen's database can be queried to obtain estimates of numbers of species and how to assess the degree of completeness of the database.

METHODS

I will illustrate the procedures using an example to estimate the number of species of birds in a transect that goes from the shoreline in the state of Guerrero, Mexico, to the mountains of the Volcanic Belt in the central part of the country (Figure 3.1). This transect has 36,300 km^2 of planimetric surface. Strictly speaking, area should be calculated by taking into account the folds of the terrain. Although this procedure can introduce very significant corrections in area estimates (Nogués and Araújo, 2006), for the purpose of this note, it is not necessary to perform this type of correction. The total transect has a length of about 3 × 1°, and was later subdivided into three regions of 12,100 km^2, and then into six regions of 6,050 km^2. The most comprehensive database existing in the world for Mexican bird species is the Atlas of the Birds of Mexico (Navarro-Sigüenza et al. 2003). This database has been assembled by experts over a period of about fifteen years and contains probably the data of nearly 90% of all the specimens that have been collected in Mexico and that are deposited in museums. A query was performed of all the specimens in localities with coordinates inside the transect or its subdivisions. This query yielded 35,261 specimens of 600 species in 830 unique "localities." "Localities" are all the different combinations of latitude and longitude at one minute of precision. The resulting database was then queried to obtain the specimens present in each of the rectangles. The results are summarized in Table 3.1.

DEBUGGING A HETEROGENEOUS-ORIGIN DATABASE

When using primary biodiversity datasets obtained from heterogeneous sources (such as GBIF), the first step should be to ensure that the taxonomy of the database is consistent. Since the data may come from collections in

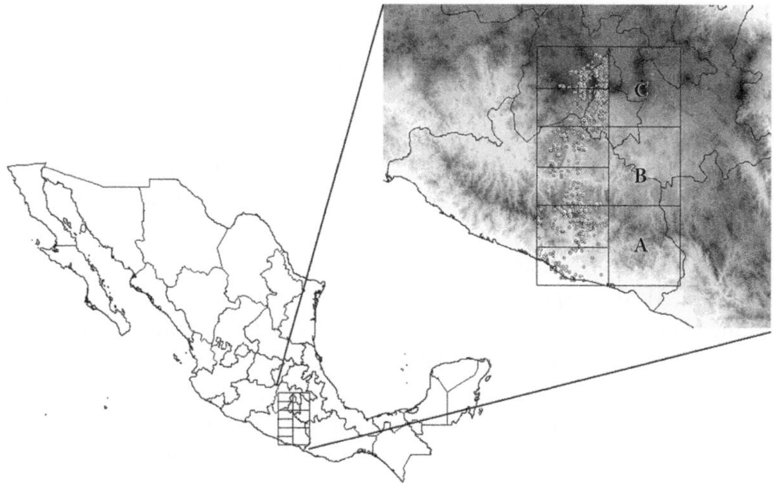

FIGURE 3.1 Transect in Central-Southern Mexico, and three subdivisions of one degree of side each (A, B, and C), and three further subdivisions of approximately _ degree X one degree each (A1 and A2, B1 and B2, and C1 and C2). Only the first subdivisions are labeled.

TABLE 3.1
Different Subdivisions of the Transect in Figure 3.1

Rectangle	Area (in km^2)	Number of Records	Number of Localities	Observed Species	C Value
Total	36,300	35,261	830	600	0.85
A	12,100	19,021	227	471	0.83
A1	6,050	16,798	148	394	0.84
A2	6,050	2,223	79	266	0.67
B	12,100	3,217	138	298	0.76
B1	6,050	1,303	54	216	0.72
B2	6,050	1,914	84	232	0.71
C	12,100	13,023	465	446	0.83
C1	6,050	6,026	212	326	0.79
C2	6,050	6,997	253	374	0.83

very different stages of the curatorial process, this means that a process of comparisons with checklists of current names and synonyms is normally needed (Soberón et al., 2002; Chapman, 2005). Georeferencing of the data by assigning coordinates to the geographical descriptions also is indispensable, since mere verbal descriptions of a locality do not allow most analysis to be performed. Although apparently very simple, georeferencing data is a procedure fraught with subtle problems. Best practices have been developed to explain and solve the many obvious and subtle problems of georeferencing (Wieczorek et al., 2004). Georeferencing is an essential procedure not only because it allows recovering of records defined by geographical queries (within a given altitudinal range, within a given biome, etc.), but because georeferenced records can be subject to a number of indispensable quality-control procedures (Chapman, 2005). If the georeferencing is performed without rigorous protocols, and it is not subject to quality-control routines, the result may look like the example in Figure 3.2, which is a database of Mexican specimens (obtained from one of the largest collections in the developed world) that displays all the localities spotted by the quality-control procedures as doubtful, for one or more reasons (terrestrial species with a georeference in the sea; Mexican specimen with a georeference outside of the boundaries of Mexico; specimen with a locality in a state but coordinates in another, etc.). In the case of the birds of Mexico that is being used in this example, the database has been maintained by experts over almost fifteen years, and the names and georeferences have been checked very thoroughly.

When the taxonomy and georeferencing of the database are ready, it is straightforward to query it to obtain the number of species names that the database contains in any geographical region. The obtained results are displayed in Table 3.1. However, it is obvious that taxonomic databases, almost universally, represent insufficient and biased collecting efforts. It is well-known that sampling localities often concentrate around roads, cities, field stations, and other regions of high collecting intensity (Nelson et al., 1990; Prendergast et al., 1993; Bojórquez-Tapia et al., 1995). It also is known that the intensity of collecting efforts, when done for taxonomic purposes, tends to be almost always insufficient as to provide estimates of some reasonable "true number" of species (Petersen et al., 2003; Walther and Moore, 2005). In fact, it is known that the observed number of species often is the most negatively biased estimator of the true value (Walther and Moore, 2005). A solution to this problem is to estimate the "true value" by any of a number of statistical procedures that have been described to the purpose (Soberón and Llorente, 1993; Colwell and Coddington, 1994). The essence of the idea is to use the history of collecting within a region as a measure of effort, and then use this history, which is presented in terms of an accumulation curve, or in terms of a distribution of number of observations per species, to obtain estimates of the number of species expected under scenarios of increased effort, or around some theoretical asymptote of total number of species that an infinite effort would reveal (Gotelli and Colwell, 2001; Colwell et al., 2004). A quotient C of the "observed number of species" to the "estimated number of species" provides an index of how close the database is to providing an answer to the question of how many species exist in a given polygon (Soberón et al., 2007).

In this case, the algorithm called ICE (Colwell, 2005) was used to obtain estimates of the true number of species in the entire transect, in three subregions of A, B, and C,

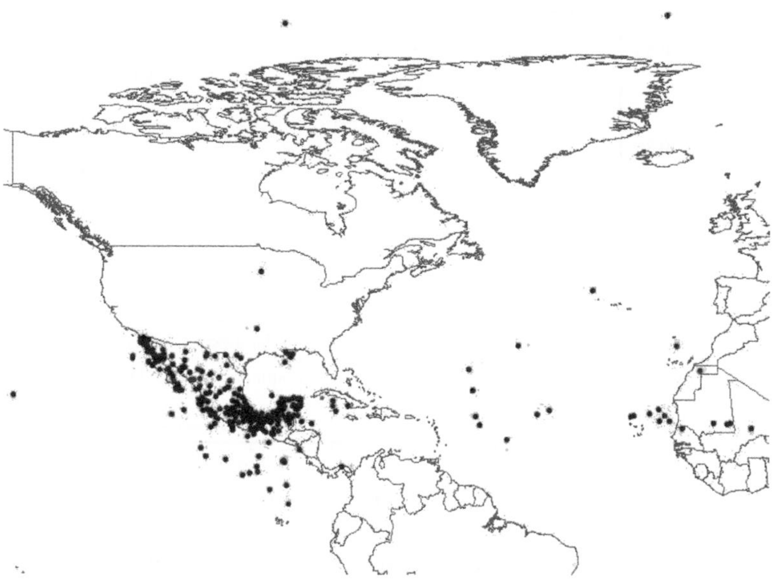

FIGURE 3.2 Examples of doubtful georeference in a database of terrestrial organisms before quality-control procedures are applied. Every point in the database has been identified because its coordinates are inconsistent with the description in its label. For example, all specimens are terrestrial organisms collected in Mexico. It is apparent that many points have georeferences outside Mexico or even in the sea. The points inside the polygon of Mexico are still inconsistent in relation to state or to municipality. Every one of these points has to be checked by referring to the original label.

and in the six subregions of A1, A2, B1, B2, C1, and C2. The results appear in Table 3.1. Instead of ICE, other algorithms could have been used. For example, Chao2, which has been shown by Walther and Moore (2005) to have less bias than ICE for certain kinds of data, may be used. However, for specimen data at regional extents, C indices based on estimators ICE, Chao2, Jack1, and Jack2 are highly correlated (Soberón et al., 2007). Therefore, using ICE as an example does not affect the results.

RESULTS AND DISCUSSION

The first purpose of the above exercise consists of displaying in a graphical way how suitable a database, at a given resolution, is in estimating the number of species in a region. A plot of the density of points in a subregion plotted against the value of the index C has been proposed as a simple but informative way of estimating the degree of completeness of a database (Soberón et al., 2000). The graph in Figure 3.3 displays the information. The value of C shows what fraction of the hypothetical "true value" of the number of species within a polygon the database contains. The density of points is simply the number of records per km^2. The results show a nonlinear relation, where at higher point densities there is a tendency for polygons to be better known, relative to the database of the polygon in question. By establishing a $C = 0.8$ as a cutting point of proportion of completeness, we see that a density of at least 1 record/km^2 is necessary to get to this degree of knowledge, and many of the subpolygons simply do not have this amount of sampling. Of course, other cutoff values could be used, or none at all. Figure 3.3 is a mere description without theoretical implications.

In terms of the geographical location of poorly known and better known segments of the transect, Figure 3.4 displays its elevation profile together with labels of its subdivisions. It can be seen that in the middle part of the transect, the database is not capable of giving an estimate of the number of species beyond 80%. At a resolution of 6,050 km^2, the only two regions in which the database provides at least 80% of the species list are the Pacific Coast plain and the highlands around Mexico City. Attempts to use this database to document or understand important biodiversity patterns associated with elevation gradients, such as humps in species richness or peaks of endemism (Grytnes and McCain, 2007), would be hindered by insufficient data. The highlands of central Mexico appear to be the best-known region, which is probably not strange, since the well-collected areas include the Valley of Mexico that traditionally has concentrated scientific efforts, including bird collections. However, much of these highlands consist of plateaus lacking steep gradients and perhaps should not be considered mountainous areas, *sensu stricto* (Körner, 2007). By decreasing the resolution (increasing the size of

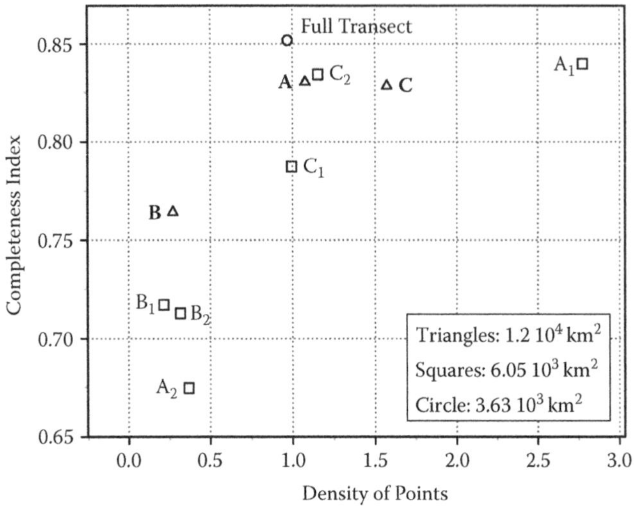

FIGURE 3.3 Completeness of databases plot. The circle represents the entire transect, the triangles the first one degree of side subdivisions, and the squares the 1 degree subdivisions.

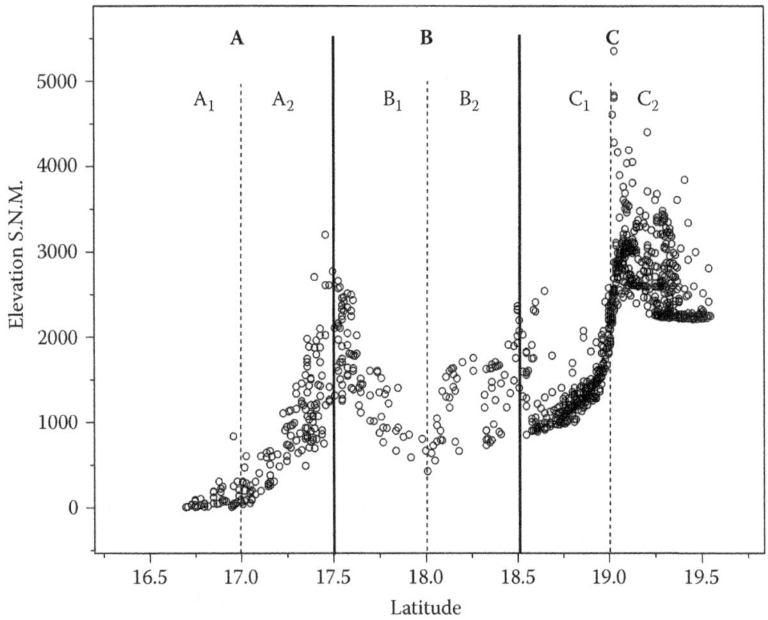

FIGURE 3.4 Elevation profile of the transect and its subdivisions. The circles represent every locality in the database. The letters in bold type label the parts of the transect with $C > 0.8$. The solid lines represent the first subdivision, and the broken line the second subdivisions.

the segments), the reliability of the database increases, and now only one region remains in the poorly known region of Graph 4. For the entire transect, a list of species from the database probably represents about 85% of the true total. Decreasing the resolution then increases the reliability at which the database can be consulted, but the capacity to inspect the details of patterns is lost.

The example presented here shows the power of the use of databases of specimens but also some of its most important limitations. Primary biodiversity databases are becoming available at a very fast rate. More and more museums and herbaria are realizing the enormous value added to their data by georeferencing, databasing specimens, and making these accessible to the public

(Krishtalka and Humphrey, 2000; Soberón and Peterson, 2004). One of the main uses of such databases is compiling species lists for arbitrary geographical regions. As we have shown in this paper, the reliability of such operations cannot be taken for granted. We describe a general technique to deal with the problem of calculating and displaying an index of exploration and completeness of collection effort that is an integral part of a particular primary biodiversity database covering a given region or set of regions. This technique is exemplified using a completeness index (C) applied to a particular spatial grid and database. One striking result of this exercise is that, for a gradient in central Mexico, using the best available database, the crucial areas of high steepness and middle elevation also are those where the completeness of the database appears to be poorest.

The method explained in this work has good potential to assess completeness of databases. It currently is being used in the Mexican National Commission on Biodiversity (CONABIO) to design future biological explorations, assess the status of collections, and optimize resources. It also can be used to look for well-collected regions that can then be reliably used for correlative studies. Although in the example we presented here the country was subdivided using a regular grid, the same method can be applied to irregular partitions (e.g., ecoregions or biotic provinces). A recent exercise, using irregular grids, was used to estimate the completeness of the lists of flora and fauna for the pine–oak forest ecoregion of Mexico (Koleff et al., 2004).

SUMMARY

Pilot initiatives such as the North America Biodiversity Information Network (NABIN) and the Red Mexicana de Biodiversidad (REMIB) demonstrated in the late '90s the feasibility and usefulness of sharing online, primary biodiversity data. Nowadays, there are significantly more than one hundred million registers of biological species publicly available, the major Web portal being the Global Biodiversity Information Network (GBIN). In this chapter, I demonstrate how a specimen's database can be queried to obtain estimates of numbers of species and how to assess the degree of completeness of the database using an example with birds of Mexico in a transect from the shoreline to the mountains of the Volcanic Belt in the central part of the country. When using primary biodiversity data sets obtained from heterogeneous sources, it is important to ensure that the taxonomy of the database is consistent and that the data is georeferenced. Taxonomic databases almost universally represent insufficient and biased collecting efforts, and sampling localities often concentrate around roads, cities, field stations, and other regions of high collecting intensity. The observed number of species is often the most negatively biased estimator of the true value. A quotient C of the *observed number of species* to the *estimated number of* species provides an index of completeness of how close the database is to providing an answer to the question of how many species exist in a given area. The example presented here shows the power of the use of databases of specimens but also some of its most important limitations.

ACKNOWLEDGMENTS

I am grateful to A.T. Peterson and Adolfo Navarro for allowing me to use part of their database of the Birds of Mexico. The Mexican biodiversity agency CONABIO provided the stimulus and chance to develop several of the ideas presented here. I am very grateful to DIVERSITAS and its Global Biodiversity Mountains Assessment for organizing and supporting the workshop of Databases of Mountain Ecosystems that took place in Kazbegi, Georgia, in 2006. In particular, I am grateful to Christian Körner, Silvia Martinez, and Eva Spehn for their knowledge, help, and enthusiasm, and to the local organizers Otar Abdaladze and Giorgi Nakhutsrishvili of the Georgian Academy of Sciences for their hospitality and generosity in sharing the knowledge of their country's very rich flora.

REFERENCES

Bojórquez-Tapia, L.A., I. Azuara, E. Ezcurra, and O. Flores-Villela. 1995. Identifying conservation priorities in Mexico through geographic information systems and modeling. *Ecological Applications* 5:215–31.

Chalmers, N.R. 1996. Monitoring and inventorying biodiversity: Collections, data and training. In *Biodiversity, Science and Development: Towards a New Partnership*, pp.171–79. di Castri, F., and T. Younes, eds. Wallingford: CAB International.

Chapman, A. 2005. *Principles and Methods of Data Cleaning Primary Species and Species-Occurrence Data.* Copenhagen: Global Biodiversity Information Facility.

Colwell, R.K. 2005. EstimateS: Statistical estimation of species richness and shared species from samples. Version 7.5. Available at http://purl.oclc.org/estimates

Colwell, R.K., and J.A. Coddington. 1994. Estimating terrestrial biodiversity through extrapolation. *Philosophical Transactions of the Royal Society B* 345:101–18.

Colwell, R.K., X.C. Mao, and J. Chang. 2004. Interpolating, extrapolating and comparing incidence-based species accumulation curves. *Ecology* 85:2717–27.

Edwards, J. 2004. Research and societal benefits of the global biodiversity information facility. *BioScience* 54:485–86.

Gotelli, N., and R.K. Colwell, R.K. 2001. Quantifying biodiversity: Procedures and pitfalls in the measurement and comparison of species richness. *Ecology Letters* 4:379–91.

Grajal, A. 1999. Biodiversity and the nation state: Regulating access to genetic resources limits biodiversity research in developing countries. *Conservation Biology* 13:6–10.

Grytnes, J.-A., and C.M. McCain. 2007. Elevation trends in biodiversity. In *Encyclopedia of Biodiversity*, pp.1–8. Levin, S., ed. Elsevier.

Jetz, W., C. Rahbek, and R.K. Colwell. 2004. The coincidence of rarity and richness and the potential signature of history in centres of endemism. *Ecology Letters* 7:1180–91.

Koleff, P., J. Soberón, and A. Smith. 2004 Bosques de Pino-Encino de la Sierra madre. In *Hotspots: Biodiversidad Amenazada II*. Mittermeier, R., P. Robles-Gil, M. Hoffman, J.D. Pilgrim, T.M. Brooks, C. Mittermeier, J. Lamoreux, and G.A.B. Fonseca, eds. Mexico: CEMEX, S. A. de C. V.

Körner, C. 2007. The use of "altitude" in ecological research. *Trends in Ecology & Evolution* 22:569–74.

Körner, C., and M. Ohsawa. 2004 Mountain systems. In *Ecosystems and Human Well Being*, pp.683–716. Watson, R.T., and A.H. Zakri, eds. Washington: Island Press.

Krishtalka, L., and P.S. Humphrey. 2000. Can natural history museums capture the future? *BioScience* 50:611–17.

May, R.M. 1990. How many species? *Phil. Trans. R. Soc. London B* 330:293.

Navarro-Sigüenza, A., A.T. Peterson, and A. Gordillo-Martinez. 2003. Museums working together: The atlas of the birds of Mexico. *Bulletin of the British Ornithologists' Club* 123A:207–25.

Nelson, B.W., C.A.C. Ferreira, M. da Silva, and M. Kawasaki. 1990. Endemism centres, refugia and botanical collection density in Brazilian Amazonia. *Nature* 345:714–16.

Nogués, D., and M.B. Araújo. 2006. Species richness, area and climate correlates. Global *Ecology and Biogeography* 15:452–60.

Petersen, F.T., R. Meier, and M. Nykjaer. 2003. Testing species richness estimation methods using museum label data on the Danish Asilidae. *Biodiversity and Conservation* 12:687–701.

Peterson, A.T., V.O. Flores, P.L.S. Leon, J.E. Llorente A.M. Luis, A.G. Navarro, C.M.G. Torres, and I. Vargas. 1993. Conservation priorities in Northern Middle America: Moving up in the world. *Biodiversity Letters* 1:33–38.

Prendergast, J.R., S.N. Wood, J.H. Lawton, and B.C. Eversham. 1993. Correcting for variation in recording effort in analyses of diversity hotspots. *Biodiversity Letters* 1:39–53.

Soberón, J. 1999. Linking biodiversity information resources. *Trends in Ecology & Evolution* 14:291.

Soberón, J., L. Arriaga, and L. Lara. 2002 Issues of quality control in large, mixed-origin entomological databases. In *Towards a Global Biological Information Infrastructure*, pp.15–22. Saarenmaa, H., and E. Nielsen, eds. Copenhagen: European Environment Agency.

Soberón, J., R. Jiménez, J. Golubov, and P. Koleff. 2007. Assessing completeness of biodiversity databases at different spatial scales. *Ecography* 30:152–60.

Soberón, J., and J. Llorente. 1993. The use of species accumulation functions for the prediction of species richness. *Conservation Biology* 7:480–88.

Soberón, J., J. Llorente, and H. Benítez. 1996. An international view of national biological surveys. *Annals of the Missouri Botanical Garden* 83:562–73.

Soberón, J., J. Llorente, and L. Oñate. 2000. The use of specimen label databases for conservation purposes: An example using Mexican Papilionid and Pierid butterflies. *Biodiversity and Conservation* 9:1441–66.

Soberón, J., and A.T. Peterson. 2004. Biodiversity informatics: Managing and applying primary biodiversity data. *Philosophical Transactions of the Royal Society B* 35:689–98.

Walther, B.A., and J. Moore. 2005. The concepts of bias, precision and accuracy, and their use in testing the performance of species richness estimators, with a literature review of estimator performance. *Ecography* 28:815–29.

Wieczorek, J., Q. Guo, and R.J. Hijmans. 2004. The point-radius method for georeferencing locality descriptions and calculating associated uncertainty. International *Journal of Geographic Information Science* 18:745–67.

4 The Global Need for, and Appreciation of, High-Quality Metadata in Biodiversity Database Work

Falk Huettmann

CONTENTS

The Need for Metadata 25
What Are Metadata? 25
References 28

THE NEED FOR METADATA

Due to an increase in computing and Internet capability worldwide, it is obvious that we will soon drown in data. For instance, the International Polar Year (IPY 2007–08) alone, a Mega-Science project with more than 60 nations and 60,000 investigators, and also dealing with mountain areas, will easily produce more than 50,000 data sets, all to be placed online. The Global Biodiversity Information Facility (GBIF), the global biodiversity Web portal covering data for virtually all mountain regions of the world, serves already "174,082,706 occurrence records from 285 data providers" (http://www.gbif.org). Ocean Biogeographic Information System (OBIS), the ocean counterpart, states it serves "17 million records of 104000 species from 519 databases" (http://www.iobis.org). The American National Research Council of the National Academies (2003) promotes any of those concepts for data from its funded projects, too. One should keep in mind that such countries as China, India, and Brazil are just now coming online, too.

So how can the general user—a global citizen—keep track of and find the best available and most suitable data set in the international data jungle, and fully understand its complexities for the most appropriate use? Metadata are tools that help, and then can further aid to overcome the "digital divide" (Stiglitz, 2006), and reach global transparency for decision making, resource distribution, and sustainability.

It is well known that data only have value when accompanied by metadata (Gunther, 1998; Braun, 2005; Gladney, 2007); it represents another form of scientific bookkeeping to stay clear of the coming information flood (Huettmann, 2005). As will be shown, without metadata, science and management lack transparency, trustworthiness, and liability; projects ignoring metadata should not be funded by public money. It simply is not professional to ignore metadata. Only with metadata can the published data translate into precious information, e.g., as required for a science-based management schema such as adaptive management (Huettmann, 2007).

WHAT ARE METADATA?

Metadata describe the data as such, e.g., title of the dataset, its abstract, methods used to assemble the data, geographic projections, the date when data were collected, and whom to contact for obtaining the data. However, the Federal Geographic Data Committee (FGDC) metadata include more than 400 specific fields. Wikipedia (http://en.wikipedia.org/wiki/Metadata) describes metadata as: "Metadata (meta data, or sometimes metainformation) is 'data about other data,' of any sort in any media. An item of metadata may describe an individual datum, or content item, or a collection of data including multiple content items and hierarchical levels, for example a database schema. In data processing, metadata is definitional data that provides information about or documentation of other data managed within an application or environment."

Metadata are not new, though, and the library sciences have spearheaded much of this development for more than four decades, e.g., in Australia and in some colonial libraries. It was United States President Bill Clinton who signed the executive order 12906 in 1994, basically requiring all federal government agencies in the United States to create and to publish metadata. This concept was further supported by the U.S. Data Quality Act (DQA) to "provide policy and procedural guidance to Federal agencies for ensuring and maximizing the quality, objectivity, utility, and integrity of information (including statistical information) disseminated by Federal agencies." By then, the FGDC had developed a standard titled "Content Standard for Digital Geospatial Metadata." Due to these U.S. initiatives, a global movement grew even bigger. FGDC became the required standard for many types of data, and has been adopted by more than 200 national and international catalogs and clearinghouses worldwide. It presents another foundation for the global certifications done by the International Organization for Standardization (ISO), which also promotes the use of metadata, and in Extensible Markup Language (XML) to achieve added online functionality. The ISO standard is widely supported by the Organisation for the Economic and Commercial Development (OECD), which consists of ca. thirty countries. The data elements included in the ISO 19115 standard (Table 4.1) are primarily based on the U.S. FGDC Content Standard for Digital Geospatial Metadata, as well as on the Australian ANZLIC Spatial Land Information Council standard.

The actual creation of metadata involves a data flow model, a business model, and expertise on metadata and the actual data set. Globally available tools for writing metadata are shown in Table 4.2. For specific application, there also are several and more focused metadata profiles, and the biological one is titled Biological Data Profile (BDP), which was developed by the National Biological Information Infrastructure (NBII). The FGDC–NBII standard matters for most biodiversity databases that carry spatial information. Further, it is probably important here to emphasize the link with ITIS (Integrated Taxonomic Information System; http://www.itis.gov/) for taxonomic names and information. Huge online functionalities can be achieved that way, including links with GenBank (http://www.ncbi.nlm.nih.gov/Genbank/) and Mapping Services, for instance.

Although a few other metadata philosophies and standards exist, e.g., Directory Interchange Format (DIF; ISO compatible for initial online discovery), Dublin Core (not really ISO compatible and for general discovery and library purposes), ABCD (Access to Biological Collection Data using BioCase; not ISO compatible but meant for global online data exchange), and Ecological Metadata

TABLE 4.1
ISO 19115 Core Metadata Elements

Mandatory Elements:	Conditional Elements:
Data set title	Data set responsible party
Data set reference date	Geographic location by coordinates
Data set language	Data set character set
Data set topic category	Spatial resolution
Abstract	Distribution format
Metadata point of contact	Spatial representation type
Metadata date stamp	Reference system
	Lineage statement
	Online Resource
	Metadata file identifier
	Metadata standard name
	Metadata standard version
	Metadata language
	Metadata character set

Source: http://www.fgdc.gov/metadata/geospatial-metadata-standards

TABLE 4.2
A Small Selection of Metadata Editors

Tool Name	Designed by (Availability & URL)	Comments
Metavist	USDA Forest Service (Free: http://ncrs.fs.fed.us/pubs/viewpub.asp?key=2737)	Stand-alone software for Windows only. Biological Data Profile (BDP) included.
NPS Metadata Tools and Editor	National Park Service (Free: http://science.nature.nps.gov/nrdata/tools/)	Standalone or with ESRI products; includes great tools. BDP included.
Morpho	Knowledge Network for Biocomplexity (Free: http://knb.ecoinformatics.org/morphoportal.jsp)	For Ecological Metadata Language (EML) metadata records.
ArcCatalog	ESRI (Commercial: http://www.esri.com/)	Automatically fills metadata fields from Arc products (profiles not included).
Spatial Metadata Management System (SMMS)	Intergraph (Commercial: http://www.intergraph.com/)	Stand-alone software with a solid and fast database. Includes BDP.

Language (EML; ISO compatible for advanced databases and sensors and adopted by GBIF), it is obvious that the better the actual data description is, the better information can be derived from the data provided. It is important to understand that metadata standards are in flux due to ongoing technical developments. However, staying with high-quality metadata such as NBII FGDC/ISO 19115 always presents a safe investment. This is because high-quality metadata can easily be crosswalked across standards because all relevant fields exist; this is not the case for incomplete, discovery, or low-quality metadata, though. Lacking information in data fields cannot be suddenly filled with "quality content." An initial ignorance of, or investment in, discovery and low-quality metadata has proven fatal in online and Mega-Science projects. These decisions ended up costing much more money than originally assumed to save. This is because of the global use of data, and the large maintenance costs, and lost opportunity costs of poorly described data. The global user will appreciate well-documented data, and the reputation of a data set and of a Web portal depends on good metadata! Well-written metadata allow for efficient data use.

Metadata can be found in the global public clearinghouses: Web sites that are globally linked and harvested. For instance, the Geospatial One Stop (GOS; http://gos2.geodata.gov/wps/portal/gos) is a major repository for a wide variety of metadata records; it aims to become "Your One Stop for Federal, State & Local Geographic Data."

Other clearinghouses specialize in certain topical areas. NBII, for example, hosts a metadata clearinghouse (http://mercdev3.ornl.gov/nbii/) that focuses on biological metadata records (together with the U.S. Geological Survey). It further provides a subset of forestry records to the Global Forest Information Service (GFIS; http://www.gfis.net/).

Metadata can help to minimize the global digital divide by making information available to all citizens of this world. It helps to achieve a transparent government, making transparent decisions for the entire world. The costs for metadata are low, and it is well set up for the new Web 2.0. Developing specific stand-alone and national metadata formats is not helpful; it fragments the existing approach and makes no use of opportunities out there.

Despite its huge importance, there currently are no "metadata police" enforcing the creation of metadata or carrying out quality control, even. Although most metadata are in English, the delivery language question is important, e.g., for the majority of global users that are only familiar with Chinese, Hindi, Suaheli, Portuguese, and Pidgin. Online translation services, such as Google Translate (http://translate.google.com/translate), for instance, can change the traditional impact of online publications into one of global impact.

Metadata are still a bottom-up process, which needs more support through agencies and their budget processes to reach all citizens of the world. Awareness, expertise related to metadata, and a metadata culture are still widely lacking, though. Therefore, we propose that metadata should also become a global requirement, taught in schools, be a graduation requirement (Huettmann, 2007), and become part of "best professional practices" worldwide. This is because a couple of things are certain: (1) the Internet will grow; (2) its role will increase; (3) we will have more data digitized; (4) more data sets are placed online (e.g., cloud computing); and (5) the relevance of metadata will further rise and become even more crucial for a successful stewardship of the earth and its resources.

Biological archive data make no exception, and their value is based on metadata quality. This is because users from all over the world will appreciate obtaining easily findable, well-described data. A mountain biodiversity information system will heavily draw on metadata, such as FGDC NBII, ISO, and GBIF standards (EML and ABCD). More specifically, it will be essential that such metadata include not only the georeferencing, taxonomy, and sampling methods, but also information about altitude and slopes; they ought to be globally compatible. Database managers of such archives are strongly encouraged to meet a minimum of metadata support, e.g., following the tasks outlined in the Kazbegi Research Agenda for mountain biodiversity by GMBA (Chapter 16). It is only then that the best use of online and computer technologies can be achieved for global sustainability (Czech, 2008).

REFERENCES

Braun, C.E. 2005. *Techniques for Wildlife Investigations and Management*. Bethesda, MD: The Wildlife Society.

Curry, G.B., and C.J. Humphries. 2007. *Biodiversity Databases: Techniques, Politics, and Applications*. New York: CRC Press.

Czech, B. 2008. Prospects for reconciling the conflict between economic growth and biodiversity conservation with technological progress. *Conservation Biology* 22:1389–98.

Esanu, J.M., and P.F. Uhlir, eds. 2004. *Open Access and the Public Domain in Digital Data and Information for Science: Proceedings of an International Symposium*. U.S. National Committee for CODATA, National Research Council.

Gladney, H.M. 2007. *Preserving Digital Information*. New York: Springer.

Guenther, O. 1998. *Environmental Information Systems*. New York: Springer.

Huettmann, F. 2005. Databases and science-based management in the context of wildlife and habitat: Towards a certified ISO standard for objective decision-making for the global community by using the Internet. *Journal of Wildlife Management* 69:466–72.

Huettmann, F. 2007. The digital teaching legacy of the International Polar Year (IPY): Details of a present to the global village for achieving sustainability. In *Proceedings 18th International Workshop on Database and Expert Systems Applications (DEXA)*, Sept. 3–7, 2007, Regensburg, Germany, pp. 673–77. Tjoa, M., and R.R. Wagner, eds. Los Alamitos, CA: IEEE Computer Society.

Huettmann, F. 2007. Modern adaptive management: Adding digital opportunities towards a sustainable world with new values. Forum *on Public Policy: Climate Change and Sustainable Development* 3:337–42.

National Research Council of the National Academies. 2003. *Sharing Publication-Related Data and Materials*. Washington, D.C.: The National Academies Press.

Stiglitz, J. 2006. *Making Globalization Work: The Next Steps to Global Justice*. New York: Penguin Books.

5 A Possible Correlation between the Altitudinal and Latitudinal Ranges of Species in the High Elevation Flora of the Andes

Mary T. Kalin Arroyo, Leah S. Dudley, Patricio Pliscoff, Lohengrin A. Cavieres, Francisco A. Squeo, Clodomiro Marticorena, and Ricardo Rozzi

CONTENTS

The Alpine Life Zone as a Template for Testing Biogeographic Theory 29
The Alpine Life Zone in the South American Andes 30
The *Senecio* Database to Show Ecological Trends across the High Andes 31
Problems Associated with the Use of Databases in Fine-Tuned Macroecological Work 32
Trends in Altitude versus Latitude in *Senecio* 33
Discussion 34
Dealing with the Problems 34
Underlying Causes of the Putative Pattern 35
Summary 36
Acknowledgments 37
References 37

THE ALPINE LIFE ZONE AS A TEMPLATE FOR TESTING BIOGEOGRAPHIC THEORY

It is increasingly becoming recognized that the alpine life zone, defined as that vegetation occurring above the upper natural treeline on mountains (Körner, 2003), provides an impressive replicated, large-scale natural experiment, and thus an ideal system for studying macroecological patterns, and ecological and evolutionary processes. Although covering a relatively small proportion of the earth's terrestrial area (ca. 3%) (Körner, 2003), alpine vegetation is amply represented in both hemispheres, where it is found on all continents, and globally extends from subpolar to equatorial latitudes. Alpine vegetation in many parts of the world, unlike much subtending lowland vegetation, is still relatively well conserved (cf. Nogués-Bravo et al., 2008), thus providing greater assurance that any broad patterns detected in the alpine will reflect nonanthropogenic processes.

Characterized by the compression of the equivalent temperature conditions found along large distances of the latitudinal gradient into relatively short distances along steep altitudinal gradients, many macroecological patterns can be profitably investigated in the alpine life zone. For example, the physically compact alpine gradient provides an excellent system for detecting the effect of land area on patterns of species richness (Körner, 2000; Gorelick, 2008), and thus could go a long way in illuminating our understanding of the latitudinal species gradient, where differences in land area between the tropical and temperate zones becomes a major confounding factor. Because the mean growing season temperature theoretically should not vary significantly for alpine surfaces at different latitudes, comparative studies of species richness in mountain ranges that span tropical to subpolar latitudes, such as the South American Andes, should enable sorting out the relative contributions of historical, evolutionary, and ecological causes of the latitudinal species gradient. The alpine gradient has already proven to be an excellent medium for unraveling the relationship between certain plant traits (e.g., breeding systems) and their evolutionary drivers (e.g., pollinators), as they are affected by the temperature regime and local weather conditions (Arroyo et al., 1982; Arroyo and Squeo, 1990) and for studying phenomena such as facilitation (Callaway et al., 2002; Arroyo et al., 2003; Cavieres et al., 2006).

Most large-scale macroecological work to date focuses either on latitudinal patterns or on altitudinal patterns, there being no attempts to link these two environmental gradients. Favorite themes currently are the latitudinal and altitudinal species gradient (cf. Lyons and Willig, 2002; Romdal and Grytnes, 2007; Chown et al., 2008) and Rapoport's rule (Rapoport, 1982; Stevens, 1989). Yet there are reasons to expect that the altitudinal range size is correlated with geographical range size. Predictably, the wider a species altitudinal range is, the greater are the species' chances for successful dispersal beyond its present geographical range. This stems from the following: Considering an adiabatic lapse rate of 6 k per 1,000 m elevation, the individual populations of a species distributed over 2,000 m of elevation straddle (and are adapted to) a range of temperatures differing by 12 k. By contrast, the populations of a species distributed over 500 m elevation under this same adiabatic lapse rate will span a mean annual temperature range of 3 k. Consequently, seeds dispersed by a species that covers 2,000 m elevation will be adapted to a much greater range of temperature conditions than those belonging to a species distributed over 500 m elevation.

Because of the wider array of potential temperature niches represented among the dispersing seeds of the altitudinal widely distributed species, the probability that some of the dispersed seeds will encounter an adequate temperature niche outside the species' present geographical range should be higher than for a species with a narrow altitudinal range. Thus, altitudinally widely distributed species should be able to expand their geographical distributions (latitudinally and longitudinally) at a faster rate, and in a young mountain flora, come to occupy larger geographical areas than altitudinally narrowly distributed species. Consequently, in a young flora, size of altitudinal range and size of geographical range should be correlated. Apart from evident theoretical interest, this question has important connotations for understanding how successfully plant and animal species may respond to climate change in high mountain floras, although on this occasion we will not dwell more on this theme.

Our primary aim here is to discuss the usefulness of electronic databases to preliminarily investigate whether altitudinal range and geographical range size are correlated in the north–south trending South American Andes, using the genus *Senecio* as a model. Data was obtained from electronic, specimen-based floristic databases, a source of information that is being increasingly employed in macroecological research when large geographical scales are considered. At this stage, we use latitudinal range as a surrogate for geographical range. In the north–south trending Andes, this finds some justification in a preliminary study such as this, although clearly is not as precise as one would like. For this and other reasons that will eventually be expounded, in the present paper, it is not our intention to come to any definitive scientific conclusions. Rather, having carried out a preliminary exercise for exploring the power of database work in a mountain ecological context (Körner et al., 2007), our main objective is to highlight some of the problems associated with using georeferenced databases for studying macroecological trends, especially when altitude is the variable of interest, as well as offering some suggestions as to how the problems might be overcome.

THE ALPINE LIFE ZONE IN THE SOUTH AMERICAN ANDES

The alpine life zone of the South American Andes, encompassing the páramo, puna, and southern temperate alpine, runs from approximately 11° N in Venezuela to the extreme tip of southern South America (55° S), with outlying patches of alpine vegetation found on the Cape Horn Islands close to 56° S. The páramo extends from 11° N to 8° S, followed by the puna to around 26 to 27° S, and from thereon south, the southern temperate alpine. A precise definition of the lower limit of the alpine life zone in the South American Andes is hindered by the lack of a treeline over a considerable extension of the puna and over part of the eastern side of the southern temperate alpine. In our work, we rely on the criterion of homologous vegetation belts. In the northern central Andes (northern and central Peru), the vegetation belt immediately above treeline corresponds to the puna, found at about (3,000) 3,200 to 3,500 m elevation. This vegetation type shows continuity with the páramo north of 8° S, which again occurs immediately above the treeline. Consequently, in the drier Andes where a treeline is absent, we use the lower limit of the puna as the lower bound of the alpine life zone. This limit is somewhat lower than the sporadic occurrence of *Polylepis* found within a matrix of typical high altitude puna vegetation from 18 to 22° S on the western side of the Andes. Using a 90 m resolution digital elevation model, we estimate the South American alpine life zone (including bare ground and ice above the upper vegetation limit), defined previously, to cover some 913,500 km^2, the equivalent of 5.1% of the land area of the continent.

Although the global trend for the alpine belt is to decrease in altitude with latitude, the altitude of the lower limit of the Andean alpine life zone experiences little elevational change from the tropics to 25 to 27° S on account of aridity in the central Andes. Aridity counteracts the global latitudinal decrease in temperature and at the same time forces vegetation belts upwards beyond the altitude expected for their latitude (Arroyo et al., 1988). Indeed, the elevation of the lower limit of the puna at its driest point on the western side of the Andes in Chile at 25° S is higher than further north (ca. 3,700–3,800 m compared with the more typical 3,200 to 3,500 limit). We specifically elaborate on this trend here, because, as will be seen later, it will be taken advantage of in order to avoid registering a spurious relationship between altitudinal and latitudinal range. The tendency for the vegetation, including tree species, in the arid area of the Andes to seek more adequate moisture at higher elevations has been pointed out by Braun et al. (2002) in their study of Andean treelines. From 27° S southward, as aridity ameliorates, the lower limit of the alpine zone descends gradually to around 2,000 to 2,200 m at 33° S to reach 300 to 400 m elevation on Hermit Island in the Cape Horn Islands.

Ongoing work by us, consisting of developing and refining a high altitude checklist, suggests that the South American alpine could house as many as 6,700 species in around 870 genera. It should be noted that many of these species are not restricted to the alpine zone per se, and some species records included at this point on the basis of a general altitudinal criterion could eventually be eliminated because of local variation in the elevation of the treeline. The alpine life zone is characterized by seven genera, with close to, or more than, one hundred species: *Calceolaria*, *Gentianella*, *Lupinus*, *Nototriche*, *Senecio*, *Solanum*, and *Valeriana*. Many additional genera are represented by close to, or more than, fifty species (*Adesmia*, *Astragalus*, *Baccharis*, *Calamagrostis*, *Carex*, *Diphlostephium*, *Draba*, *Elaphoglossum*, *Espeletia*, *Festuca*, *Geranium*, *Gynoxys*, *Halenia*, *Hypericum*, *Huperzia*, *Miconia*, *Oxalis*, *Pentacalia*, *Poa*, *Puya*, and *Viola*). Thus, the South American alpine flora is ideal for studies that require large numbers of species. *Senecio* is by far the largest genus at high elevations in the South American Andes, with more than 300 species occurring totally or partially above the treeline. Species of *Senecio* are found in the páramo, puna, and southern temperate alpine.

THE *SENECIO* DATABASE TO SHOW ECOLOGICAL TRENDS ACROSS THE HIGH ANDES

Several subgeneric taxa are recognized in the South America species of *Senecio*, but here we have considered all species independently of their sectional placement in the absence of comprehensive phylogenetic work. We carried out separate analyses for species of *Senecio* occurring in the páramo and puna, so as to ascertain the generality of any emerging patterns. Toward this purpose, we extracted 1,209 puna and páramo *Senecio* records from a database of 95,000 specimens representing the entire South American alpine flora that we have downloaded from the open access TROPICOS electronic specimen-based database housed at the Missouri Botanical Garden (http://www.tropicos.org/) and 726 from Chilean sources. Twenty-two additional records were downloaded from the C.V. Starr Virtual Herbarium (http://sciweb.nybg.org/Science2/VirtualHerbarium.asp), an electronic gateway to the collections of the William and Lynda Steere Herbarium at the New York Botanical Garden (NYBG), to give a grand total of 1,957 records. TROPICOS is particularly well endowed with specimens from high elevations in Colombia, Ecuador, Peru, and Bolivia, and thus a valuable resource. Specimens from Argentina are now being actively incorporated in TROPICOS. However, most of the Argentine material presently found in TROPICOS lacks georeferences.

Many more nondatabased specimens of *Senecio* are found in Argentina's major herbaria, to the extent that the present information for Argentina (i.e., the eastern side of the southern part of the puna) is incomplete. Still more specimens are to be found in national herbaria in Venezuela, Colombia, Peru, Ecuador, and Bolivia. Overall, around 77.8% of the *Senecio* records downloaded from TROPICOS and NYBG contained either latitude and longitude coordinates or altitude data (including a few georeferences furnished by us) and thus are informative for the purposes here. However, only 58.6% of the records simultaneously contain both latitude and altitude in the same record. The Chilean data includes mostly specimens from Chile and is particularly rich in records for the southern half of the puna. Most herbarium specimens collected in Chile from the 1950s onward contain georeferences and altitude data taken from topographical maps or with GPS and field altimeters. The major Chilean herbaria have retroactively provided georeferences for many older specimens, which, in any case, are few for the puna. All puna *Senecio* records contain latitude, and 98.6% contain both latitude and altitude. The records for each *Senecio* species include above and below treeline occurrences—i.e., their total altitudinal range.

PROBLEMS ASSOCIATED WITH THE USE OF DATABASES IN FINE-TUNED MACROECOLOGICAL WORK

To investigate the relationship between latitudinal range and altitudinal range, the first task was to obtain latitudinal and altitudinal limits for each *Senecio* species. This is easier said than done when using floristic databases and will always involve sources of error. Altitude on many herbarium specimens is given as a range (e.g., 3,500 to 3,800 m). Any field botanist knows that such ranges can refer to the general range over which a series of specimens representing diverse species would have been collected on a particular collecting trip or day, and not necessarily the altitudinal range over which the particular species occurs at that collecting site. Unfortunately, determining when ranges on specimens in electronic databases refer to a species' real range versus a general collecting range is virtually impossible without going back to original field notes, and then might not work anyway. Secondly, in order to obtain the altitudinal range of a species from database records, one usually has no other choice than to use the data obtained from several specimens collected from different localities.

To estimate a species' altitudinal range, we employed two criteria: Criterion (1) The range was considered the number of 100 m bands calculated from the difference between the highest point elevation represented among the collections of the species under consideration (or highest upper limit among the altitudinal ranges), and the lowest point elevation represented among the collections (or lower limit among the ranges); Criterion (2) Prior to searching for the upper and lower elevation recorded for the species, all range data on individual specimens was averaged so as to provide one altitude point per specimen. Under both criteria, when a species had been collected once, its altitudinal range was considered to be 100 m. To be consistent, 100 m was added to the difference between the maximum and minimum altitude at which each species was collected. In other words, when a species was reported for, say, 3,400 m, it was assumed that it occurred at least between 3,400 to 3,500 m. The first criterion effectively assumes that the ranges given in individual database records are meaningful for the particular species, which, of course, we know is not always true. The second measure of altitudinal range is more conservative. The truth will lie somewhere between. The latitudinal range of each species was expressed as the number of 1° latitudinal bands obtained from lowest and highest latitudes registered for the species. When a species had only been collected once, it was considered to be distributed in one latitudinal band.

A second and more serious problem concerns spurious effects, which will arise if the localities from which an altitudinal range is deduced cover a very large latitudinal range. In a north–south trending mountain range such the Andes, a species distributed, say, from the latitude of the páramo into the southern temperate alpine will naturally be found at much lower elevations at the southern extreme of its latitudinal range, on account of the latitudinal reduction in temperature. The further south such a species extends (therein increasing its latitudinal range), the wider will be its altitudinal range based on the upper and lower altitudinal limits taken from database records throughout its latitudinal range. We attempted to avoid this kind of spurious effect in terms of the hypothesis we wish to test in the present study, as far as possible, by limiting our analysis to those species of *Senecio* found in the páramo and the puna, and whose maximum latitudinal distribution terminated in the 25th parallel south. In this way, we were able to consider around 36° of latitude in the Andes, where there is little latitudinal variation in the lower limit of the alpine zone. The altitudinal increase in the lower limit of the puna (ca. 200 to 300 m) at 25° S is fairly inconsequential, because the upper limit of the alpine vegetation does not increase accordingly.

Under these restrictions, the altitudinal range calculated on the basis of the highest and lowest records available should be fairly representative of a species' altitudinal range through the study area. Discarding 16 puna species whose distributions extend beyond the 25° latitudinal band limit and numerous other species for which the available data was noninformative, we obtained distribution data for 112 *Senecio* species, of which 14 are found in both the páramo and puna. For the purposes of the present analysis, these last species were assigned to that sector of the Andes corresponding to where the midpoint of their respective latitudinal ranges is found (< 8° S = páramo; ≥ 8° S = puna). The total number of species of *Senecio* in the páramo and puna, including those species that are distributed south of the 25° latitudinal band, is around 220.

Thus, the electronic databases available to us allowed consideration of around 50% of the species.

TRENDS IN ALTITUDE VERSUS LATITUDE IN *SENECIO*

The correlation coefficients between altitudinal range and latitudinal range for *Senecio* species are summarized in Table 5.1. Results obtained under Criterion 1 are shown graphically in Figure 5.1. High altitude *Senecios* may be distributed over a very narrow altitudinal range (100 m) to as much as 3,890 m elevation, and over a single, 1° latitudinal band to up to twenty-four latitudinal bands (Figure 5.1). Our original prediction of a relationship between altitudinal and latitudinal range is borne out for both the páramo and puna *Senecios* (Table 5.1, Figure 5.1). In both high Andean vegetation zones, significant correlations were obtained between these variables independently of the criterion used to represent altitudinal range. The median altitudinal and latitudinal ranges, respectively, tend to be larger in the páramo than in the puna, although the differences are not significant (Mann Whitney U tests) (Table 5.2).

While these results are auspicious, they need to be considered critically. Aside from the problem of correctly depicting the exact altitude at which a particular database entry was collected, rare species are possibly influencing the results. In general floristic surveys, the source of most database records, rare species will tend to be collected on fewer occasions than common species. Nevertheless, there will be cases where botanists studying a particular genus will seek to collect rare species. Given these caveats, the true altitudinal and latitudinal ranges of rare species in geographical regions still at the phase of general exploration, as is the case in the high Andes, will tend to be underrepresented in relation to more common species. As a consequence, rare species will tend to cluster relatively more closely to the origin on both axes of a plot of latitudinal versus altitudinal range than their true distributions dictate, in comparison with more common species. When rare species are very numerous in the data set, they could conceivably force a positive correlation between altitudinal and latitudinal range. Variation in the level of exploration (in different parts of the páramo and puna, in this particular case) also will affect results. That is, when the level of exploration is more limited, the probability of representing the full altitudinal and latitudinal ranges of a species will be diminished. One way of assessing whether rare species are influencing the analysis is to eliminate them entirely and repeat the analysis

TABLE 5.1
Correlation Coefficients for Altitudinal Versus Latitudinal Range of Species of *Senecio* in the Puna and Páramo, South American Andes, as well as for the Subset of Puna Species Occurring in Chile

		Criterion 1		Criterion 2	
	Number of species	Correlation coefficient (r)	P	Correlation coefficient (r)	P
Full set of species					
Puna (Chile)	33	0.516	0.01	0.444	0.05
Puna	88	0.540	0.01	0.522	0.01
Páramo	24	0.433	0.05	0.436	0.05
Restricted set of species					
Puna (Chile)	31	0.508	0.01	0.434	0.05
Puna	56	0.408	0.01	0.289	0.05
Páramo	15	0.097	NS	0.063	NS

Note: Criterion 1: Absolute limits obtained from all range data given on all database records; Criterion 2: When the altitude on an individual database record was given as a range, the midpoint was calculated as the representative altitude for the specimen (see text for further details); Full set of species: All species included regardless of the size of their latitudinal ranges; Restricted set of species: Species found in only one altitudinal band eliminated.

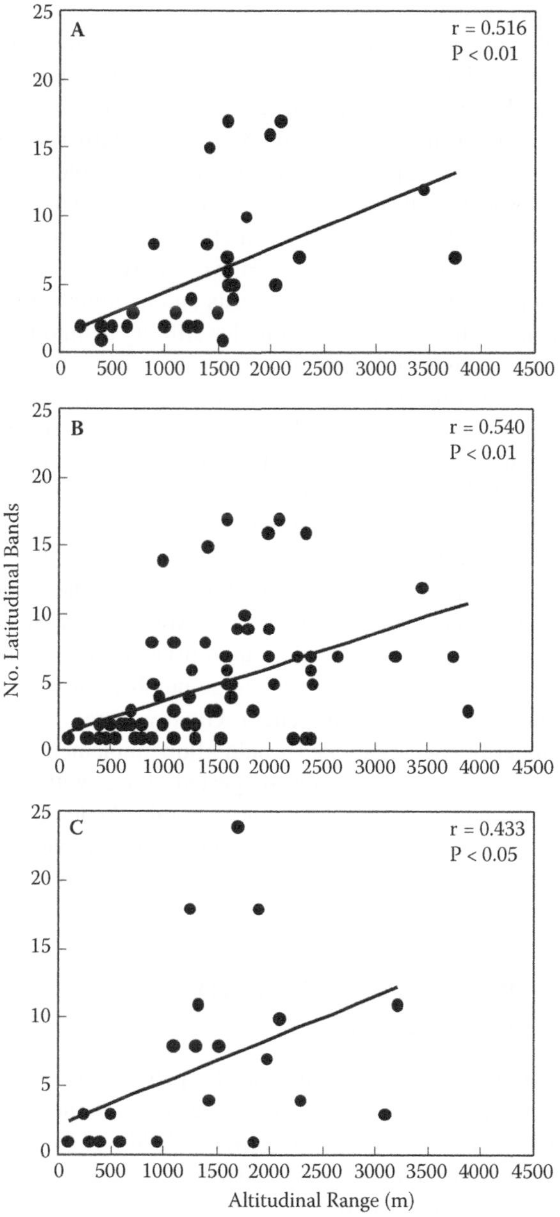

FIGURE 5.1 Relationship between altitudinal range and latitudinal range in species of Senecio (Asteraceae), South American Andes. (A) Puna of Chile only (n = 33 species); (B) puna of Chile, Peru, Argentina (n = 88 species); (C) páramo (n = 24 species). The lines show the tendencies in the data. r = product–moment correlation coefficient.

so as to determine if the correlation between altitudinal and latitudinal range still holds up. We performed the analyses again, eliminating all species recorded from a single latitudinal band (n = 9 in the páramo; n = 32 in the puna) (Table 5.1, restricted set of species). The correlation between altitudinal range and latitudinal range continued to be significant for the puna, but not for the páramo, where the number of species was now very low (Table 5.2). With respect to differing levels of exploration, the Chilean puna has been extensively collected, there being only two species restricted to a single latitudinal band, according to the data at hand. It can be seen that the relationship between altitudinal range and latitudinal range holds up in this well-collected area (Figure 5.1, Table 5.1).

TABLE 5.2
Mean and Median Altitudinal and Latitudinal Ranges for Species of *Senecio* in the Puna and Páramo, South American Andes

	Number of species	Mean altitude (m)	Median altitude (m)	Mean latitude (No. bands)	Median latitude (No. bands)
Criterion 1					
Puna	88	1194	1100	4.2	2
Páramo	24	1270	1275	6.2	3.5
Criterion 2					
Puna	88	980	835	4.2	2
Páramo	24	1127	1158	6.2	3.5

Note: Differences between species in the puna and páramo not significant (Mann Whitney U Test); Mann Whitney U Test. Altitude: Criterion 1: p = 0.499; Criterion 2: p = 0.745; Latitude: p = 0.128.

DISCUSSION

The study of large-scale macroecological patterns is dependent on the existence of sufficient and accurate plant and animal data. Our preliminary results in *Senecio* are in line with the hypothesis that altitudinal range and geographical range, as proxied by latitudinal range, are correlated. However, numerous uncertainties have appeared with the use of raw georeferenced database records, in particular because of the need of precise altitudinal data. Consequently, we feel that it is too early to come to any definitive conclusions.

DEALING WITH THE PROBLEMS

How to deal with altitudinal ranges in electronic database records constitutes a major challenge for macroecologists. If abundant specimens are available, those for which altitude is cited as a range could be eliminated. For the *Senecio* data set, this would be tantamount to eliminating 27.5% of the informative specimens obtained from TROPICOS and NYBG, signifying a huge loss of information. In order for such valuable resources as TROPICOS to reap maximum value, field botanists might endeavor to report the precise altitude where a collection is made. If a species is sighted beyond that point on an altitudinal gradient, its additional range could be explicitly noted on the herbarium specimen label and be included in the general habitat data that the best databases included. At the same time, ecologists should be explicit regarding the limitations of any databases employed.

Controlling for spurious effects between altitudinal and latitudinal ranges insofar as our interests are concerned is even more challenging. The situation in the Andes, whereby the lower limit of the alpine life zone does not vary much over its northern 36 degrees of latitude, and if anything, increases with latitude, would seem unique worldwide. Notwithstanding taking advantage of this fortuitous situation, the *Senecio* altitudinal ranges will not be totally accurate because of local differences in the lower limit of the alpine zone.

We suggest that the best way of getting around this problem, in terms of the hypothesis of interest here, is to convert the altitudinal ranges of each species into an equivalent mean growing season temperature range, thereby abandoning altitude, per se (cf. Körner, 2007). This procedure, besides leading to greater accuracy in depicting the breath of a species' temperature niche in the present study, would have enabled the inclusion of many more species of *Senecio* found in the southern temperate alpine, thereby increasing statistical power. However, to be able to reach this point, regional climate models capable of discriminating between temperatures over small spatial scales on altitudinal gradients must be available. Even then, some difficulties will remain. At any given altitude, significant small-scale variation in temperature and other environmental variables, associated with opposite facing slopes, exposed versus windy slopes, and different substrate types, is typical at high elevations (Squeo et al., 1983; Rozzi et al., 1997; Körner, 2007). Additionally, in large-scale studies involving many degrees of latitude, latitudinal ranges should be converted into linear distances, thereby overcoming the problem that latitudinal bands are of different widths.

A procedure is needed to deal with the distributions of rare species, which by their very nature will tend to be undersampled in most floristic databases. It should be born in mind that databases of museum records reflect the accumulated effort of generations of plant and animal collectors, such that unequal sampling of rare and common species and between different geographical areas will always be a problem when using this kind of data. Resampling the database to obtain a standard number of collections per species provides a possible solution to this problem, always that a sufficient number of species in the group of organisms under study is represented by a large number of collections. However, except for a few groups of organisms, such as birds and mammals, reality has it that the tropical latitudes still lag behind temperate latitudes in terms of collection density. Another uncertainty that tends to be disregarded in these kinds of studies concerns the misidentification of closely related allopatric species, which will lead to an exaggeration of the latitudinal or latitudinal range of a species. Of course, the best test of the hypothesis would consist of systematically sampling the altitudinal gradient at 100 m intervals and 1° intervals along both sides of the entire Andean chain. Until more formal work is carried on these various issues, the most appropriate avenue would seem to be to work with very large numbers of species drawn from many plant genera. Greatly increasing the number of species studied would allow implementation of some of the procedures suggested above, without a significant loss of statistical power. For the Andes, needless to say, inclusion of herbarium records in national herbaria could lead to significant changes in the distributions of many *Senecio* species and thus is critical.

UNDERLYING CAUSES OF THE PUTATIVE PATTERN

Assuming for the moment that the pattern revealed in this preliminary study of a single large taxon in the South American Andes could conceivably hold up with the inclusion of additional data and better altitudinal range control, to our knowledge, such a pattern has not been previously revealed for any other plant or, for that matter, animal group. The closest parallel we have been able to find is Brown and Mauer's (1987) demonstration of a strong correlation between the length and width of geographical ranges expressed as km north–south versus km east–west in North American and European land birds, which, of course, is not exactly the same. At some level, the underlying process leading to the macroecological pattern of interest here is akin to the notion that species with larger geographical distributions will tend to provide larger numbers of dispersing propagules and hence be more likely to succeed better in long-distance dispersal. However, here we are talking not only about more diaspores, but about a set of diaspores that are adapted to many different temperature conditions. Thompson et al. (1999) showed that a measure of niche breadth is the best predictor of range size in the herbaceous flora of England.

Species with wide altitudinal distributions effectively have wide temperature niche breadths. As a corollary, among species with equal-sized geographical distributions, those species spread out over a steep altitudinal gradient should be more successful as long-distance colonizers than species with narrow altitudinal ranges. In passing, it is worthwhile pointing out that some of the most striking cases of long-distance dispersal concern dispersal from the northern hemisphere into the South American Andes (e.g., *Halenia*, *Valeriana*; Von Hagen and Kadereit, 2000; Bell and Donoghue, 2005). Dispersal into alpine areas is usually considered easy because of the less competitive, open nature of the alpine habitat. However, the high probability of temperature niche matching could also be a factor here.

Evolutionary history and differences in dispersal capacity also may influence the putative correlation between altitudinal range and latitudinal extension in the genus *Senecio*. All other things being equal, the more recently evolved species in a genus could be expected to have smaller altitudinal and latitudinal distributions, whereas earlier branching species would have had more time to colonize into the alpine, both altitudinally and latitudinally.

Assessing the effect of evolutionary history on the size of latitudinal and altitudinal ranges requires access to appropriate phylogenetic information, as yet not available for *Senecio*. Although we are not totally versed in the relevant literature, we note that much recent phylogeographic research on alpine plants focuses on their glacial history (e.g., Holderegger and Abbott, 2003; Schoenswetter et al., 2003). Insofar as dispersal is concerned, poorly dispersing species could be expected to have smaller altitudinal and latitudinal ranges, simply because they are not well-adapted for getting around. The achenes of *Senecio* bear a pappus and are wind dispersed. However, considerable variation in achene size is seen among species, which could make for differences in dispersal capacity among species. In the genus *Celmisia* (Asteraceae) in New Zealand, altitude is significantly and negatively correlated to seed dry weight (Fenner et al., 2001), suggesting that establishment at higher elevations favors smaller and, presumably, more easily dispersed seeds. A comparison of the latitudinal and altitudinal ranges of Andean taxa characterized by different dispersal types would be very enlightening in this respect.

Finally, it should be noted that our hypothesis presupposes that new evolved species will be able to colonize altitudinal gradients relatively rapidly, after which time lateral spread will occur. Very little seems to be known about how alpine species extend their geographic ranges. The unwritten dogma is that tracking the same temperature latitudinally constitutes an easier option than the spawning of new populations adapted to increasingly colder temperatures at the higher altitudes. Nevertheless, clinal variation in seed longevity over a compact alpine gradient in a fairly continuously distributed species has been reported by Cavieres and Arroyo (2001). For species distributed at high elevations, tracking the same temperature latitudinally (and longitudinally) will often involve crossing significant barriers from one mountain peak to another. Thus, lateral temperature tracking may not be as easy as intuitively assumed. But, of course, in reality, both processes are likely to occur simultaneously.

In conclusion, the use of electronic databases to detect large-scale patterns in alpine ecosystems poses numerous challenges, yet at the same time opens the door for taking further advantage of one of the best replicated natural experiments on Earth for detecting and understanding fundamental macroecological patterns. Just as altitude compresses large temperature differences into short distances, imprecise altitude data will signify magnified errors in macroecological analyses. Yet with caution, ecologists should use the highly valuable information contained in electronic databases while ensuring maximum rigor.

SUMMARY

Understanding the determinants of large-scale biodiversity patterns depends on access to sufficient and accurate georeferenced collection data. Macroecological patterns can be profitably studied in the alpine zone, a replicated, large-scale natural experiment that extends over large regions and exists at all vegetated latitudes of the globe. In our study, we take a look at the relationship between latitudinal range and altitudinal range for species of the genus *Senecio* occurring in the páramo and puna of the South American Andes. We compared latitudinal ranges of species measured as the number of 1° latitudinal bands and altitudinal ranges depicted as the number of 100 m elevational bands, over which a species is distributed, along an area of the Andes where temperature does not become markedly depressed with latitude, and found that these two variables to be positively correlated. We discuss the caveats of the method: the problem of rare species; variation in levels of exploration; the effect of misidentification of closely related allopatric species. We highlight several problems associated with the use of georeferenced data when altitude is the variable of interest, pointing out that use of mean growing temperature range instead of altitudinal range is biologically more realistic. Comparisons of altitudinal and latitudinal ranges will help to disentangle historical, evolutionary, and ecological causes. With appropriate phylogenetic information (not yet available for *Senecio*) one could assess the effect of evolutionary history on macroecological patterns. Differences in dispersal capacities on the size of latitudinal and altitudinal ranges also needs to be considered. The pattern uncovered in the genus *Senecio* in high elevation habitats in the South American Andes needs to be tested across a wide range of taxa in order to determine whether it is a general macroecological pattern.

ACKNOWLEDGMENTS

Research supported by Contracts ICM P05-002 and PFB-23, Conicyt-Chile to the Instituto de Ecología y Biodiversidad (IEB) and an IEB Postdoctoral fellowship to LD. We thank Ana María Humaña, Soledad Muñoz, Daniela Dominquez, and José Montalve for their help in downloading records from TROPICOS.

REFERENCES

Arroyo, M.T.K., L.A. Cavieres, A. Peñaloza, and M. Arroyo-Kalin. 2003. Positive association between the cushion plant *Azorella monantha* (Apiaceae) and alpine plant species in the Chilean Patagonian Andes. *Plant Ecology* 169:121–29.

Arroyo, M.T.K., R. Primack, and J.J. Armesto. 1982. Community studies in pollination ecology in the high temperate Andes of Central Chile. I. Pollination mechanisms and altitudinal variation. *American Journal of Botany* 69:82–97.

Arroyo, M.T.K., and F.A. Squeo. 1990. Relationship between plant breeding systems and pollination. In *Biological Approaches and Evolutionary Trends in Plants*, pp. 205–227. Kawano, S., ed. London: Academic Press.

Arroyo, M.T.K., F.A. Squeo, J.J. Armesto, and C. Villagrán. 1988. Effects of aridity on plant diversity in the northern Chile Andes: Results of a natural experiment. *Annals of the Missouri Botanical Garden* 75:55–78.

Bell, C.D., and M.J. Donoghue. 2005. Phylogeny and biogeography of Valerianaceae (Dipsacales) with special reference to the South American valerians. *Organisms, Diversity and Evolution* 5:147–59.

Braun, G., J. Mutke, A. Reder, and W. Barthlott, W. 2002. Biotope patterns, phytodiversity and forestline in the Andes, based on GIS and remote sensing data. In *Mountain Biodiversity A Global Assessment*, pp. 75–101. Körner, C., and E.M. Spehn, eds. New York, Parthenon Publishing.

Brown, J.H., and B.A. Mauer. 1989. Macroecology: The division of food and space among species and communities. *Science* 243:1145–50.

Callaway, R.M., et al. 2002. Positive interactions among alpine plants increase with stress. *Nature* 417:844–48.

Cavieres, L.A., and M.T.K. Arroyo. 2001. Persistent soil seed banks in *Phacelia secunda* (Hydrophyllaceae): Experimental detection of variation along an altitudinal gradient in the Andes of Central Chile. *Journal of Ecology* 89:31–39.

Cavieres, L.A., E.I. Badano, A. Sierra-Almeida, S. Gómez-González, and M.A. Molina-Montenegro. 2006. Positive interactions between alpine plants species and the nurse cushion plant *Laretia acaulis* do not increase with elevation in the Andes of central Chile. *New Phytologist* 169:59–69.

Chown, S.L., B.J. Sinclair, H.P. Leinaas, and K.J. Gaston. 2008. Hemispheric asymmetries in biodiversity—a serious matter in ecology. *PloS Biology* 2:11 (published online).

Fenner, M., W.G. Lee, and E.H. Pinn. 2001. Reproductive features of *Celmisia* species (Asteraceae) in relation to altitude and geographical range in New Zealand. *Biological Journal of the Linnean Society* 74:51–58.

Gorelick, R. 2008. Species richness and the analytic geometry of latitudinal and altitudinal gradients. *Acta Biotheoretica*. Published online March 18, 2008.

Holderegger, R., and R.J. Abbott. 2003. Phylogeography of the Arctic-alpine *Saxifraga oppositifolia* (Saxifragaceae) and some related taxa based on cpDNA and ITS sequence variation. *American Journal of Botany* 90:931–36.

Körner, C. 2000. Why are there global gradients in species richness? Mountains might hold the answer. *Trends in Ecology & Evolution* 15:513–14.

Körner, C. 2003. *Alpine Plant Life: Functional Plant Ecology of High Mountain Ecosystems*, 2nd ed. Berlin: Springer.

Körner, C. 2007. The use of "altitude" in ecological research. *Trends in Ecology & Evolution* 22:569–74.

Körner C., M. Donoghue, T. Fabbro, C. Häuser, D. Nogués-Bravo, M.T.K. Arroyo, J. Soberon, L. Speers, E.M. Spehn, H. Sun, A. Tribsch, P. Tykarski, and N. Zbinden. 2007. Creative use of mountain biodiversity databases: The Kazbegi Research Agenda of GMBA-DIVERSITAS. *Mountain Research and Development* 27:276–81.

Nogúes-Bravo D., M.B. Araújo, T. Romdal, and C. Rahbek. 2008. Scale effects and human impact on the elevational species richness gradient. *Nature* 453:216–19.

Romdal, T. S., and J.A. Grytnes. 2007. An indirect area effect on elevational species richness patterns. *Ecography* 30:440–48.

Rozzi, R., M.T.K. Arroyo, and J.J. Armesto. 1997. Ecological factors affecting gene flow between populations of *Anarthrophyllum cumingii* (Papilionaceae) growing on equatorial and polar-facing slopes in the Andes of Central Chile. *Plant Ecology* 132:171–79.

Schönswetter, P., A. Tribsch and H. Niklfeld. 2003. Phylogeography of the high alpine cushion plant *Andosace alpina* (Primulaceae) in the European Alps. *Plant Biology* 5:623–30.

Squeo, F.A., H. Veit, G. Arancio, J.R. Gutiérrez, M.T.K. Arroyo, and N. Olivares. 1993. Spatial heterogeneity of high mountain vegetation in the Andean desert zone of Chile (30°ES). *Mountain Research and Development* 13:203–09.

Stevens, G.C. 1989. The latitudinal gradient in geographical range: How so many species coexist in the tropics. *American Naturalist* 133:240–56.

Thompson, K., K.J. Gaston, and S.T. Band. 1999. Range size, dispersal and niche breadth in the herbaceous flora of central England. *Journal of Ecology* 97:150–55.

Von Hagen, K.B., and J.W. Kadereit. 2000. The diversification of *Halenia* (Gentianaceae): Ecological opportunity versus key innovation. *Evolution* 57:2507–18.

6 Exploring Patterns of Plant Diversity in China's Mountains

Jingyun Fang, Xiangping Wang, Zhiyao Tang, Zehao Shen, and Chengyang Zheng

CONTENTS

Introduction.....39
Aims of the Plant Species Diversity Survey for China's Mountain Project.....40
Methods.....42
Study Sites and Design.....42
Nested Sampling Design.....42
Data Collection.....42
Database.....43
Data Analysis.....43
First Results of the Project.....44
Latitudinal Pattern of Species Richness at Plot-Level.....44
Altitudinal Pattern of Species Richness.....44
Relationship between Species Richness and Climate.....45
Conclusions.....46
Summary.....46
Acknowledgments.....46
References.....46

INTRODUCTION

China, one of the world's "megabiodiversity countries" (McNeely et al., 1990), is home to more than 30,000 vascular plants and 6,300 vertebrate species (Chen, 1998; Tang et al., 2006). The mountain areas, which account for ca. two-thirds of the total country area and span from tropical to boreal zones and from perhumid to hyper-arid climate (Figure 6.1), are the key for maintaining this rich diversity. They have not only acted as essential refuges to numerous species during historical glaciations, as evolution centers for some taxa, and paths for tropic species dispersal, but also provided great habitat heterogeneity for the coexistence of a variety of species. These features have been suggested as major causes for the significantly higher diversity in East Asia than in Europe and North America (Guo, 1999; Qian et al., 2007), and these regions also are recognized as a critical field to quantify the relative roles of history and contemporary climate in determining broad-scale biodiversity patterns (e.g., Ricklefs et al., 2004).

Many authors have studied the patterns of plant diversity in mountain areas in China (e.g., Fang et al., 1996; Liu, 1997; Tang and Ohsawa, 1997; Jiang et al., 2000; Wang et al., 2002). However, most of these studies focused on the altitudinal pattern of species diversity for either a single or a few mountains, and often did not apply consistent methods and sampling designs for their surveys. Therefore, the data obtained from these previous studies are often difficult to compare and to upscale to quantify large-scale diversity patterns of plants.

In order to explore geographic patterns of plant species diversity and the underlying mechanisms for China's mountains, the Department of Ecology, Peking University, has launched a multiyear project, Peking University Plant Species Diversity Survey for China's Mountains

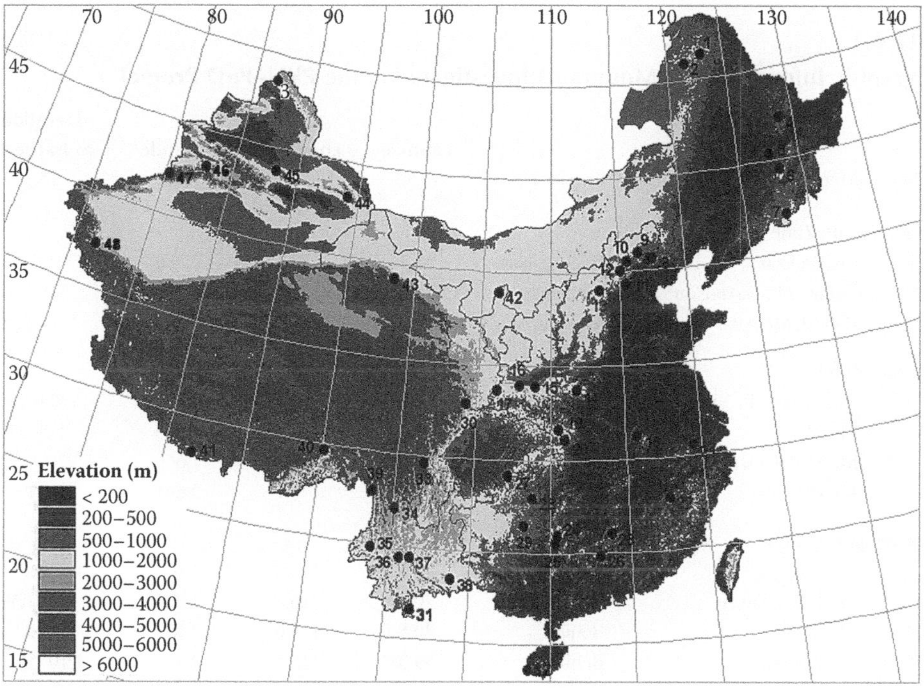

FIGURE 6.1 Study sites of the PKU–PSD project. For names and geographic information of the mountains, see Table 6.1.

(abbreviated as PKU–PSD), commencing in the mid-1990s. Over the past fifteen years, the PKU–PSD project has surveyed plant communities in more than fifty mountains across the country. In this chapter, we present a brief overview on the aims, methods, sampling sites, and database developments of the project, and provide three examples of the preliminary findings from the project using the electronic archive data.

AIMS OF THE PLANT SPECIES DIVERSITY SURVEY FOR CHINA'S MOUNTAIN PROJECT

The PKU–PSD project is designed to investigate species composition, structures, growth, and environment of plant communities and ecological traits of dominant species along altitudinal gradients in typical mountains across the country, using consistent sampling protocols. At the same time, information on flora, climate and topography, and remote sensing data for the mountain areas also are collected. Using these data sets at plot, local, and regional levels, the project is exploring the following issues:

1. Altitudinal patterns of species composition, structure, and species diversity of plant communities in each mountain.
2. Geographical differentiations in these altitudinal patterns across the country.
3. Latitudinal and altitudinal patterns of plant diversity and their environmental controls.
4. Relationship between community diversity and regional species pool.
5. Changes in species–area relationship and their possible connections with regional species pools, geological history, regional processes, and human dimensions.
6. Numerical characteristics of major vegetation types (e.g., boreal, temperate, and subtropic forests) in China's mountains.
7. Ecophysiological traits of dominant species for the representative mountains.
8. A test of several hypotheses on the broad-scale biodiversity patterns, e.g., the Rapoport's rule (Stevens, 1989; Stevens, 1992), the water–energy dynamic hypothesis (O'Brien 2006), the metabolic theory (Allen et al., 2002; Brown et al., 2004), the species pool hypothesis (Zobel, 1997), and the middle domain effect (Colwell and Hurtt, 1994; Colwell and Lees, 2000).

TABLE 6.1
Geographic Information of Mountains Investigated in the PKU–PSD Project

Number	Mountain	Province	Latitude (° N)	Longitude (° E)	Altitude (m)	Elevation of Transects (m)
Cool-Temperate Zone						
1	Mt. Baikalu	Heilongjiang	51.3	123.1	1,460	701–1,114
2	Genhe, Mt. Daxing'an	Neimenggu	50.9	121.5	1,142	815–839
3	Kanasi, Mt. Altai	Xinjiang	48.5	86.8	4,374	970–2,538
Temperate Zone						
4	Liangshui, Mt. Xiaoxin'an	Heilongjiang	47.2	128.9	707	344–466
5	Mt. Mao'er	Heilongjiang	45.3	127.6	805	350–420
6	Mt. Datudingzi	Jilin	44.4	128.2	1,669	970–1,524
7	Mt. Changbai	Jilin	42.0	128.1	2,691	530–1,945
Warm-Temperate Zone						
8	Mt. Wuling	Hebei	40.6	117.6	2,116	1,070–2,095
9	Mt. Labagoumen	Beijing	40.9	116.5	1,935	876–1,695
10	Mt. Song	Beijing	40.5	115.6	2,199	592–1,468
11	Mt. Dongling	Beijing	39.3	115.5	2,303	1,107–1,302
12	Mt. Xiaowutai	Hebei	40.0	115.2	2,882	1,250–2,850
13	Mt. Wutai	Shanxi	39.0	113.5	3,058	2,520–2,751
14	Mt. Laojun	Henan	33.7	111.7	2,192	800–1,800
15	Mt. Niubeiliang	Shannxi	33.9	108.9	2,802	1,430–2,800
16	Mt. Taibai	Shannxi	34.0	107.8	3,767	1,200–3,650
17	Mt. Xiaolong	Gansu	33.7	106.3	2,446	700–2,440
Subtropic Zone						
18	Mt. Dabie	Anhui	31.2	115.7	1,729	160–1,720
19	Mt. Shennongjia	Hubei	31.6	110.5	3,105	650–3,100
20	Mt. Dalao	Hubei	31.1	110.9	2,005	1,320–2,000
21	Mt. Xitainmu	Zhejiang	30.5	119.4	1,787	610–1,495
22	Mt. Wuyi	Fujian	27.8	117.6	2,158	600–2,150
23	Mt. Bamian	Hunan	26.0	113.8	2,042	650–2,040
24	Mt. Maoer	Guangxi	25.9	110.4	2,142	660–2,107
25	Mt. Dupang	Hunan	25.5	110.3	1,856	617–1,856
26	Mt. Mang	Hunan	24.8	113.1	1,902	610–1,895
27	Mt. Jinfo	Sichuan	29.1	107.2	2,251	650–2,240
28	Mt. Fanjing	Guizhou	27.9	108.8	2,494	1,000–1,850
29	Mt. Leigong	Guizhou	26.4	108.3	2,179	1,300
30	Mt. Motian	Gansu	32.9	104.3	4,072	670–3,400
Tropic Zone						
31	Xishuangbanna	Yunnan	21.6	101.6	1,250	660–1,250
32	Mt. Jianfeng	Hainan	18.5	108.7	1,412	790
The Tibet and Hengduan Mountains Zone						
33	Mt. Gongga	Sichuan	29.6	101.8	7,556	1,200–4,500
34	Mt. Yulong	Yunnan	27.0	100.1	5,596	2,722–4,190
35	Mt. Gaoligong	Yunnan	24.8	98.8	3,374	1,840–2,370
36	Mt. Wuliang	Yunnan	24.4	100.7	3,306	1,389–2,800

(continued)

TABLE 6.1
Geographic Information of Mountains Investigated in the PKU–PSD Project (continued)

Number	Mountain	Province	Latitude (° N)	Longitude (° E)	Altitude (m)	Elevation of Transects (m)
37	Mt. Ailao	Yunnan	24.4	101.3	3,166	1,830–2,520
38	Mt. Bozhu	Yunnan	23.4	103.9	2,409	2,312–2,997
39	Mt. Gong	Yunnan	27.8	98.6	5,128	2,025–3,630
40	Mt. Nanjiabawa	Xizang	29.6	95.1	7,782	750–4,650
41	Mt. Pushila	Xizang	28.1	86.7	5,425	5,176–5,390
The Arid Zone of Northwest China						
42	Mt. Helan	Ningxia	38.9	106.2	3,554	1,236–3,402
43	Mt. Qilian	Gansu	39.2	98.5	5,547	2856–3,355
44	Mt. Tuomu'erti	Xinjiang	43.1	94.1	4,886	2,210–2,805
45	Mt. Bogeda	Xinjiang	43.8	88.3	5,445	1,730–2,430
46	Mt. Xitianshan	Xinjiang	43.0	82.9	5,068	2,050–2,520
47	Mt. Tuomu'er	Xinjiang	42.1	80.3	7,435	2,280–3,026
48	Mt. Kunlun	Xinjiang	37.3	76.6	7,719	2,773–3,418

METHODS

Study Sites and Design

More than fifty mountains across all biogeographic regions of China have been studied over the past fifteen years. They belong to seven climatic and physiographical zones, with five zones in the humid east part and two zones in the west part (Figure 6.1). Table 6.1 shows the geographic information on these mountains.

In each mountain, we established one to several transects (along different slopes) from low to high altitudes to cover all the vegetation types (see Table 6.1). Plots are located along the transects at an elevational interval of 50–100 m. The plots were chosen to include ecologically homogeneous and physiognomically representative zonal vegetation, without indications of remarkable recent disturbance. Ravines, rock faces, and heterogeneous topographic habitats were avoided.

Nested Sampling Design

The plot size for forest vegetation is 600 m^2 (20 × 30 m), consisting of six 10 × 10 m subplots. In each plot, one of the six subplots is randomly selected for the study of the shrub layer. Five 1 × 1 m quadrates located at the four corners and the center of the plots are used for herb layer investigation. For the scrub and grassland, the plot size is 10 × 10 m.

For mountains with a clear altitudinal zonation of the vegetation, the species–area relationship was studied in nested quadrates for each vegetation zone. In each plot, we recorded all vascular plant species in nested quadrates of different sizes: 0.25 m^2 (0.5 × 0.5 m), 0.5 m^2 (0.5 × 1 m), 1 m^2 (1 × 1 m), 2 m^2 (2 × 1 m), 4 m^2 (2 × 2 m), etc., up to 2,000–5,000 m^2 (the maximum area of the nested quadrates depended on the vegetation types).

Data Collection

Latitude, longitude, altitude, aspect (degree to real North), inclination, and position on slope were recorded for each plot. Indications of human disturbances also were recorded. All species of a plot (20 × 30 m) were recorded to obtain number of species per plot. Species were identified *in situ* by experts. For plants that could not be identified in the field, specimens were collected and identified by relevant experts. In forested plots, we recorded species and individual tree diameters at breast height (DBH, breast height = 1.3 m) of all stems with DBH ≥ 3 cm, and tree height was measured for at least one-third of trees. In the shrub layer species, name, abundance, cover, and mean height were recorded for each species and quadrate (for herbs in the five 1 m^2 quadrates).

Soil profiles were investigated at an altitudinal interval of 100 to 200 m in some of the mountains. For each profile, soils were described *in situ* for each horizon, and soil samples were collected for 0 to 10 cm, 10 to 20 cm,

20 to 40 cm, 40 to 60 cm, 60 to 80 cm, 80 to 100 cm, and ≥ 100 cm soil depth in order to assess soil moisture, texture, bulk density, and organic carbon and nitrogen concentration in the laboratory.

The local climate was recorded in some of the mountains with a large altitudinal range. We measured air temperature at 1 to 2 m high above ground in half-hour intervals for at least one year using HOBO microloggers (Onset Computer Corp.) at an elevation interval of 100 to 200 m. The microloggers were mounted on tree trunks in closed canopy stands (e.g., Tang and Fang, 2006).

Database

We developed a database which documents the following information for each plot, including:

1. Geographical coordinates, indicators of local topography, soil variables (soil type, soil depth and moisture, etc.), and variables describing human disturbance.
2. Climatic variables. For each plot, monthly mean temperature and precipitation were estimated with linear models, which were developed for different regions using latitude, longitude, and altitude as predictors (Fang et al., 1996). The climate data used are from thirty-year averaged climatic records from ca. 700 climate stations across China. A comparison of the predictions of the models and the observed climatic data (from climate stations or HOBO microloggers in the plots) showed a good match between the estimated and predicted temperature and precipitation (e.g., Tang and Fang, 2006; Wang et al., 2008). Several climatic indices (e.g., annual potential and actual evapotranspiration) also were calculated for each plot based on the estimated monthly mean temperature and precipitation.
3. Species diversity data sampled in the field as described previously.

Data Analysis

Focusing on the following issues, we explored plant diversity patterns for China's mountains using the data set developed in the project, together with local flora and corresponding environmental data compiled from the literature.

1. Differences in altitudinal diversity patterns among mountains and the underlying drivers. Climatic drivers, such as moisture availability and temperature, human disturbances (often more frequent in the lowlands or at mid-altitude), and the scale of analysis (grain and extent), are commonly suggested as underlying mechanisms for the differences in altitudinal patterns among mountains (e.g., Rahbek, 2005; McCain, 2007; Nogués-Bravo et al., 2008). The data collected in this project have provided an excellent opportunity for examining why altitudinal patterns of alpha diversity differ among mountains and regions. Our first results clearly indicate that vegetation structure (plant functional group) also is, on top of other drivers, an important factor affecting altitudinal gradients of species richness at the plot scale (Figure 6.3).
2. Comparison of latitudinal gradients of species richness at a similar altitude (e.g., Figure 6.2) or within the same vegetation zone. Until now, it remains unclear whether alpha diversity changes along latitudinal gradients in the same way at different altitudes or different vegetation zones (e.g., alpine and subalpine zones). Another unrevealed question is if changes in alpha diversity along latitudinal gradients are different between humid and arid parts of China. Our data can be used to quantify these questions, and the potential abiotic and biotic drivers of these relationships.
3. Patterns of mountain diversity in China across both latitudinal and altitudinal gradients. Many studies have analyzed the large-scale patterns of species richness; however, most of these studies used gamma diversity (i.e., species richness at a regional scale) data within large grain sizes. Our data set offers detailed information on alpha diversity, which is one of very few data sets of this type worldwide (e.g., Gentry, 1988), and thus provides a unique opportunity for exploring geographic diversity gradients in the hyperdiverse ecoregion of East Asia. Our data set also can be used to examine the relative role of history versus climate in shaping the diversity patterns and to compare species richness between East Asia and North America at different scales (Ricklefs et al., 2004).
4. Changes in species–area relationship along gradients of latitude and altitude. The nested plot design yields a specific species–area relationship for each vertical vegetation zone per mountain. These nested plots, together with data of the regional species pool, can be used to analyze the species–area relationship in relation to

possible environmental drivers, and assess the effects of species pool and geological history on community diversity.

FIRST RESULTS OF THE PROJECT

Latitudinal Pattern of Species Richness at Plot-Level

The latitudinal pattern of species richness has long been a focus in biodiversity study (e.g., Gaston, 2000; Hillebrand, 2004), and many mechanisms, such as explanations based on climate (Brown et al., 2004; O'Brien, 2006), history (Ricklefs, 2006), and geometric constraints (Colwell and Lees, 2000), have been proposed. Nevertheless, the latitudinal gradients in plant richness have rarely been examined with plot data in East Asia. In this chapter, to avoid the nonelevation specific effect of water availability (Körner, 2007), we used data collected from the humid eastern part of China to explore the latitudinal patterns of species diversity. Our result showed that woody species richness decreases with increasing latitude at similar altitudes (800–1400 m), with a rate of 1.2 species/plot (20 x 30 m) per degree of latitude (Figure 6.2). This negative relationship between species richness and latitude is widely reported from other regions of the world (e.g., Stevens, 1989; Gaston, 2000; Hillebrand, 2004), but most of these previous studies were based on gamma diversity within geographic grids (but see Gentry, 1988). Our study sites span a latitudinal range from 18° N (Mt. Jianfengling, Hainan) to 53° N (Mt. Baikalu, Heilengjiang) and provide new support for the latitudinal gradient of species diversity at community level. However, this is just a preliminary result, and further examinations are needed for different elevations and for the arid west region of China.

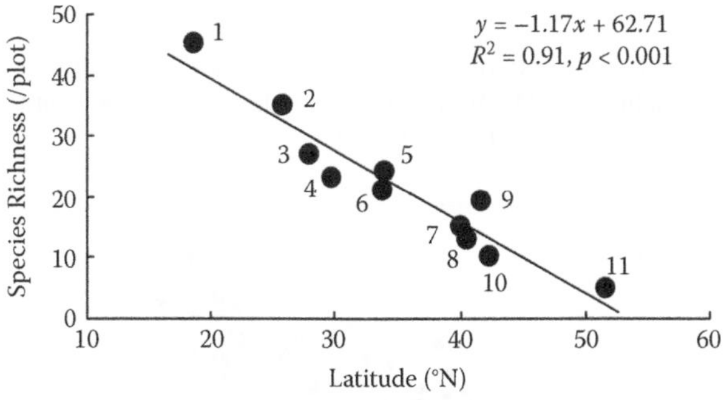

FIGURE 6.2 Latitudinal change in woody species richness for major mountains in humid east part of China. Data showed are mean values for plots (20 x 30 m) between 800 to 1,400 m in each mountain (1, Mt. Jianfengling; 2, Mt. Maoer; 3, Mt. Wuyi; 4, Mt. Gongga; 5, Mt. Taibai; 6, Mt. Niubeiliang; 7, Mt. Xiaowutai; 8, Mt. Songshan; 9, Mt. Changbai; 10, Mt. Datuding; 11, Mt. Baikalu). For detailed information on these mountains, see Table 6.1.

Altitudinal Pattern of Species Richness

Altitudinal gradients have been suggested to mirror latitudinal gradient, which provides another excellent way to test various hypotheses on geographic diversity patterns and to quantify the effects of spatial scales (Rahbek, 2005; Nogués-Bravo et al., 2008). Our plot data showed that woody species richness decreases with increasing altitude for mountains in the humid eastern part of China, e.g., Mt. Changbai (Figure 6.3) and Mt. Taibai (Figure 6.4),

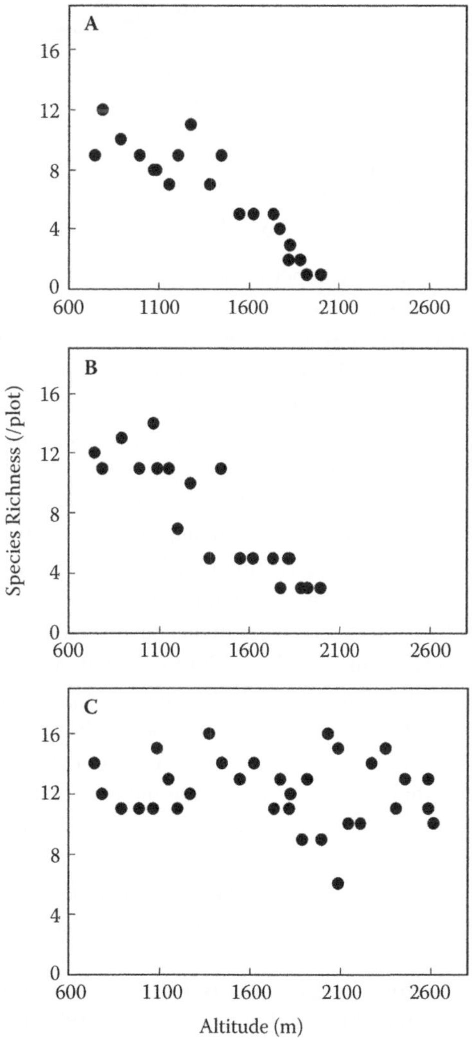

FIGURE 6.3 Patterns of species richness along altitudinal gradient on the northern slope of Mt. Changbai, Jilin, northeast China. (A) tree layer, (B) shrub layer, and (C) herb layer.

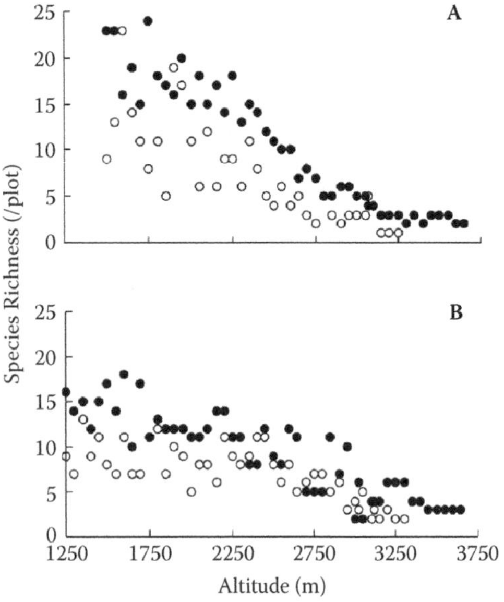

FIGURE 6.4 Patterns of woody species richness along altitudinal gradients of Mt. Taibai, central China. (A) southern slope, (B) northern slope. ○: tree layer, ●: shrub layer.

consistent with that reported for other humid mountains of the world (e.g., Stevens, 1992; Vazquez and Givnish, 1998; Aiba and Kitayama, 1999; Odland and Birks, 1999). However, in Mt. Wuyi (Fujian, southeast China), which underwent significant transformation by human disturbance at low altitude, not unexpectedly, tree species richness shows a mid-elevation peak (Figure 6.5). In addition, the northern slope of Mt. Taibai, Qingling Range, showed a much less pronounced elevational gradient of diversity than the southern slope (Figure 6.4). This difference could not be accounted for by temperature, because the temperature in the former is actually higher than the latter at the low altitudes (Tang and Fang, 2006). This may either be associated with differences in the regional species pools or be related to differences in land use on the two slopes (Tang et al., 2004), awaiting a more detailed analysis.

Altitudinal patterns of plant community richness may be quite different among different plant growth forms (Figures 6.3 and 6.4). For instance, species richness of herbaceous plants does not show a clear altitudinal pattern in Mt. Changbai (Figure 6.3C), which may be more related to local environmental factors, such as canopy openness. However, gamma diversity in Mt. Gaoligong showed similar altitudinal trends for tree, shrub, and herb species (Wang et al., 2007). These two cases suggest that the alpha and gamma diversity may be controlled by different processes, and thus comparisons among ecological groups at different scales (and among mountains) are required for better understanding of altitudinal diversity patterns (Lomolino, 2001; Grytnes, 2003; Rahbek, 2005).

Relationship between Species Richness and Climate

Many authors have pointed out that species richness is often closely correlated with climate, and that the relative importance of energy versus water availability varies geographically, and water limitation is more important in the low than the high latitudes (O'Brien, 1993; Hawkins et al., 2003; Kreft and Jetz, 2007). Because water availability gradients are not confined to altitudinal (or latitudinal) gradients, but can occur anywhere, they may regionally confound environmental gradients typical for all mountains, such as temperature, atmospheric pressure, or available land area (Körner, 2007). Under the cold and humid conditions, the woody plant species richness of northeast China is closely associated with energy availability (another expression of warmth, Figure 6.6). This is consistent with the species–energy hypothesis (Wright,

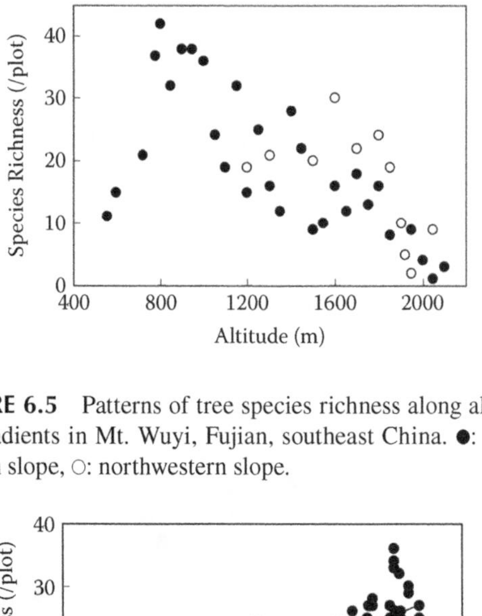

FIGURE 6.5 Patterns of tree species richness along altitudinal gradients in Mt. Wuyi, Fujian, southeast China. ●: southeastern slope, ○: northwestern slope.

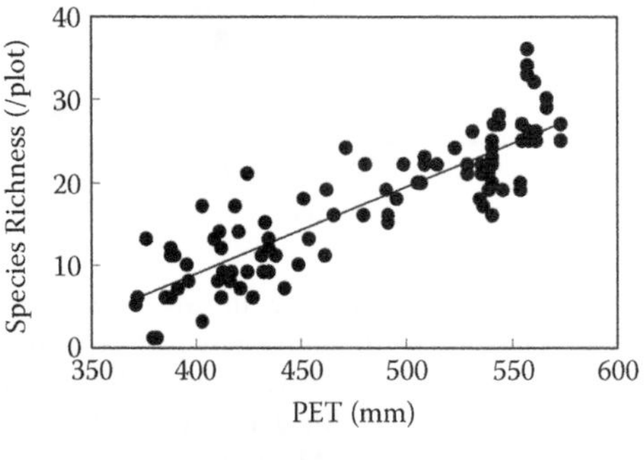

FIGURE 6.6 Relationship between woody species richness and annual potential evapotranspiration (PET) in forests in northeast China.

1983) and supports that energy availability constrains species richness in the high latitudes (Hawkins et al., 2003), although it is unresolved what the actual mechanism is, most likely temperature via its influence on season length and meristematic activity (Körner, 2003).

CONCLUSIONS

The database developed by the PKU–PSD project is a huge data set for the broad-scale plant diversity patterns in East Asia. Our first results have suggested the great potential of this database in exploring the effects of various factors (e.g., climate, functional groups, geometric constraints, and history) on altitudinal richness patterns at different scales, and to compare the latitudinal diversity gradients among regions and vertical vegetation zones. By combining the data sets for community diversity and regional diversity, as well as the species–area relationships obtained in this project, it also is possible for us to examine how community diversity was limited by regional species pool and to examine the possible factors regulating this process (e.g., productivity). At the same time, the database provides a good opportunity to compare diversity patterns between East Asia and North America at plot-scale, and to test several important hypotheses on geographic diversity patterns, which have rarely been tested with data in mountains across China.

SUMMARY

Understanding the mechanisms underlying geographic diversity patterns is a central issue in broad-scale ecology and biogeography. In this chapter, we described the experimental design, sampling protocol, and the database developed for the PKU–PSD project, which aimed to examine the overall patterns of vascular plant diversity in mountains across China. Over the past fifteen years, the project has surveyed plant communities in more than fifty mountains across the country using consistent sampling protocols. The first results of the project showed that species richness at community scale decreased with increasing latitude in humid East China, which may be largely a result of decreasing energy availability. However, the altitudinal richness patterns differed remarkably among mountains (and different slopes within a specific mountain), functional groups, and spatial scales. These differences suggested that systematical comparisons of diversity gradients among regions and functional groups at different scales are necessary for a better understanding of the underlying mechanisms. Consequently, a database including diversity data, environmental factors, biotic factors (e.g., functional groups), and variables describing evolution history (e.g., species pool) and human disturbance is the base for such analyses.

ACKNOWLEDGMENTS

Many students in the Terrestrial Ecology Group of Peking University have participated in field work for this project. We thank A.P. Chen, T. Gu, W.X. Han, Z.L. Liu, S.L. Piao, H.H. Shen, Z.H. Wang, X.P. Wu, L. Zhang, S.Q. Zhao, B. Zhu, and many others for assistance in data collection. The PKU–PSD project has been supported by the National Natural Science Foundation of China (#40638039, 40228001, 90711002, 90211016, 49971002, 39425003).

REFERENCES

Aiba, S.I., and K. Kitayama. 1999. Structure, composition and species diversity in an altitude–substrate matrix of rain forest tree communities on Mount Kinabalu, Borneo. *Plant Ecology* 140:139–57.

Allen, A.P., J.H. Brown, and J.F. Gillooly. 2002. Global biodiversity, biochemical kinetics, and the energetic-equivalence rule. *Science* 297:1545–48.

Brown, J.H., J.F. Gillooly, A.P. Allen, V.M. Savage, and G.B. West. 2004. Toward a metabolic theory of ecology. *Ecology* 85:1771–89.

Chen, C.D., ed. 1998. *China's Biodiversity: A Country Study Organized by State Environmental Protection Administration*. Beijing: China Environmental Science Press.

Colwell, R.K., and G.C. Hurtt. 1994. Nonbiological gradients in species richness and a spurious Rapoport effect. *American Naturalist* 144:570–95.

Colwell, R.K., and D.C. Lees. 2000. The mid-domain effect: Geometric constraints on the geography of species richness. *Trends in Ecology & Evolution* 15:70–76.

Fang, J.Y., M. Ohsawa, and T. Kira. 1996. Vertical vegetation zones along 30° N latitude in humid East Asia. *Vegetatio* 126:135–49.

Gaston, K.J. 2000. Global patterns in biodiversity. *Nature* 405:220–27.

Gentry, A.H. 1988. Changes in plant community diversity and floristic composition on environmental and geographical gradients. *Annals of the Missouri Botanical Garden* 75:1–34.

Grytnes, J.A. 2003. Species-richness patterns of vascular plants along seven altitudinal transects in Norway. *Ecography* 26:291–300.

Guo, Q.F. 1999. Ecological comparisons between eastern Asia and North America: Historical and geographical perspectives. *Journal of Biogeography* 26:199–206.

Hawkins, B.A., R. Field, H.V. Cornell, D.J. Currie, J.F. Guegan, D.M. Kaufman, J.T. Kerr, G.G. Mittelbach, T. Oberdorff, E.M. O'Brien, E.E. Porter, and J.R.G. Turner. 2003. Energy, water, and broad-scale geographic patterns of species richness. *Ecology* 84:3105–17.

Hillebrand, H. 2004. On the generality of the latitudinal diversity gradient. *American Naturalist* 163:192–211.

Jiang, Y., M. Kang, S. Liu, L. Tian, and M. Lei. 2000. A study on the vegetation in the east side of Helan Mountain. *Plant Ecology* 149:119–130.

Körner, C. 2000. Why are there global gradients in species richness? Mountains might hold the answer. *Trends in Ecology & Evolution* 15:513–14.

Körner, C. 2003. *Alpine Plant Life*, 2nd ed. Heidelberg, Germany: Springer.

Körner, C. 2007. The use of "altitude" in ecological research. *Trends in Ecology & Evolution* 22:569–74.

Kreft, H., and W. Jetz. 2007. Global patterns and determinants of vascular plant diversity. *PNAS* 104:5925–30.

Liu, Q.J. 1997. Structure and dynamics of the subalpine coniferous forest on Changbai mountain, China. *Plant Ecology* 132:97–105.

Lomolino, M.V. 2001. Elevation gradients of species-density: Historical and prospective views. *Global Ecology & Biogeography* 10:3–13.

McCain, C.M. 2007. Could temperature and water availability drive elevational species richness patterns? A global case study for bats. *Global Ecology & Biogeography* 16:1–13.

McNeely, J.A., K.R. Miller, W.V. Reid, R.A. Mittermeir, and T.B. Werner. 1990. *Conserving the World's Biological Diversity*. IUCN, Gland, Switzerland.

Nogués-Bravo, D., M.B. Araújo, T. Romdal, and C. Rahbek. 2008. Scale effects and human impact on the elevational species richness gradients. *Nature* 453:216–20.

O'Brien, E.M. 1993. Climatic gradients in woody plant species richness: Towards an explanation based on an analysis of Southern Africa's woody flora. *Journal of Biogeography* 20:181–98.

O'Brien, E.M. 2006. Biological relativity to water–energy dynamics. *Journal of Biogeography* 33:1868–88.

Odland, A., and H.J.B. Birks. 1999. The altitudinal gradient of vascular plant richness in Aurland, western Norway. *Ecography* 22:548–66.

Qian, H., P.S. White, and J.-S. Song, 2007. Effects of regional vs. ecological factors on plant species richness: An intercontinental analysis. *Ecology* 88:1440–53.

Rahbek, C. 2005. The role of spatial scale and the perception of large-scale species-richness patterns. *Ecology Letters* 8:224–39.

Ricklefs, R.E. 2006. Evolutionary diversification and the origin of the diversity–environment relationship. *Ecology* 87:S3–S13.

Ricklefs, R.E., H. Qian, and P.S. White. 2004. The region effect on mesoscale plant species richness between eastern Asia and eastern North America. *Ecography* 27:129–36.

Stevens, G.C. 1989. The latitudinal gradient in geographical range: How so many species coexist in the tropics. *American Naturalist* 133:240–56.

Stevens, G.C. 1992. The elevational gradient in altitudinal range: An extension of Rapoport's latitudinal rule to altitude. *American Naturalist* 140:893–911.

Tang, C.Q., and M. Ohsawa. 1997. Zonal transition of evergreen, deciduous, and coniferous forests along the altitudinal gradient on a humid subtropical mountain, Mt. Emei, Sichuan, China. *Plant Ecology* 133:63–78.

Tang, Z., and J. Fang. 2006. Temperature variation along the northern and southern slopes of Mt. Taibai, China. *Agricultural and Forest Meteorology* 139:200–207.

Tang, Z., J. Fang, and L. Zhang. 2004. Patterns of woody plant species diversity along environmental gradients on Mt. Taibai, Qinling Mountains. *Biodiversity Science* 12:115–22.

Tang, Z.Y., Z.H. Wang, C.Y. Zheng, and J.Y. Fang. 2006. Biodiversity in China's mountains. *Frontiers in Ecology and the Environment* 4:347–52.

Vazquez, J.A.G., and T.J. Givnish. 1998. Altitudinal gradients in tropical forest composition, structure, and diversity in Sierra de Manantlan. *Journal of Ecology* 86:999–1020.

Wang, G., G. Zhou, L. Yang, and Z. Li. 2002. Distribution, species diversity and life-form spectra of plant communities along an altitudinal gradient in the northern slopes of Qilianshan Mountains, Gansu, China. *Plant Ecology* 165:169–81.

Wang, X.P., J.Y. Fang, and B. Zhu. 2008. Forest biomass and root-shoot allocation in northeast China. *Forest Ecology and Management* 255:4007–20.

Wang, Z., Z. Tang, and J. Fang. 2007. Altitudinal patterns of seed plant richness in the Gaoligong Mountains, south-east Tibet. China. Diversity and Distributions: 13:845–54.

Wright, D.H. 1983. Species–energy theory: An extension of species–area theory. *Oikos* 41:496–506.

Zobel, M. 1997. The relative role of species pools in determining plant species richness: An alternative explanation of species coexistence? *Trends in Ecology & Evolution* 12:266–69.

7 Elevational Pattern of Seed Plant Species Richness in the Hengduan Mountains, Southwest China

Area and Climate

Da-Cai Zhang and Hang Sun

CONTENTS

Introduction....49
Materials and Methods....50
  Study Area....50
  Source of Data....50
  Elevational Belts....50
  Area Data....50
  Climate Data....50
  Analysis of Data....51
Results....52
Discussion....52
Summary....54
Acknowledgments....54
References....55

INTRODUCTION

Environmental and climatic variables change greatly along an elevational gradient within a short distance, and since mountains often are among the last wilderness areas, they permit unbiased approaches toward an understanding of natural gradients of species richness (MacArthur, 1972; Walter, 1979; Körner, 2000; Fang, 2004). Many different elevational patterns of species richness have been reported (see Sanders et al., Chapter 10), often with a species richness peak at mid-elevations (Rahbek, 1995, 2005). However, such gradients often are confounded with region specific moisture (Körner, 2007) or land use gradients (Nagués-Bravo, 2008), leading to a reduction of species richness at lower elevations, unrelated to elevation, per se. On top of such nonstrictly elevation-related drivers, the reduction of land area per elevational belt exerts an additional, often unaccounted, driver of species richness (MacArthur, 1972; Bachman et al., 2004; Körner, 2000, 2007). Land area may even show a mid-elevation peak as well (Fu et al., 2006; Wang et al., 2007). Therefore, the elevational patterns should account for all drivers and separate physically elevation-related from particular regional patterns on top of the thermal gradient (Bhattarai et al., 2004; Fu et al., 2006). We argue, because other information often is limited, that elevational gradients of species richness should at least be adjusted for area (Bachman et al., 2004). Here we ask how this land area adjustment will affect species richness patterns by using electronic databases of species distribution and topography (geographical information systems, or GIS). The relationship between land area and species richness is one of the few well-established ecological rules, with larger areas supporting more species (Schoener, 1976), so area is a good predictor of species richness and can explain a large proportion of variation in species richness along an elevational gradient (Körner, 2000, 2007; Sanders, 2002; Bachman et al., 2004; Jankowski and Weyhenmeyer, 2006). Specifically, we will examine whether area is a good predictor for the elevational patterns of species richness in the Hengduan Mountains using a large floristic database and geostatistics.

Climatic variables, such as potential evapotranspiration and rainfall, are important explanatory variables for elevational pattern of species richness (Bhattarai et al., 2004; Fu et al., 2006). The relationship between climate and elevational pattern of species richness, however, greatly varies among regions, and different climatic variables have different explanatory power. In this study, we also will examine how climatic variables influence the elevational pattern of species richness in the Hengduan Mountains.

The Hengduan Mountains region is recognized as being one of the biodiversity hot spots in the world (Boufford and Dijk, 2000; Myers et al., 2000; Boufford et al., 2004), and it offers one of the widest bioclimatic gradients, so this region is ideal to analyze elevational patterns of species richness. In this present study, the main aims are to describe the elevational patterns of total and endemic seed plant richness in the Hengduan Mountains, and analyze the effect of land area and climate on these patterns using GIS and a digital floristic database.

MATERIALS AND METHODS

Study Area

The Hengduan Mountains are located in southwest China, an area of 364,000 km^2 stretching from 24°40′–34°00′ N, and 96°20′–104°30′ E (Li, 1987; Figure 7.1). This region comprises mostly steep mountains, with narrow gorges and an alpine plain (Li, 1989). The vegetation can be broadly categorized into two types: to the south, the vegetation is subtropical, evergreen, broad-leaved forest with a mixture of evergreen and deciduous species; to the north, the vegetation is Qinghai–Tibetan plateau alpine vegetation (Yu et al., 1989; Figure 7.1). The coniferous forest covers an elevational range from 3,100 to 4,100 m, with treeline at ca. 4,100 m (Liu et al., 1984). The Hengduan Mountains are located entirely within subtropical latitudes, ranging from 25° to 32° latitude. We selected four weather stations to show climatic variation with latitude (two stations in the South at 25° to 27° latitude, and two in the North at 30° latitude) and also with altitude, as each of the pair of stations varied in altitude (South: 1,991 m and 3,276 m; North: 3,080 m and 4,014 m; Figure 7.1).

Source of Data

Our main source of data was a floristic database, "The Vascular Plants of the Hengduan Mountains" (Wang, 1993, 1994), available online data from biodiversity of the Hengduan Mountains, adjacent areas of south-central China (http://hengduan.huh.harvard.edu/fieldnotes), and additional collections from field trips by the authors.

Elevational Belts

The full range of elevations from 800 to 5,500 m was divided into forty-seven 100 m elevational belts. A species range was defined by records in every 100 m belt between its upper and lower elevation limits (Vetaas and Grytnes, 2002; Bhattarai et al., 2004; Bachman et al., 2004). The number of species in each 100 m elevational belt was used to measure species richness.

Area Data

We used a digital elevational model (DEM) to calculate the area at each 100 m elevational belt in the region using ESRI's Arcview 3.1. We interpolated the DEM in each 100 m elevational belt and rasterized at 1 × 1 km grid cells, and then counted the number of grid cells to arrive at total area per belt.

Climate Data

We used climate data of ninety weather stations from the China Meteorological Data Sharing Service System (http://cdc.cma.gov.cn) and the Chinese Natural Resources Database (http://www.data.ac.cn/zrzy/g03.asp). Mean annual biotemperature (MAT; biotemperature refers to all temperatures above freezing, with all temperatures below freezing adjusted to 0°C) showed a strong linear relationship with elevation ($R^2 = 0.87$, $P < 0.001$), and adiabatic lapse rate = 0.44°C/100 – m with increasing elevation, so linear regression can be used to estimate MAT for elevational belt lacking a weather station. We used the formula of Holdrige (1976) to calculate potential evapotranspiration (PET), PET = mean annual biotemperature × 58.93. Variation in mean annual rainfall (MAR) along the elevational gradient was complicated, and showed a weak relationship with elevation ($R^2 = 0.23$, $P < 0.01$). The

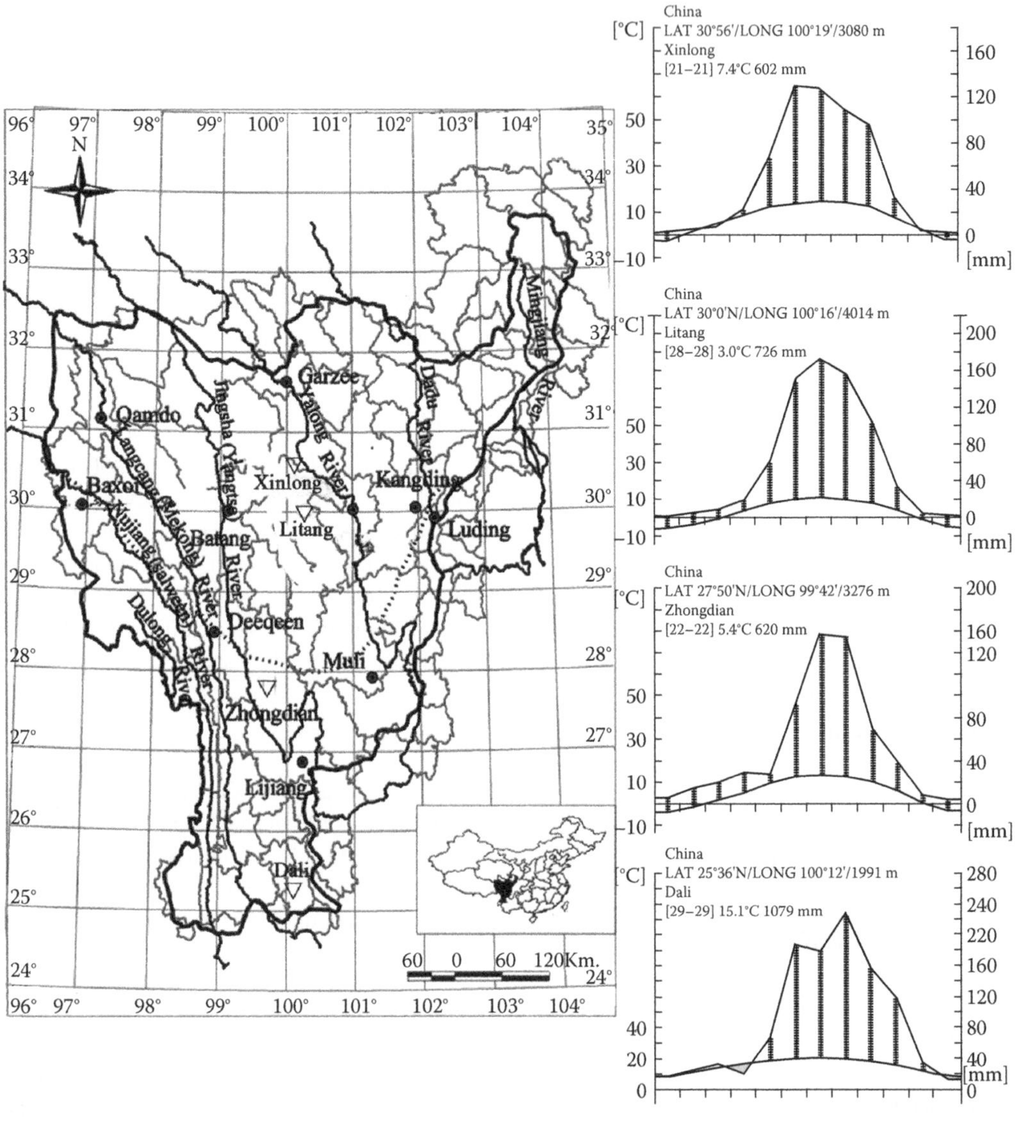

FIGURE 7.1 Location and climate of the Hengduan Mountains in southwest China. The broken line indicates the division of vegetation. Four weather stations were selected to show the variation of climate with latitude and elevation in this region.

Kriging algorithm was used to determine spatial interpolation of MAR using ESRI's Arcview 3.1 (Karnieli, 1990; Hofierka et al., 2002). Variation in PET and MAR along this elevational gradient was shown in Figure 7.2 (a,b).

Analysis of Data

Species density was used to eliminate effects of area on species richness by the following equation: Species density (D) = S / ln (A), where S was number of species, and A was area (Qian, 1998; Vetaas and Grytnes, 2002; Wang et al., 2007), and we used the relation S – ln (A) to analyze the relationship between species richness and area.

Ordinary least squares (OLS) and conditional autoregressive (CAR) models, performing in package SAM 2.0 (Rangel et al., 2006), were used to relate species richness and independent variables, and to analyze the multiple relationships between species richness and independent variables. Akaike's information criteria (AIC) were used to compare the fitness of models, with the smaller value of AIC indicating a better fit.

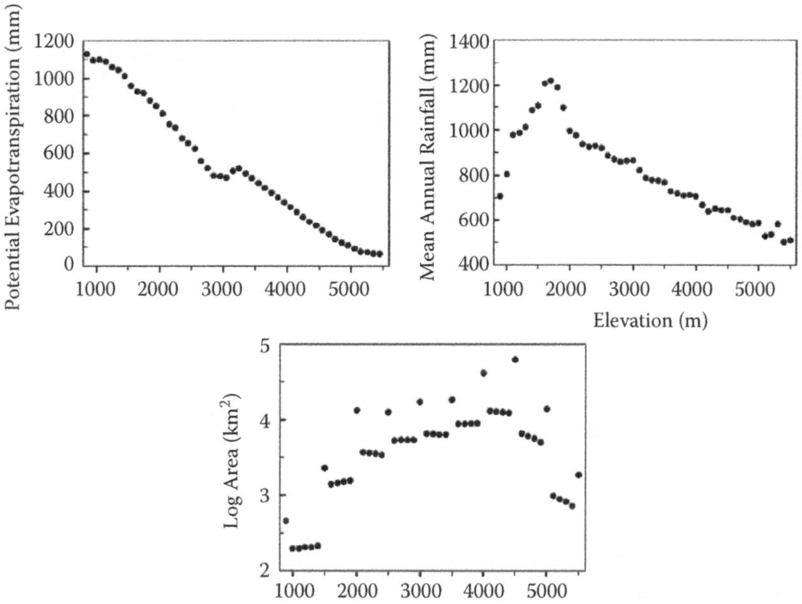

FIGURE 7.2 Topography and climate along an elevational gradient in the Hengduan Mountains. (a) Mean annual potential evapotranspiration (PET) obtained from the linear regression between biotemperature (PET = biotemperature × 58.93; Holdrige, 1976) and elevations. (b) Mean annual rainfall obtained from the Kriging interpolation of the observed values in the climatic stations. (c) Land area (km^2) per 100 m elevation belt (note the peak around 4,000 m).

A species was defined as endemic if its distribution was limited to the Hengduan Mountains region or extended only slightly into neighboring regions.

RESULTS

Variation of land area along the elevational gradient in the Hengduan Mountains turned out to be hump-shaped, with elevations between 4,000 and 4,500 m contributing the largest area (Figure 7.2c). The total number of seed plants in the Hengduan Mountains is 8,590, with 2,783 of these species (32.4%) being endemic to this region. Disregarding land area variation, the elevational pattern of total species richness is unimodal with elevations from 2,500 to 3,200 m showing the highest richness (Figure 7.3a). The distribution is skewed toward lower elevations, and there is a steep trend in species richness at both ends of the elevational gradient, namely a sharply increasing rate in species richness between 800 and 2,000 m (19.8 ± 13.8%) and reduction between 3,800 and 5,500 m (24.6 ± 11.0%).

The pattern of endemic species richness along the elevational gradient also is unimodal, with the frequency of endemics increasing toward higher elevations (Figure 7.3a). The maximum of endemic species richness occurs at elevations between 3,000 and 3,800 m, i.e., above the peak of total species richness. Endemic species richness changes at both ends of the elevational gradient at a faster rate than total richness, and it increases with elevation at a rate 23 ± 24% below 3,000 m, and then decreases at a rate 29 ± 15% above 3,900 m.

Variation of species density also is unimodal along the elevational gradient for both total and endemic species, and peaks at similar elevations as do species richness (Figure 7.3b).

Species richness is significantly correlated with land area per 100 m belt, and endemic richness shows an even stronger relationship with area than total richness; species richness, however, shows a weak relationship with PET and MAR (Table 7.1). According to the OLS model, all explanatory variables together explain more than half of the variation in species richness, and they explain more variation in total richness than endemic richness (Table 7.2). Among all explanatory variables, land area is the most important variable explaining the elevational pattern of species richness (Table 7.2).

DISCUSSION

Our database analysis illustrates that the elevational patterns of total and endemic richness in the Hengduan Mountains is unimodal (Figure 7.3a), which is a common

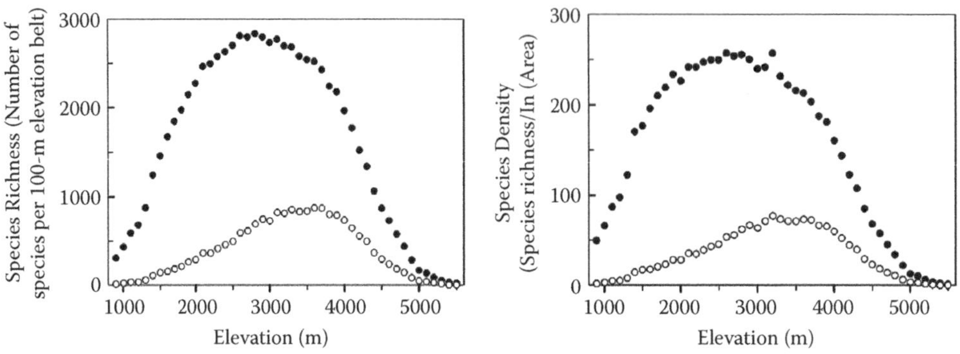

FIGURE 7.3 Elevational patterns of species richness (a) and species density (b) both presented to be unimodal in the Hengduan Mountains, and peaked at same elevations. ● = total species; ○ = endemic species.

TABLE 7.1
Relationship between Elevational Patterns of Species Richness and Explanatory Variables

		OLS Model			CAR Model		
		Area	PET	MAR	Area	PET	MAR
Total species	R^2	0.23*	0.09	0.24*	0.37	0.15	0.14
	AIC	645	653	644	639	653	654
Endemic species	R^2	0.45*	0.04	0.02	0.45**	0.39	0.26
	AIC	519	545	546	522	527	536

Note: *$p < 0.05$; **$p < 0.01$; AIC, Akaike's information criteria; OLS, ordinary least squares; CAR, conditional autoregressive; PET, potential evapotranspiration; MAR, mean annual rainfall.

TABLE 7.2
Multirelationship between Elevational Patterns of Species Richness and Explanatory Variables

	OLS model					CAR model				
			Standardized coefficient					Standardized coefficient		
	R^2	AIC	Area	MAR	MAT	R^2	AIC	Area	MAR	MAT
Total species	0.66	611	0.74*	0.29	0.46**	0.48	634	0.45*	0.19	0.29
Endemic species	0.52	516	0.81*	−0.22	0.50	0.44	529	0.60*	−0.31	0.35

Note: *$p < 0.05$; **$p < 0.01$; abbreviations expressed as in Table 7.1.

pattern of species richness along elevational gradients if covariables, such as land area availability and regional climatic gradients unspecific to elevation, remain unaccounted (Vetaas and Grytnes, 2002; Bhattarai et al., 2004; Krömer et al., 2005; Sánchez-González and López-Mata, 2005; Wang et al., 2007). Higher elevations are expected to have more endemic species because of the fragmented topography (Ibisch et al., 1996; Kessler et al., 2001), more isolated habitats and allopatric speciation (McGlone et al., 2001; Vetaas and Grytnes, 2002), low immigration rates of species (Lomolino, 2001), and younger evolutionary history (McGlone et al., 2001), so that endemic richness should become higher at higher elevations than total richness (Figure 7.3a). Two reasons may result in the steep variation in species richness at both ends of the elevational gradient. First, interpolation may lead to a steeper decrease in species richness toward both elevational ends (Grytnes and Vetaas, 2002), and second, the steeper clines in land area per 100 m steps of elevation at both high and low elevations (Figure 7.2c) leads to higher species–area ratios at a given diversity.

The relationship between land area and species richness is one of the few ecological rules that is well-established (MacArthur, 1972; Schoener, 1976), and the land areas in each 100 m elevational belt are not equal, therefore, the effect of area on species richness should be eliminated when comparing the elevational pattern of species richness per belt (Rahbek, 2005). In this study, elevational patterns of species richness were not changed when expressed per belt area, still showing peak at the same elevation as absolute diversity per belt (Figure 7.3b), which also was observed in the Gaoligong Mountains (Wang et al., 2007) and the Himalayan (Vetaas and Grytnes, 2002). In these mountains, the elevational variation of land area per belt also shows a hump-shaped pattern paralleling the species richness pattern. The coincident variations of area and species richness along the elevational gradient in the Hengduan Mountains make area a strong, potentially explanatory variable of the elevational patterns of species richness per 100 m belt (Tables 7.1 and 7.2). The fact that endemic species richness peaks at higher elevations is consistent with the largest area found at higher elevations, so that endemic richness shows an even stronger relationship with area than total richness (Table 7.1), and this phenomenon has been observed in other taxa, such as frogs in the Hengduan Mountains (Fu et al., 2006).

Because PET shows a near to linear change with elevation (Figure 7.2a), it is no surprise that the species diversity response to PET parallels that of elevation (Table 7.1). However, MAR does not match this pattern, with a strong increase from an 800 m level of only 700 mm at subtropical temperatures to a 1,300 mm peak at ca. 1,700 m followed by a steep linear decline toward the upper limit of plant life. In other words, drought constrains species richness in low altitude habitats and extremely low temperature at very high elevation (Ricklefs and Schluter, 1993), offering a second avenue toward explaining the mid-elevation diversity peak (Hegazy et al., 1998; Körner, 2007). The extreme climate, e.g., the drought at valley and low temperature above timberline, may be a limit of species richness, and the condition of climate at mid-elevations contributes to the highest species richness. Therefore, reduction of land area and extreme climate at both ends of the elevational gradient seem to be the main drivers of the hump-shaped pattern of species richness in the Hengduan Mountains.

The multirelationship between elevational pattern of species richness and explanatory variables supports the importance of land area and moisture regimes in determining the elevational pattern of species richness (Körner, 2007; Table 7.2). Combining GIS data on topography, meteorological databases, and species richness data permits distilling functional relationships between biodiversity and elevation. The confounding of several parallel or autocorrelated environmental gradients with elevation does, however, call for studies along elevational gradients that differ in those nonelevation-tied variables, such as land area and moisture. The detailed records of species distribution in publications and online database are the basis for this analysis, but more field investigations are needed to compare the differences of species richness at local scale across larger areas of contrasting mountain topography and water relations (Körner, 2007).

SUMMARY

Patterns of seed plant species richness along elevational gradients in mountains are often confounded with other factors, such as climate and land area. In our study, we adjusted for land area and tested how climatic variables affect the elevational pattern in the Hengduan Mountains in southwest China. Hengduan Mountains, one of the biodiversity hotspots of the world, are especially well-suited for such a study, as they cover a very wide amplitude of bioclimate, measured by more than 90 weather stations. Land area itself (calculated in each 100 m elevational belts using a Digital Elevation Model) along the elevational gradient was hump-shaped, with a peak at 4,000 to 4,500 m. The 8,590 different plant species also showed a hump-shaped pattern along the elevational gradient, with a peak at 2,500 to 3,000 m. The extreme climate, e.g., the drought in the valley, and the low temperatures above timberline and the reduction of land area at both ends of the elevational gradient seem to be the main drivers of the hump-shaped pattern of species richness in Hengduan Mountains. Confounding of environmental gradients calls for replications of elevational gradients differing in land area, moisture, and other, nontemperature-related gradients.

ACKNOWLEDGMENTS

This study was supported by grants-in-aid from the Natural Science Foundation of China (NSFC, 30625004, 40771073 to H. Sun).

REFERENCES

Bachman, S., W.J. Baker, N. Brummitt, J. Dransfield, and J. Moat. 2004. Elevation gradient, area and tropical island biodiversity: An example from the palms of New Guinea. *Ecography* 27:299–310.

Bhattarai, K.R., O.R. Vetaas, and J.A. Grytnes. 2004. Fern species richness along a central Himalayan elevation gradient, Nepal. *Journal of Biogeography* 31:389–400.

Boufford, D.E., and P.P.V. Dijk. 2000. South-Central China. In *Hotspots: Earth's Biologically Richest and Most Endangered Terrestrial Ecoregions*, pp. 338–51. Mittermeier, R.A., N. Myers, and C.G. Mittermeier, eds. Mexico: Cemex.

Boufford, D.E., P.P.P. Dijk, and L. Zhi. 2004. Mountains of Southwest China. In *Hotspots Revisited: Earth's Biologically Richest and Most Endangered Ecoregions*, 2nd ed., pp. 159–64. Mittermeier, R.A, P. Robles-Gil, M. Hoffmann, J.D. Pilgrim, T.M. Brooks, C.G. Mittermeier, J.L. Lamoreux, and G. Fonseca, eds. Mexico: Cemex.

Fang, J.Y. 2004. Exploring altitudinal patterns of plant species diversity of China's mountains. *Biodiversity Science* 12:1– 4.

Fu, C.Z., X. Hua, J. Li, Z. Chang, Z.C. Pu, and J.K. Chen. 2006. Elevational patterns of frog species richness and endemic richness in the Hengduan Mountains, China: Geometric constraints, area and climate effects. *Ecography* 29:919–27.

Grytnes, J.A., and O.R. Vetaas. 2002. Species richness and altitude: A comparison between null models and interpolated plant species richness along the Himalayan altitudinal gradient, Nepal. *The American Naturalist* 159:294–304.

Hegazy, A.K., M.A. El-Demerdash, and H.A. Hosni. 1998. Vegetation, species diversity and floristic relations along an altitudinal gradient in south-west Saudi Arabia. *Journal of Arid Environments* 38:3–13.

Hofierka, J., J. Parajka, and H. Mitasova. 2002. Multivariate interpolation of precipitation using regularized spline with tension. *Transaction in GIS* 6(2):135–50.

Holdrige, L.R. 1976. *Life Zone Ecology*. San Jose: Tropical Science Center.

Ibisch, P.L., A. Boegner, J. Nieder, and W. Barthlott. 1996. How diverse are neotropical epiphytes? An analysis based on the "Catalogue of the flowering plants and gymnosperms of Peru." *Ecotropica* 1:13–28.

Jankowski, T., and G.A. Weyhenmeyer. 2006. The role of spatial scale and area in determining richness-altitude gradients in Swedish lake phytoplankton communities. *Oikos* 115:433–42.

Karnieli, A. 1990. Application of kriging technique to areal precipitation mapping in Arizona. *GeoJournal* 22:391–98.

Kessler, M., S.K. Herzog, J. Fjeldså, and K. Bach, K. 2001. Species richness and endemism of plant and bird communities along two gradients of elevational, humidity, and land use in the Bolivian Andes. *Diversity and Distributions* 7:61–77.

Körner, Ch. 2000. Why are there global gradients in species richness? Mountains might hold the answer. *TREE* 15:513.

Körner, Ch. 2007. The use of "altitude" in ecological research. *TREE* 22:569–74.

Krömer, T., M. Kessler, S.R. Gradstein, and A. Acebey. 2005. Diversity patterns of vascular epiphytes along an elevational gradient in the Andes. *Journal of Biogeography* 32:1799–1809.

Li, B.Y. 1987. On the boundaries of the Hengduan Mountains. *Mountain Research* 5(2):74–82.

Li, B.Y. 1989. Geomorphologic regionalization of the Hengduan mountainous region. *Mountain Research* 7(1):13–20.

Liu, L.H., Y.D. Yu, and J.H. Zhang. 1984. The division of vertical vegetation zone in Hengduanshan. *Acta Botanica Yunnanica* 6:205–16.

Lomolino, M.V. 2001. Elevation gradients of species-density: Historical and prospective views. *Global Ecology & Biogeography* 10:3–13.

MacArthur, R.H. 1972. *Geographical Ecology: Patterns of the Distribution of Species*. New York: Harper and Row.

McGlone, M.S., R.P. Duncan, and P.B. Heenan. 2001. Endemism, species selection and the origin and distribution of the vascular plant flora of New Zealand. *Journal of Biogeography* 28:199–216.

Myers, N., R.A. Mittermeier, C.G. Mittermeier, G.A.B. Fonseca, and J. Kent. 2000. Biodiversity hotspots for conservation priorities. *Nature* 403:853–58.

Nogués-Bravo, D., M.B. Araújo, T. Romdal, and C. Rahbek. 2008. Scale effects and human impact on the elevational species richness gradients. *Nature* 453:216–19.

Qian, H. 1998. Large-scale biogeographic patterns of vascular plant richness in North America: An analysis at the genera level. *Journal of Biogeography* 25:829–36.

Rahbek, C. 1995. The elevational gradient of species richness: A uniform pattern? *Ecography* 18:200–205.

Rahbek, C. 2005. The role of spatial scale and the perception of large-scale species richness patterns. *Ecology Letters* 8:224–39.

Rangel, T.F.L.V.B., J.A.F. Diniz-Filho, and L.M. Bini. 2006. Towards an integrated computational tool for spatial analysis in macroecology and biogeography. *Global Ecology & Biogeography* 15:321–27.

Ricklefs, R.E., and Schluter, D. 1993. Species diversity in ecological communities: Historical and geographical perspectives. Chicago: University of Chicago.

Sánchez-González, A., and L. López-Mata. 2005. Plant species richness and diversity along an altitude gradient in the Sierra Nevada, Mexico. *Diversity and Distribution* 11:567–75.

Sanders, N.J. 2002. Elevational gradients in ant species richness: Area, geometry, and Rapoport's rule. *Ecography* 25:25–32.

Schoener, T.W. 1976. The species-area relationship within archipelagoes: Models and evidence from island land birds. *Proceedings of the 16th International Ornithological Congress* 6:629–42.

Vetaas, O.R., and J.A. Grytnes. 2002. Distribution of vascular plant species richness and endemic richness along the Himalayan elevational gradient in Nepal. *Global Ecology & Biogeography* 11:291–301.

Walter, H. 1979. *Vegetation of the Earth Ecological System of the Geobiosphere*. 2nd ed. New York: Springer-Verlag.
Wang, Z.H., Z.Y. Tang, and J.Y. Fang. 2007. Altitudinal patterns of seed plant richness in the Gaoligong Mountains, south-east Tibet, China. *Diversity and Distribution* 13:845–54.
Wang, W.S. 1993, 1994. *Vascular Plants of the Hengduan Mountains*. Vol. 1 and Vol. 2. Beijing: Science Press.
Yu, Y.D., L.H. Liu, and J.H. Zhang. 1989. Vegetation regionalization of the Hengduan mountainous region. *Mountain Research* 7(1):47–55.

8 Elevational Gradients of Species Richness Derived from Local Field Surveys versus "Mining" of Archive Data

Michael Kessler, Thorsten Krömer, Jürgen Kluge, Dirk N. Karger, Amparo Acebey, Andreas Hemp, Sebastian K. Herzog, and Marcus Lehnert

CONTENTS

Introduction 57
Methods 58
Results 59
Discussion 59
Summary 61
Acknowledgments 62
References 62

INTRODUCTION

The documentation and explanation of global and regional gradients of species richness is one of the major challenges of ecological and biogeographical research. In addition to the classical latitudinal richness gradient, patterns of species richness along elevational gradients also have received considerable attention in the last decade (e.g., Rahbek, 1995, 2005; Lomolino, 2001; Körner, 2007). Compared to the latitudinal gradient, elevational gradients have a number of advantages that make them appealing study objects (Lomolino, 2001; Körner, 2007; Grytnes and McCain, 2007). They are replicated on many mountain ranges with different topographical, climatic, geological, and historical conditions, allowing for statistical separation of potential explanatory factors. Also, individual transects are spatially concise and located in specific biogeographic regions. However, conducting field surveys along elevational gradients is challenging, as especially at higher elevations a rugged topography makes many areas hard to access. Furthermore, along most extensive elevational transects at least part of the transect is usually heavily impacted by humans, mostly at low, but also at high, elevations (Nogués-Bravo et al., 2008). As a consequence, the assessment of a complete gradient up to the upper vegetation limit is, if at all possible, time consuming and, therefore, expensive. As an alternative to intensive field surveys, elevational patterns of species richness can be obtained from data compilations from the literature or scientific collections (data mining) (Körner et al., 2007). This approach has a great advantage: Despite the fact that these data sources are built up on a huge amount of effort often spanning over decades, now these data are existing and mostly easily accessible, and form a treasure of valuable information for species diversity assessment. Data mining generally documents regional patterns of diversity, because it typically covers larger areas than local surveys and because location data, especially from older sources, often is less precise. Combining the data from numerous collecting efforts spread over longer time spans, species lists obtained from data mining are more complete than local surveys and include a larger fraction of the total biota. However, data mining is biased by three main factors:

(i) Field collecting is usually spatially uneven, as collectors prefer easily accessible sites and areas that look particularly interesting (Nelson et al., 1990).

(ii) Surface area usually declines with elevation (Körner, 2000), and area is well known to influence regional species richness (Rosenzweig and Ziv, 1999; Lomolino, 2001).

(iii) Actual species records are always patchy, and it is contentious whether species distributions should be interpolated between these records. Along elevational gradients, some authors have suggested that species should be considered present in a given elevational belt if they have been recorded at both higher and lower elevations (Williams et al., 1996; Lees et al., 1999), whereas others claim that this leads to artificially elevated species numbers at mid-elevations, because such interpolation is not possible at the extremes of the gradient (Grytnes and Vetaas, 2002).

In contrast, field surveys, hereafter called local studies, have the advantage of usually employing a standardized sampling method, thus eliminating the influence of varying sampling intensity and sampling area on the observed richness patterns (e.g., Lomolino, 2001). Further, the obtained species records correspond to taxa unambiguously present at the study sites. On the other hand, field surveys are often labor and time consuming, and, especially in highly diverse tropical ecosystems, they are typically time constrained and record only part of the local flora or fauna, with a certain proportion of mostly rare species evading detection. To make things worse, this bias potentially varies with elevation due to differences in, e.g., overall richness (complete samples are more readily achieved in species-poor assemblages), abundance distributions (more common taxa are more readily sampled), and vegetation structure (canopy-living organisms are more easily sampled in high-elevation forests with low canopy than in tall lowland forests). Therefore, local surveys along elevational gradients may suffer from systematic biases. Both of these methodological approaches thus have specific shortcomings and potential biases, and may be expected to recover different elevational patterns of diversity. However, to date there have been no attempts to compare these different methods. In the present study, we therefore compiled regional data for five groups of organisms (birds, four plant groups) along six tropical elevational gradients (Carrasco in Bolivia, Pichincha in Ecuador, Braulio Carrillo in Costa Rica, Los Tuxtlas in Mexico, Kilimanjaro in Tanzania, Kinabalu in Borneo) for which we already had local field data. Our main aim was to assess the degree of congruence between the different sampling approaches and of how potential biases in the different methods might influence our perception of diversity patterns. We purposefully refrained from linking the observed patterns to environmental variables or other potential determining factors, because this has been done both in our own previous publications (e.g., Hemp, 2001, 2005, 2006; Kessler, 2001; Kessler et al., 2001; Herzog et al., 2002, 2005; Kluge et al., 2006; Krömer et al., 2006) in numerous other studies (e.g., McCain 2005, 2007; Oommen and Shanker, 2005; Brehm et al., 2007), and because the focus of this study was on sampling methods and not on the interpretation of the patterns.

METHODS

Regional richness patterns were generated by adding species numbers for different elevational steps from electronic databases and published species lists. Sources and data type are listed in Table 8.1. Because sampling is typically uneven at different elevations, in some cases, interpolation between range records and rarefaction techniques were applied to correct for incomplete sampling records (Colwell and Coddington, 1995).

Local data for the present study were derived from five previously conducted field studies. Except for the Kilimanjaro transect, plant studies all used the same sampling method with plots of 400 m^2 each. These plots were usually of square shape and placed in natural forest on slopes, i.e., avoiding secondary and special habitats, such as ravines or ridges (Kessler and Bach, 1999; Kessler, 2001). The Carrasco transect in Bolivia was surveyed by Amparo Arabey (AA), Sebastian K. Herzog (SKH), Michael Kessler (MK), and Thorsten Krömer (TK) for plants and birds in 1996 and 1997 (Kessler, 2001; Herzog et al., 2005; Krömer et al., 2006). Birds were surveyed along a continuous elevational gradient using the species-list method, where temporally consecutive visual and acoustical (including extensive use of sound recordings) bird observations are subdivided into lists of ten species that are subsequently analyzed using rarefaction curves and species richness estimators (Herzog et al., 2002). The Pichincha transect in Ecuador was sampled in 2005 by Marcus Lehnert (ML) and MK. The Braulio Carrillo transect in Costa Rica was surveyed in 2002 and 2003 by Jürgen Kluge (JK) in 156 plots (Kluge et al., 2006). The Los Tuxtlas transect in Mexico was sampled in 2005 and 2006 by TK and AA for aroids, bromeliads, and ferns (TK and AA, unpublished data). Mt. Kilimanjaro in Tanzania was extensively sampled by Andreas Hemp (AH), with several hundred plots of 1,000 m^2 each (Hemp, 2001, 2005, 2006). Finally, along

TABLE 8.1
Organisms, Source, and Data Types of Twelve Regional Data Sets of Species Richness Along Elevational Gradients

Organism	Transect Location	No. of Records	Record Type	Source	Data Correction
Araceae	Bolivia—Carrasco	n.a.	country list	Kessler and Croat, 1999	no
Birds	Bolivia—Carrasco	> 60,000	museum, local lists, field reports	Hennessey et al., 2003	no
Bromeliaceae	Bolivia—Carrasco	n.a.	country list	Krömer et al., 1999	no
Palms	Bolivia—Carrasco	n.a.	country list	Moraes, 1996	no
Pteridophytes	Bolivia—Carrasco	> 27,000	herbarium records	Kessler and Smith, unpublished data	no
Pteridophytes	Ecuador—Pichincha	n.a.	country list	Jørgensen and León-Yánez, 1999	no
Pteridophytes	Costa Rica	n.a.	regional flora	Moran and Riba, 1995	interpolation
Araceae	Mexico—Los Tuxtlas	n.a.	regional list	Acebey and Krömer, in press	no
Bromeliaceae	Mexico—Los Tuxtlas	n.a.	regional flora, regional list	Espejo et al., 2005; Krömer and Acebey, 2007	no
Pteridophytes	Mexico—Los Tuxtlas	n.a.	country flora	Mickel and Smith, 2004	no
Pteridophytes	Tanzania—Mt. Kilimanjaro	n.a.	regional list	Hemp, 2002	interpolation
Pteridophytes	Borneo—Mt. Kinabalu	n.a.	herbarium records	Grytnes and Beaman, 2006	rarefaction

Note: For the Carrasco transect, we defined the regional area as the forested northeastern part of the department of Cochabamba (i.e., excluding the arid inter-Andean valleys); n.a. = not available.

the Mt. Kinabalu transect in Borneo, ferns were sampled in fifteen plots by MK in 2000 (Kessler et al., 2001). Local and regional richness patterns were compared visually and using correlation analysis.

RESULTS

Our twelve case studies showed a wide range of regional and local richness patterns. Most patterns (nineteen) were hump shaped, but there also were three monotonically decreasing patterns, one with roughly constant values to mid-elevations followed by a decline, and one with increasing richness (Figure 8.1). Correlation values between local and regional patterns were significantly positive in ten out of twelve cases with R-values between 0.46 and 0.92, but there also were one positive (fern gradient in Mexico) and one negative, nonsignificant (Bromeliaceae in Mexico) relationship each. These mostly high positive correlation values corresponded to the visual impression, which showed fairly good concordance between the local and regional patterns. In six cases, the peak of regional richness was located at somewhat lower elevations than that of local richness, while the inverse was observed in four cases. In two cases, local and regional richness peaks coincided.

DISCUSSION

Our study recovered a wide range of different elevational species richness patterns with a prevalence of hump-shaped patterns (see also Rahbek, 1995, 2005). An evaluation of potential explanations for these patterns is beyond the scope of our study, but it is evident that there are taxon-specific differences. For example, along all of our study transects, fern species richness peaked at 1,000 to 2,000 m, whereas richness of Araceae peaked at 0 to 1,000 m, both in Mexico and Bolivia. These differences cannot be explained by factors that affect all taxa in similar ways (e.g., area) but clearly point to taxon-specific physiological or evolutionary causes. Thus, Araceae are well known to be highly sensitive to low temperatures (Mayo et al., 1997), whereas many ferns require high levels of environmental humidity (Bhattarai et al., 2004; Kluge et al., 2006). Therefore, in order to explain such taxon-specific patterns, the actual environmental drivers and organismic requirements both need to be accounted for (Körner, 2007). In most of our study cases, species richness patterns along elevational gradients recovered from the data mining and field studies were in good agreement, with positive correlation values in the range of 0.46 to 0.92. Therefore, compiling elevational richness

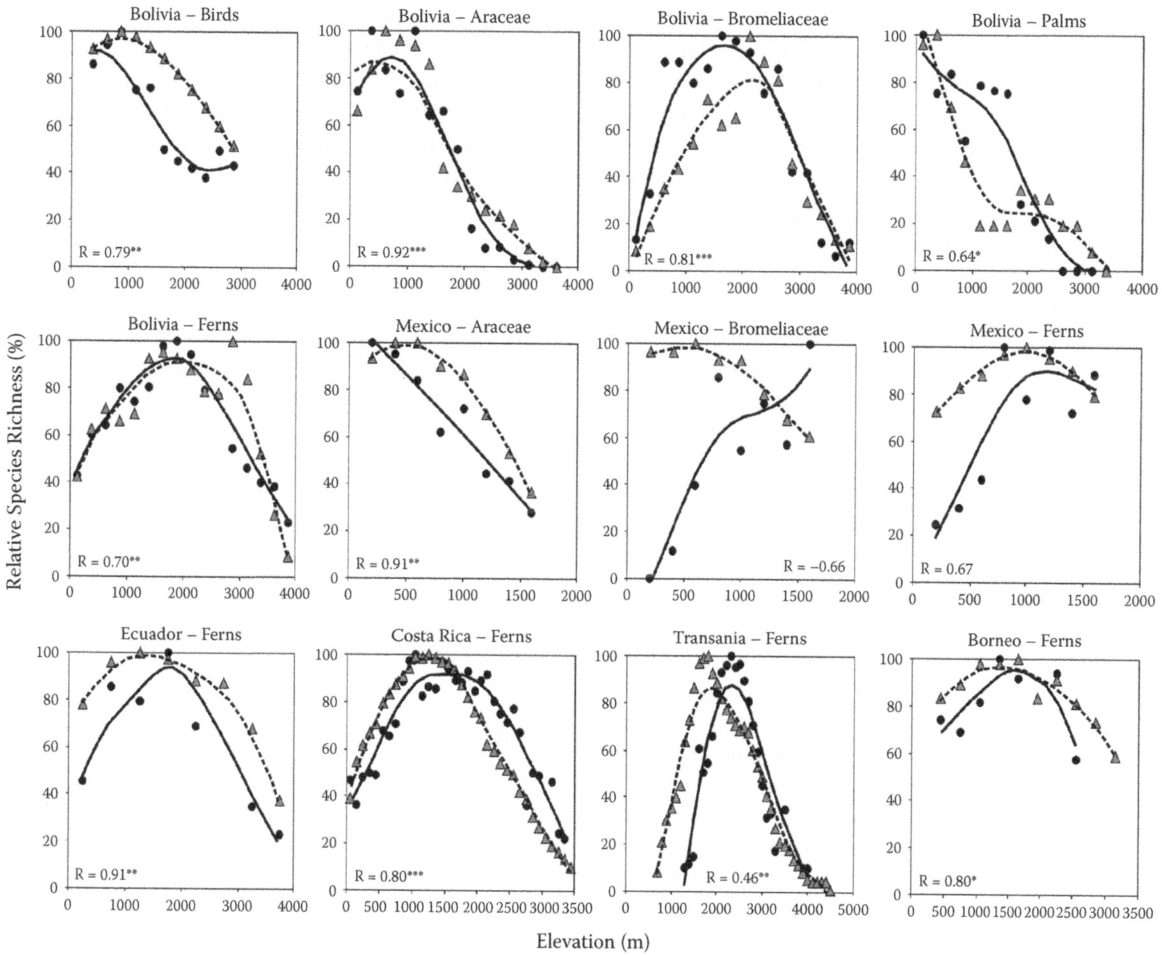

FIGURE 8.1 Elevational patterns of local (black circles, solid line) and regional (gray triangles, broken line) species richness for five groups of organisms along six study transects. The correlation values between local and regional patterns are indicated in each graph. $^{*}p < 0.05$; $^{**}p < 0.01$; $^{***}p < 0.001$. Trend lines were fitted with distance weighted least square smoothing, using SYSTAT 7.0.

data by data mining can act as a fairly good proxy for local richness patterns, and vice versa.

However, it is precisely the differences between the two approaches that are ecologically informative and scientifically intriguing. The causes for deviations between local and regional patterns fall into two main categories. First, every sampling regime suffers to some degree from statistical noise, and a comparison of two data sets is therefore highly unlikely to achieve perfect concordance. While annoying, this source of variation between local and regional studies is not overly troubling because it involves no systematic biases.

The second category of causes, involving systematic variations due to methodological sampling differences or due to natural factors, is much more interesting. Area is probably one of the most general factors potentially leading to systematic differences between local and regional studies. Because land surface area typically declines with increasing elevation (Körner, 2000; Lomolino, 2001), and because regional area directly influences the regional species pool (MacArthur and Wilson, 1967; Romdal and Grytnes, 2007), regional species richness may be expected to be relatively higher at lower elevations than local richness. Indeed, in our sample, this was the case in six (50%) of the studies, but the opposite was true in four (33%) studies. Furthermore, larger areas may have a higher diversity of ecological conditions and habitats, further increasing regional species numbers (Rahbek and Graves, 2001). As already outlined in the introduction, other systematic natural changes with elevation may involve overall richness, abundance distributions, and vegetation structure. The impact of these factors on individual studies is likely to be dependent on the overall context of the elevational transect (topography, climate, etc.) and the taxon under consideration.

As a result of the scale dependence of the influence of different ecological and geographical parameters along elevational gradients, local and regional patterns of species richness are to be expected to deviate somehow (Rahbek and Graves, 2001; Oommen and Shanker, 2005). For this reason, it is not unexpected that data mining at the regional scale may uncover different patterns than local field studies. And rather than attempting to decide which approach is "better," both should be seen as complementary in jointly providing a more complete picture of patterns of diversity at different spatial scales. A comparison of local and regional data sets may be instrumental in uncovering the processes determining the co-occurrence of species (alpha diversity) and shifts of species composition (beta diversity) along ecological and geographical gradients (Oommen and Shanker, 2005).

As an example, the striking negative relationship between local and regional richness of bromeliads at Los Tuxtlas in Mexico shows opposing patterns, caused by a much higher variability of habitats at low elevations and by a strong shift in the local density of bromeliad individuals along the elevational gradient (T. Krömer, unpublished data). At low elevations, bromeliad diversity is high, but the species are represented by few, scattered individuals, resulting in a low species representation in randomly placed local plots. At higher elevations at Los Tuxtlas, overall bromeliad species richness declines, but the abundance of individuals increases, resulting in a higher representation of bromeliads on individual local plots. Hence, we observe low alpha and high beta diversity at low elevations, and the reverse (high alpha and low beta diversity) at high elevations. For the other elevational case studies, the general patterns recovered were the same for the local and regional studies (as documented by the positive correlation values), but when looking in detail, there were many differences. In particular, the elevational peaks of local and regional richness patterns coincided only in two cases, one of which (palms in Bolivia) involved highest richness at the lowest elevation. Further, the relative differences of species richness between the levels of highest and lowest diversity differed at local and regional levels. For example, whereas fern species richness along the Los Tuxtlas transect in Mexico peaked at about 1,100 m at both local and regional scales, regional species richness declined by only about 30% to the lowlands, whereas local richness declined by about 80% (Figure 8.1). This suggests relatively higher beta diversity at low elevations. In other cases, e.g., ferns at high elevations in Bolivia and Costa Rica, regional richness declined more strongly than local richness, which may be interpreted as low beta diversity at high elevations. In combinations, these studies suggest that there is a decline of beta diversity with elevation among ferns, something that has not been possible to document with local field data alone, despite several attempts (Kessler, 2001; J. Kluge and M. Kessler, unpublished data). In conclusion, our study shows that data mining and local field studies are complementary in their approaches, biases, and information contained. None can replace the other, and in combination they can provide insights that cannot be gained by either of them taken alone. E-mining of archive data is a promising tool for areas where local data do not exist and some diversity estimates are needed for research, conservation, or management purposes, and for areas where local data are available, but where the addition of regional data leads to a more complete picture of the spatial distribution of biodiversity.

SUMMARY

Studies of elevational species richness patterns involve data compilations from the literature or scientific collections (regional studies: "data mining") or field surveys (local studies). Both approaches have potential methodological biases, but so far no comparison has been made to assess the degree of congruence between the different sampling approaches and of how potential differences might influence our perception of diversity patterns. We compared local and regional diversity patterns for five groups of organisms (birds, four plant groups) along six tropical elevational gradients (Carrasco in Bolivia, Pichincha in Ecuador, Braulio Carrillo in Costa Rica, Los Tuxtlas in Mexico, Kilimanjaro in Tanzania, Kinabalu in Borneo). In ten cases, local and regional patterns were significantly correlated ($R = 0.46$ to 0.92), but in detail, there were many deviations in the location of maximum diversity and the slopes of species number increases and declines. The perception of diversity patterns strongly depends on scale, and therefore it is not surprising that the compiled data mined from regional lists do not necessarily perfectly reflect local patterns. These differences were at least partly determined by ecological and geographical shifts along elevational gradients (e.g., reduced surface area and ecological heterogeneity at higher elevations). As a result, e-mining data do not mimic local data, but rather both give insight to species assemblage-structuring processes from different points of view. Both data sources combined shape our understanding about the mechanisms that drive richness patterns, and therefore are of strong complementary and synergistic value.

ACKNOWLEDGMENTS

For invaluable support of our field work, we thank L. Salazar, H. Navarette, N. Krabbe, Bellavista Lodge, H. Brieschke, Fundación Jocotoco, Tandeyapa Lodge, and C. Vits and B. de Roguer in Ecuador; the Herbario Nacional de Bolivia, I. Jimenez, J. Gonzales, K. Bach, S.G. Beck, M. Cusicanqui, A. de Lima, R. de Michel, M. Moraes, the Dirección Nacional de Conservación de la Biodiversidad, the park guards of Carrasco National Park, the mayor of Villa Tunari, Hotel El Puente, and the Facultad de Agricultura, Universidad Mayor de San Simón, Cochabamba, in Bolivia; F. Corrales and the staff of Biological Station La Selva and the Organization for Tropical Studies OTS, the National Herbarium in San José and the park rangers of the Sistema Nacional de Area de Conservaciones (SINAC) and the Area de Conservación Cordillera Volcánica Central (ACCVC) in Costa Rica; and the Chief Park Wardens of Kilimanjaro National Park, L. Moirana, and N. Mafuru in Tansania. For valuable comments on the manuscript, we thank C. Körner and E. Spehn. Field work was partly funded by the Deutsche Forschungsgemeinschaft DFG (grants to MK, AH), the Deutscher Akademischer Austauschdienst DAAD (to JK, SKH), and the Universidad Nacional Autónoma de México (to TK).

REFERENCES

Acebey, A., and T. Krömer. In press. Diversidad y distribución de las aráceas de la Reserva de la Biosfera "Los Tuxtlas," Veracruz, México. Revista Mexicana de Biodiversidad.

Bhattarai, K.R., O.R. Vetaas, and J.A. Grytnes. 2004. Fern species richness along a central Himalayan elevational gradient, Nepal. *Journal of Biogeography* 31:389–400.

Brehm, G., R.K. Colwell, and J. Kluge. 2007. The role of environment and mid-domain effect on moth species richness along a tropical elevational gradient. *Global Ecology and Biogeography* 16:205–19.

Colwell, R.K., and J.A. Coddington. 1995. Estimating terrestrial biodiversity through extrapolation. In *Biodiversity: Measurement and Estimation*, pp. 101–118. Hawksworth, D.L., ed. London: Chapman & Hall.

Espejo, A., A.R. López-Ferrari, and M.I. Ramírez. 2005. Bromeliaceae. *Flora de Veracruz*. Fascículo 136. Xalapa: Instituto de Ecología, A.C., and University of California.

Grytnes, J.A., and J.H. Beaman. 2006. Elevational species richness patterns for vascular plants on Mount Kinabalu, Borneo. *Journal of Biogeography* 33:1838–49.

Grytnes, J-A., and C.M. McCain. 2007. Elevational trends in biodiversity. In *Encyclopedia of Biodiversity*, pp. 1–8. Levin, S., ed. Amsterdam: Elsevier.

Grytnes, J.A., and O.R. Vetaas. 2002. Species richness and altitude: A comparison between null models and interpolated plant species richness along the Himalayan altitudinal gradient, Nepal. *American Naturalist* 159:294–304.

Hemp, A. 2001. Ecology of the pteridophytes on the southern slopes of Mt. Kilimanjaro. Part II: Habitat selection. *Plant Biology* 3:493–523.

Hemp, A. 2002. Ecology of the pteridophytes on the southern slopes of Mt. Kilimanjaro. Part I. Altitudinal distribution. *Plant Ecology* 159:211–39.

Hemp, A. 2005. The impact of fire on diversity, structure and composition of Mt. Kilimanjaro's vegetation. In *Land Use Change and Mountain Biodiversity*, pp. 51–68. Spehn, E., M. Liberman, and C. Körner, eds. Boca Raton, Fla.: CRC Press.

Hemp, A. 2006. Ecology and altitudinal zonation of pteridophytes on Mt. Kilimanjaro. In *Taxonomy and Ecology of African Plants, their Conservation and Sustainable Use*, pp. 309–37. Proceedings of the 17th AETFAT Congress, Addis Ababa, Ethiopia. Ghazanfar, S.A., and H.J. Beentje, eds. Kew: Royal Botanic Gardens.

Hennessey, A.B., S.K. Herzog, and F. Sagot. 2003. *Lista anotada de las aves de Bolivia*. 5th ed. Santa Cruz de la Sierra, Bolivia: Asociación Armonía/BirdLife International.

Herzog, S.K., M. Kessler, and K. Bach. 2005. The elevational gradient in Andean bird species richness at the local scale: A foothill peak and a high-elevation plateau. *Ecography* 28:209–22.

Herzog, S.K., M. Kessler, and T.M. Cahill. 2002. Evaluation of a new rapid assessment method for estimating avian diversity in tropical forests. *The Auk* 199:749–69.

Jørgensen, P.M., and S. León-Yánez. 1999. *Catalogue of the Vascular Plants of Ecuador*. St. Louis: Missouri Botanical Garden Press.

Kessler, M., and T. Croat. 1999. State of knowledge of Bolivian Araceae. *Selbyana* 20:224–34.

Kessler, M. 2001. Patterns of diversity and range size of selected plant groups along an elevational transect in the Bolivian Andes. *Biodiversity and Conservation* 10:1897–1920.

Kessler, M., and K. Bach. 1999. Using indicator groups for vegetation classification in species-rich Neotropical forests. *Phytocoenologia* 29:485–502.

Kessler, M., B.S. Parris, and E. Kessler. 2001. A comparison of the tropical montane pteridophyte communities of Mount Kinabalu, Borneo, and Parque Nacional Carrasco, Bolivia. *Journal of Biogeography* 28:611–22.

Kluge, J., M. Kessler, and R.R. Dunn. 2006. What drives elevational patterns of diversity? A test of geometric constraints, climate and species pool effects for pteridophytes on an elevational gradient in Costa Rica. *Global Ecology and Biogeography* 15:358–71.

Körner, C. 2000. Why are there global gradients in species richness? Mountains might hold the answer. *Trends in Ecology & Evolution* 15:513–14.

Körner, C. 2007. The use of "altitude" in ecological research. *Trends in Ecology & Evolution* 22:569–74.

Körner, C., M. Donoghue, T. Fabbro, C. Häuse, D. Nogués-Bravo, M.T. Kalin Arroyo, J. Soberon, L. Speers, E.M. Spehn, H. Sun, A. Tribsch, P. Tykarski, and N. Zbinden. 2007. Creative use of mountain biodiversity databases: The Kazbegi Research Agenda of GMBA-DIVERSITAS. *Mountain Research and Development* 27:276–81.

Krömer, T., and A. Acebey. 2007. The bromeliad flora of the San Martín Tuxtlas volcano, Veracruz, Mexico. *Journal of the Bromeliad Society* 57:62–69.

Krömer, T., M. Kessler, B.K. Holst, H.E. Luther, E. Gouda, W. Till, P.L. Ibisch, and R. Vásquez. 1999. Checklist of Bolivian Bromeliaceae with notes on species distribution and levels of endemism. *Selbyana* 20:201–23.

Krömer, T., M. Kessler, and S.K. Herzog. 2006. Distribution and flowering ecology of bromeliads along two climatically contrasting elevational transects in the Bolivian Andes. *Biotropica* 38:183–95.

Lees, D.C., C. Kremen, and L. Andriamampianina. 1999. A null model for species richness gradients: Bounded range overlap of butterflies and other rainforest endemics in Madagascar. *Biological Journal of the Linnean Society* 67:529–84.

Lomolino, M.V. 2001. Elevation gradients of species-density: Historical and prospective views. *Global Ecology and Biogeography* 10:3–13.

MacArthur, R.H., and E.O. Wilson. 1967. *The Theory of Island Biogeography: Monographs in Population Biology*. Vol. 1. Princeton, N.J.: Princeton University Press Princeton.

Mayo, S.J., J. Bogner, and P.C. Boyce. 1997. *The Genera of Araceae*. Kew: Royal Botanical Gardens.

McCain, C.M. 2005. Elevational gradients in diversity of small mammals. *Ecology* 86:366–72.

McCain, C.M. 2007. Cold temperature and water availability drive elevational species richness patterns? A global case study for bats. *Global Ecology and Biogeography* 16:1–13.

Mickel, J.T., and A.R. Smith. 2004. The pteridophytes of Mexico. Memoirs of the New York Botanical Garden, Vol. 88.

Moraes, M. 1996. Diversity and distribution of palms in Bolivia. *Principes* 40:75–85.

Moran, R.C., and R. Riba. 1995. *Flora Mesoamericana Vol. 1: Psilotaceae a Salviniaceae*. Mexico: Universidad Nacional Autónoma de Mexico, Mexico D.F.

Nelson, B.W., C.A.C. Ferreira, M.F. da Silva, and M.L. Kawasaki. 1990. Endemism centers, refugia and botanical collection density in Brazilian Amazonia. *Nature* 345:714–16.

Nogués-Bravo, D., M.B. Araújo, T. Romdal, and C. Rahbek. 2008. Scale effects and human impact on the elevational species richness gradients. *Nature* 453:216–19.

Oommen, M.A., and K. Shanker. 2005. Elevational species richness patterns emerge from multiple local mechanisms in Himalayan woody plants. *Ecology* 86:3039–47.

Rahbek, C. 1995. The elevational gradient of species richness: A uniform pattern? *Ecography* 18:200–205.

Rahbek, C. 2005. The role of spatial scale and the perception of large-scale species-richness patterns. *Ecology Letters* 8:224–39.

Rahbek, C., and G.R. Graves. 2001. Multiscale assessment of patterns of avian species richness. *Proceedings of the National Academy of Science* 98:4534–39.

Romdal, T.S., and J.-A. Grytnes. 2007. An indirect area effect on elevational species richness patterns. *Ecography* 30:440–48.

Rosenzweig, M.L., and Y. Ziv. 1999. The echo pattern of species diversity: Pattern and process. *Ecography* 22:614–28.

Williams, P.H., G.T. Prance, C.J. Humphries, and K.S. Edwards. 1996. Promise and problems in applying quantitative complementary areas for representing the diversity of some Neotropical plants (families Dichapetalaceae, Lecythidaceae, Caryocaraceae, Chrysobalanaceae, and Proteaceae). *Biological Journal of the Linnean Society* 58:125–57.

9 Species Richness of Breeding Birds along the Altitudinal Gradient—An Analysis of Atlas Databases from Switzerland and Catalonia (NE Spain)

Niklaus Zbinden, Marc Kéry, Verena Keller, Lluís Brotons, Sergi Herrando, and Hans Schmid

CONTENTS

Introduction 65
Methods 66
Switzerland 66
Catalonia 67
Species Set 67
Statistical Analysis 67
Results 68
Differences in Environment 68
Observed Species Richness 69
Relationships between the Observed Species Richness and Environmental Variables 69
Discussion 71
Summary 72
References 72

INTRODUCTION

The distribution of bird species is strongly determined by climatic conditions. Climate change is therefore likely to affect the ranges of many species, and a recent analysis predicts latitudinal changes of species' ranges in Europe of an average 550 km toward northeast (Huntley et al., 2007). Analogous changes also are to be expected in relation to altitude. Changes in environmental conditions along the altitudinal gradient of mountains show similarities to those related to latitude, and a decrease in species richness has been observed both with increasing altitude and increasing latitude (MacArthur, 1972). A decrease in species richness of birds with increasing altitude has been observed in different mountain regions around the globe, e.g., in the Alps (Bezzel, 1971; Wartmann and Furrer, 1977), the Sierra Nevada, Spain (Zamora-Rodríguez, 1987), the Himalayas (Thiollay, 1980), and in the Sierra Madre del Sur, Mexico (Navarro, 1992).

Several factors have been postulated to account for the decline in species richness, e.g., the reduction of primary production with decreasing temperature, at least as long as enough water is available (species–energy hypothesis; Wright, 1983; Turner et al., 1988; Currie, 1991; Lennon et al., 2000; Currie et al., 2004; Evans et al., 2005; Hawkins et al., 2003). Higher energy availability also allows larger population sizes, which in turn reduces the risk of local extinction and could therefore result in larger numbers of species surviving at a particular altitude level (Evans et al., 2005; Carnicer et al., 2007).

Temperature itself may have a direct influence on birds, in particular during the first days after hatching when thermoregulation is not yet fully developed (Podulka et al., 2004). With decreasing temperature along the altitudinal or latitudinal gradient, fewer species might have been able to develop the specific adaptations necessary to cope with these conditions.

In accordance with the well-known species–area relationship (MacArthur, 1972), it also has been postulated that a decrease in species richness on mountains might result from the fact that the area available decreases with increasing altitude, similarly to the decrease in area with increasing latitude (Rahbek, 1995; Körner, 2000; Körner, 2004; Romdal and Grytnes, 2007).

In addition to temperature and other environmental factors that vary in parallel to altitudinal gradients, ecotones have been shown to influence distribution patterns of birds in the Andes (Terborgh, 1971, 1977). Habitat diversity has been identified as a major factor determining species richness in North American mammals in the more southern regions, whereas in northern areas (Alaska, Canada) energy was determined as the main limiting factor (Kerr and Packer, 1997).

Species distribution patterns may, however, also be influenced by historical events, such as glaciation cycles, leading to local extinction and recolonization. These patterns vary between regions. Taxonomic species composition also varies between continents, making comparisons difficult. Mountains are therefore good models to study the causes of changes in species richness, since the elevational gradients are situated within the same biogeographic regions and have gone through a similar geological and evolutionary history (Körner, 2000). The comparison of different mountain massifs located within the same biogeographic region may provide further insights into the factors explaining patterns of species distribution, as well as give more generality.

The Alps and the Pyrenees are two major mountain massifs in Europe. Despite a slight difference in latitude, the bird communities in both regions share a large number of species. In Switzerland, situated in the center of the Alpine chain, and in Catalonia, covering the eastern parts of the Pyrenees, large bird distribution data sets (bird atlas) have recently been produced using similar database methodology (Schmid et al., 1998; Estrada et al., 2004), enabling scientists to tackle ecological questions on a larger scale, both in space and time. In this article, we document and compare species richness patterns on the same spatial scale, 1 km^2, along altitudinal gradients of two different mountain regions, demonstrating the power of electronic databases for such analyses and large-scale comparisons. We analyze species richness patterns of breeding birds using abiotic and biotic environment variables and discuss the factors likely to be responsible for the observed patterns.

METHODS

The analysis was based on a large set of monitoring data of breeding birds, collected for distribution atlases of birds. Such databases are available for many countries and regions in Europe and elsewhere, as birds figure among the animal groups which are best monitored. Distribution atlases from Switzerland and Catalonia are well-suited for comparative analyses, as the scale used for collecting the data is equal. Limitations arise from the fact that field methods differed.

SWITZERLAND

Data used in this analysis were collected from 1993 to 1996 for the national distribution atlas of breeding birds (Schmid et al., 1998). Territories of abundant and widespread species were mapped on 1×1 km squares distributed in a spatially representative fashion across the country (Figure 9.1). No surveys were carried out above ca. 2,500 m, where hardly any birds breed. Three visits below the treeline and three visits above the treeline were carried out and spread across the breeding season. No time limits were set for visits, but each one lasted normally between two and four hours. Birds breeding in colonies and rare species found in the same squares were recorded separately. For the analyses in this paper, we used 2,622 squares lying completely within Switzerland and for which land cover data were available. Land cover data were extracted from the 1 ha grid provided by the Federal Statistical Office (Bundesamt für Statistik, 1999). Categories were aggregated as follows: forest (including categories closed and open forest, shrubs/dwarf shrubs, and small woodlands), agricultural land (lowland grassland, arable land, orchards, vineyards), alpine pastures (meadows and pastures at high altitude, including some lower-lying areas on steep slopes), unproductive areas, lakes and rivers, and urban habitats (built-up areas).

Temperature and precipitation data correspond to the average values 1,961 to 1,990, interpolated for each square, they were provided by the Swiss Federal Institute of Forest, Snow and Landscape Research, Birmensdorf (K. Ecker). Mean daily temperature and total precipitation during the breeding season (March–July) were used.

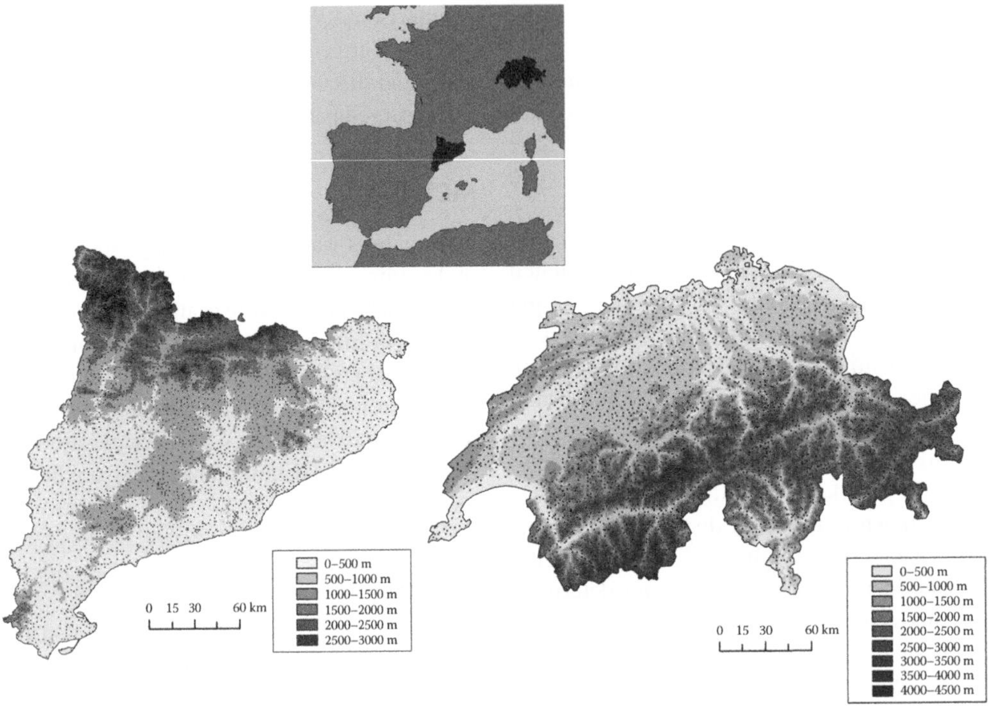

FIGURE 9.1 Location of the two study regions in Western Europe: Switzerland, top right; Catalonia, bottom left. Country maps including main topographic features and the location of the squares (1 × 1 km) surveyed in each atlas. (Maps produced by J. Guélat.)

Catalonia

Data for the Catalan Breeding Bird Atlas were collected from 1999–2002 (Estrada et al., 2004), with 1 × 1 km squares distributed in a habitat-driven, stratified manner across the country (Figure 9.1). Two visits were carried out during the breeding season, each lasting one hour. As in the case of Switzerland, we used only 3,038 squares lying completely in Catalonia and for which land cover data were available.

Land cover data come from different sources and are described in detail in Brotons et al. (2004) and Viñas and Baulies (1995). In general, categories corresponded to the ones used in Switzerland, with one exception. The category "scrub" was included in the category "forest" for our analysis, in order to be compatible with the Swiss categories.

Climate data were extracted from the Catalan Digital Atlas (CDA; Ninyerola et al., 2000), calculated in the same way as for Switzerland.

Species Set

Nocturnal and crepuscular species were excluded from the analysis, because they could not be recorded well during the standard surveys (Eurasian Stone-curlew *Burhinus oedicnemus*, Eurasian Woodcock *Scolopax rusticola*, nightjars *Caprimulgus europaeus*, *C. ruficollis*, and owls). Nonindigenous species and feral pigeon were also excluded.

Statistical Analysis

Owing to imperfect detection probability of most species (Kéry and Schmid, 2006; Dorazio et al., 2006; Kéry et al., 2008), we did not directly observe species richness. Instead, we analyzed the number of observed species. As is customary, albeit usually unstated in studies such as ours, we make the untested assumption that the proportion of observed species among the species truly present does not change along the dimensions of interest, such as altitude (Kéry et al., 2008). Data from Catalonia and Switzerland were analyzed separately. We used linear models to test for a piece-wise linear relationship between observed avian species richness and altitude. To explore the relationship between observed species richness and environmental variables, we used generalized additive models (GAMs) implemented in the mgcv library (Wood, 2006) in R both for inference (testing) and for graphical exploration.

Originally, we considered the following environmental variables as potential explanatory variables for the observed avian species richness: altitude, relief (difference between min and max altitude in sample squares), cover of alpine pasture, agricultural land, forests and urban areas, and the mean temperature and total precipitation during the breeding season (March–July). Altitude and mean temperature were almost totally collinear (Catalonia: r = –0.98; Switzerland: r = –0.99), so the latter was dropped from the list.

When fitting a GAM, we accounted for the strong richness–altitude pattern by retaining in the model the parametric form of the piece-wise linear regression with altitude and adding spline (i.e., smooth) terms of the remaining environmental variables. For model identifiability reasons, and similar to random-effect factors, effects of spline terms sum to zero (Wood, 2006), hence their effects must be interpreted as being "on top" of those of the fixed effects, here, the altitude model. In mgcv, the degree of smoothness (the converse of "wiggliness") is determined automatically by generalized cross-validation. To identify those environmental factors most likely related to the observed species richness, we used a backwards elimination scheme starting from a full model with the parametric form of altitude, as well as spline terms of the remaining variables, and dropped nonsignificant spline terms until only significant ones remained in the model. We note that we use statistical significance only as a rough guide in model selection, since inference in stepwise model selection schemes does not take into account model selection uncertainty.

RESULTS

Differences in Environment

A comparison between Switzerland and Catalonia revealed different patterns in relation to altitude for some major environmental variables (Figure 9.2). Forest cover (recall that this includes scrub) was much higher in Catalonia than in Switzerland, in particular above 1,000 m a.s.l., where it was close to 100% around 1,500 m a.s.l. Accordingly, the percentage of agricultural land was lower in Catalonia but only above ca. 400 m. The lower percentage of agricultural land in Switzerland below 400 m is due to high percentages of built-up areas and the presence of large lakes at this altitude, which makes up only a small percentage of the total area of Switzerland.

Temperature was ca. 2 to 3°C higher in Catalonia than in Switzerland, consistently along the whole altitudinal range (Figure 9.3). Precipitation, on the other hand, was higher in Switzerland and showed a different pattern. In Catalonia, precipitation shows a regular increase with altitude, while in Switzerland it remained almost constant

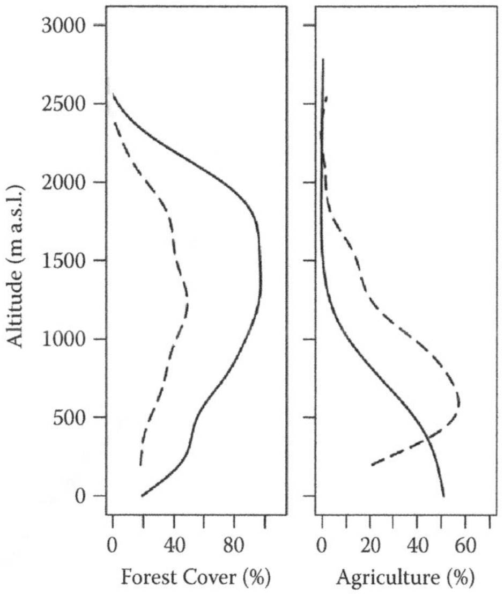

FIGURE 9.2 Forest cover and proportion of agricultural land in relation to altitude in surveyed squares in Catalonia (solid line) and Switzerland (dashed line). Lines are fitted values from a GAM.

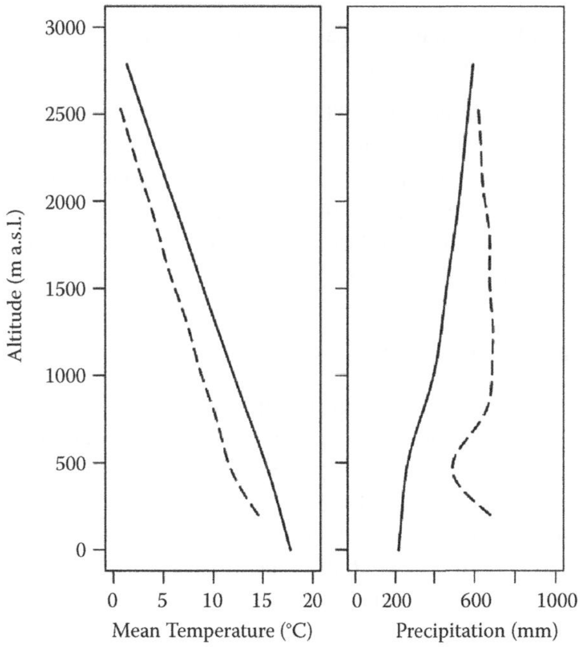

FIGURE 9.3 Mean temperature (°C) and precipitation sum during the breeding season (March–July) in relation to altitude in surveyed squares in Catalonia (solid line) and Switzerland (dashed line). Lines are fitted values from a GAM.

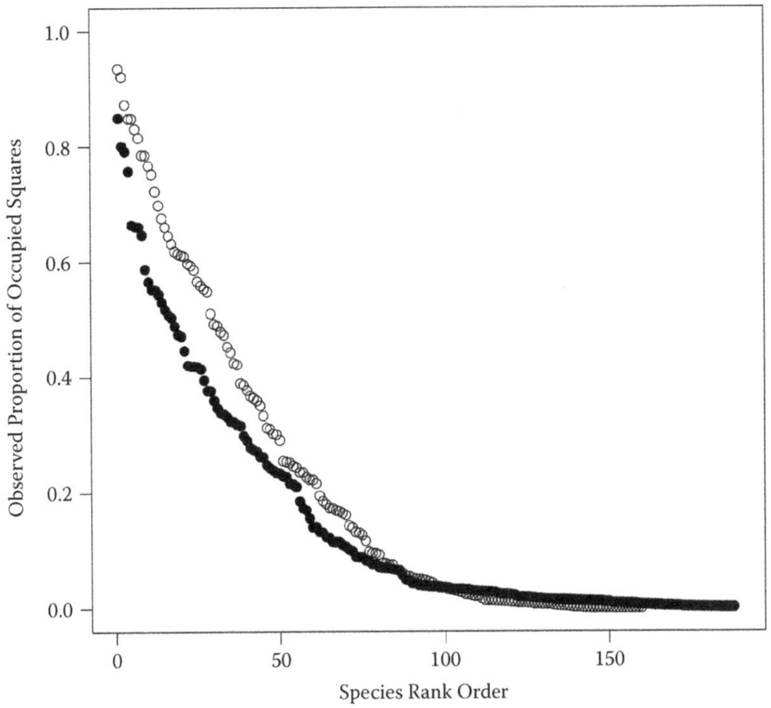

FIGURE 9.4 Rank-occupancy plot for 188 diurnal avian species in Catalonia (solid circles) and 160 diurnal avian species in Switzerland (open circles).

above an altitude of ca. 800 m a.s.l. The pattern at lower altitude is largely influenced by the local topography. Most of the lowest-lying areas in Switzerland (below ca. 300 m) are situated on the southern slope of the Alps, which is characterized by an insubrian climate with high precipitation in summer.

Observed Species Richness

In total, 216 diurnal breeding bird species were recorded in Catalonia and Switzerland together, 188 in Catalonia and 160 in Switzerland; 125 were shared species. The shape of the observed rank–occupancy curves for both countries is similar, with only about 15% of the species recorded in more than half of the squares and about 50% of the species recorded in fewer than 5% of all squares (Figure 9.4). However, for the more widespread species, occupancy was higher in Switzerland than in Catalonia; for the restricted-range species, it was higher in Catalonia. Hence, in Catalonia, there is higher evenness and therefore greater species richness.

Species richness declined with increasing altitude in both countries (Figure 9.5 and Figure 9.6). However, this decline is only clearly visible above a certain altitude. The "breakpoint," identified by piece-wise linear regression analysis, was located at 1,453 m a.s.l. in Catalonia and at 1,528 m in Switzerland. At lower altitude, observed species richness in Catalonia (ca. thirty species per square) was lower than in Switzerland (40 species per square), but higher at an altitude above about 2,400 m.

Relationships between the Observed Species Richness and Environmental Variables

Catalonia: Apart from the parametric (piece-wise linear) structure of altitude, the most parsimonious model for observed species richness in Catalonia contained all predictors that we considered *a priori* and explained 22.4% of the total variability in observed species richness. The optimal smooth structure contained 5.8 edf for relief ($p = 0.0004$), 1.8 edf for the cover of alpine pastures ($p = 0.002$), 2.1 edf for agriculture ($p = 0.003$), 7.5 for forest ($p \ll 0.001$), 6.3 for urban habitats ($p \ll 0.001$), and 8.7 for precipitation sum ($p \ll 0.001$). More species than expected at a given altitude were observed with little relief, at low values of the cover of alpine pastures, at low-medium values of the cover of agricultural habitats, at intermediate values of forest cover, at low values of urban cover, and at medium levels of precipitation.

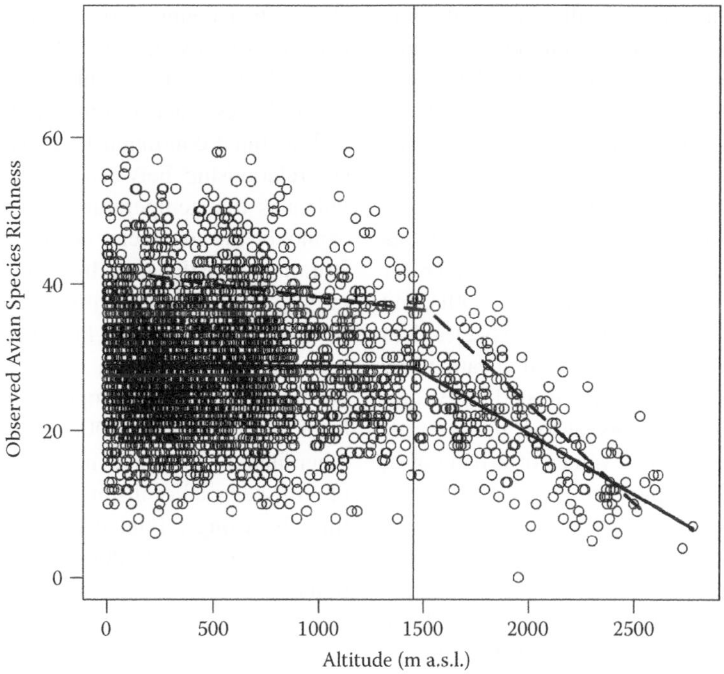

FIGURE 9.5 Observed avian species richness in relation to altitude in Catalonia. Best-fit piece-wise linear regression line (solid line) and location of the "breakpoint" at 1,453 m are superimposed. For comparison, the piece-wise regression line for the Swiss data (see Figure 9.6) also is shown (dashed line).

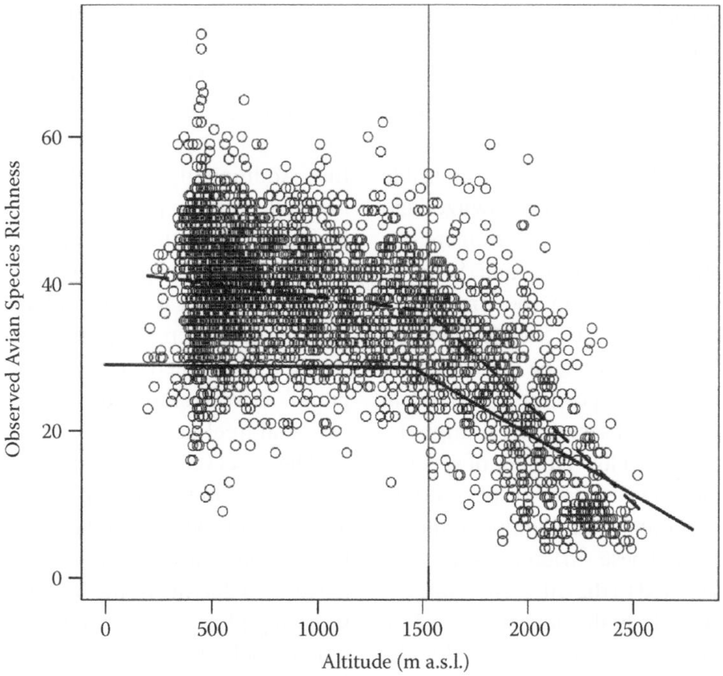

FIGURE 9.6 Observed avian species richness in relation to altitude in Switzerland. Best-fit piece-wise linear regression line and location of the "breakpoint" at 1,528 m are superimposed (dashed line). For comparison, the piece-wise regression line for the Catalonian data (see Figure 9.5) also is shown (solid line).

Switzerland: Mirroring these results from Catalonia in an almost perfect way, the most parsimonious model for the observed species richness in Switzerland contained all environmental variables other than relief (with an approximate $F_{6, 2582} = 1.765$, $p = 0.10$). This model explained a major portion of the total observed variability in the data ($R^2 = 66\%$). The optimal smooth structure contained 4.8 edf for the cover of alpine habitats ($p \ll 0.001$), 8.1 edf for agriculture ($p \ll 0.001$), 7.8 for forest ($p \ll 0.001$), 6.2 for urban habitats ($p < 0.001$), and 7.1 for precipitation sum ($p \ll 0.001$). More species than expected at a given altitude were observed at rather low values of the cover of alpine habitats, at medium values of the cover of agricultural habitats, at low, but not very low, values of forest cover, at low values of urban cover, and at medium levels of precipitation.

DISCUSSION

Our study demonstrates the usefulness of atlases, large biodiversity databases built to study the distribution of birds and other organisms. In this article, we compared observed species richness of diurnal breeding birds in 1 km^2 squares along two important altitudinal gradients in the Swiss Alps and the Catalan Pyrenees to investigate the shape of the richness-elevation relationship and possible habitat factors responsible for the altitudinal distribution of observed avian species richness. We found consistent patterns in both regions, with a decline in richness with altitude that seemed to be well described by a broken stick.

Slightly different field methods might be responsible for some of the differences found between Switzerland and Catalonia. Detection probability of species increases with increasing observation effort (Kéry and Schmid, 2004, 2006). Therefore, differences in the time spent for surveys (ca. 6 to 9 hours/square in Switzerland and only 2 hours/square in Catalonia) might be the reason for the overall lower species richness per km^2 observed in Catalonia. (However, the lower observer effort might be less important in situations with a low number of species, as is the case at high altitude.) Differences in species richness found between Switzerland and Catalonia should thus not be overemphasized. These differences in effort do not affect the findings related to the different shapes of the curves in each study area. In both countries, a decline in species richness was found beyond a certain altitude, which is consistent with many other studies so far. This decline was, however, not linear but showed a "breakpoint" below which hardly any decline was found. An analysis of the data of Terborgh (1977), taking observation effort into account, also revealed a nonlinear relationship between species richness of breeding birds and altitude (Rahbek, 1995). The number of species in the study area in the Andes reached a maximum around 1,400 m a.s.l. and declined continuously above that level. The nonlinear relationship between species richness and altitude indicates that temperature, which is declining linearly, cannot be the sole factor determining species richness. In Britain, where the altitudinal gradient is much lower than in our study regions, summer temperature was found to be the best predictor for bird diversity patterns, but there was no evidence for a mechanism operating directly through effects of temperature on thermoregulatory loads (Lennon et al., 2000). Temperature, together with other factors, may have indirect effects on the occurrence of bird species. For instance, the altitude of the upper tree limit is strongly related to temperature (Körner, 1998; Körner and Paulsen, 2004). This marked habitat transition might be responsible for the observed decline of species above a certain altitude, especially because habitat requirements of birds are often more related to structural elements in their habitats than to the presence of, e.g., certain plant species (Wiens, 1989). However, the identified "breakpoints" (1,528 m in Switzerland and 1,453 m in Catalonia) are clearly lower than the known tree limits in both countries: 2,200 to 2,400 m in Catalonia (Carreras et al., 1996); 1,900 to 2,200 m in Switzerland (Eggenberg, 1995). This pattern could be related to reduced tree density, tree height, and tree species diversity at lower altitude, thus decreasing niche availability for many bird species at elevations well below the treeline. Additionally, the treeline may have been artificially lowered in many places to create pastures, especially in Switzerland. In the case of Catalonia, the almost complete absence of agricultural areas above 1,300 m also might contribute to a loss of habitat diversity and, consequently, to a decrease in the number of bird species. All the aforementioned causes might explain why the "breakpoint" in Switzerland and in Catalonia lies far below the potential treeline.

The observed pattern also might indicate that temperature as a limiting factor becomes only apparent above a certain altitude, while below, other environmental factors and land use will be more important. In Germany, however, a decline of the number of bird species on small sampling plots (2.3 to 6.8 ha) was found within a 500 m altitude belt at low altitude (150 to 650 m a.s.l.) in similar habitats (oak–beech forests), which was attributed to a decline in plant and invertebrate diversity with increasing altitude (Wink and Wink, 1986). At this small scale, these results might, however, simply indicate changes in density of the observed species. Land use as a factor determining species richness after correcting for altitude was found in our data for both Switzerland and Catalonia. In the two countries, all habitat categories contributed to the model. In Switzerland, forest and agricultural land, however, showed the strongest relationships to species richness, with forests positively and agricultural land negatively influencing species richness. Species richness association with agricultural land differed between Switzerland and Catalonia, with a more positive relationship being observed in the latter. This difference between the two countries might, at least partially, be a consequence of the intensity in agricultural land use in Switzerland, where many species have abandoned agricultural areas in the course of the twentieth century which can still regularly be observed in Catalonia, where the landscape still contains more natural elements.

Our analysis confirmed the general relationship between altitude and species richness beyond the region of intense land use (> 1,500 m). For lower elevations, both mountain regions show no clear elevational trend, which we attribute to a confounding of natural drivers with anthropogenic drivers. Despite the fact that species composition in Catalonia and Switzerland is fairly comparable, the patterns found showed differences, which cannot be explained easily. Further analyses comparing patterns of occurrence at a species level and studies of the factors limiting the occurrence might help explain these differences.

The comparison of data from Switzerland and Catalonia showed the value of using data from distribution atlases for large-scale comparisons between regions. Similar data are widely collected across the world and have a high potential for further analyses, provided that the data are available in digital format and are at the same scale. We recommend a standard observation scheme for future atlas projects, or data stratification, in order to provide better comparisons among schemes; for instance, in an atlas scheme based on 2 × 2 km squares, the quadrants (top left, top right, bottom left, bottom right) should be recorded separately for each observation, which would allow an analysis at a scale of 1 km^2 as well.

During the last fifty years, more than 400 bird atlases were assembled worldwide (Gibbons et al., 2007). Unfortunately, only a small portion of the data already has been digitized and is available in adequate databases. Providing atlas data in databases should be a requirement for new atlas projects, and digitization of the data from older atlases is important, given the urgent need for widespread and large-scale data on the state of biodiversity (Purvis and Hector, 2000; Jenkins et al., 2003).

SUMMARY

Data collected for distribution maps (atlas) of breeding birds provide the basis for a variety of analyses of bird species richness patterns and comparisons between regions. Here, we used such atlas databases to study species richness of diurnal breeding birds along an altitude gradient in 1 × 1 km squares from Switzerland (2,622 squares) and Catalonia, Spain (3,038 squares). In total, 216 species were recorded (Switzerland, 160; Catalonia, 188), 125 were shared. As expected, species richness decreased with altitude; however, that decline was not linear. Up to a certain altitude, hardly any decrease was found, but beyond this "breakpoint," species richness declined, in Switzerland more strongly than in Catalonia. The location of that "breakpoint" was estimated at approximately 1,450 m a.s.l. in Catalonia, and at 1,530 m a.s.l. in Switzerland. We argue that this pattern is likely to be influenced by a decrease in plant biomass associated with altitude and matching processes leading to a decline in the diversity of habitat niches for birds. The availability of bird distribution data in well-managed databases has been a precondition for our study and will no doubt foster the conduct of more studies that tackle important large-scale ecological questions.

REFERENCES

Bezzel, E. 1971. Große Analyse der Verbreitung einiger Brutvögel in den Bayerischen Alpen und ihrem Vorland. *Anz. Ornithol. Ges. Bayern* 10:7–37.

Brotons, L., W. Thuiller, M.B. Araújo, and A.H. Hirzel. 2004. Presence-absence versus presence-only modelling methods for predicting bird habitat suitability. *Ecography* 27:437–448.

Bundesamt für Statistik. 1999. *GEOSTAT—Arealstatistik 1979/85 und 1992/97*. Neuchâtel: Bundesamt für Statistik.

Carnicer, J., L. Brotons, D. Sol, and P. Jordano. 2007. Community-based processes behind species richness gradients: Contrasting abundance–extinction dynamics and sampling effects in areas of low and high productivity. *Global Eco. Biogeogr.* 16:709–719.

Carreras, J., E. Carrillo, R.M. Masalles, J.M. Ninot, I. Soriano, and J. Vigo. 1996. Delimitation of the supra-forest zone in the Catalan Pyrenees. *Bull. Soc. Linn. Provence* 46:27–36.

Currie, D.J. 1991. Energy and large-scale patterns of animal- and plant-species richness. *Am. Nat.* 137:27–49.

Currie, D.J., G.G. Mittelbach, H.V. Cornell, R. Field, J.-F. Guégan, B.A. Hawkins, D.M. Kaufman, J.T. Kerr, T. Oberdorff, E. O'Brien, and J.R.G. Turner. 2004. Predictions and tests of climate-based hypotheses of broad-scale variation in taxonomic richness. *Ecol. Lett.* 7:1121–34.

Dorazio, R.M., J.A. Royle, B. Söderström, B., and A. Glimskär. 2006. Estimating species richness and accumulation by modeling species occurrence and detectability. *Ecology* 87:842–54.

Eggenberg, S. 1995. Ein biogeographischer Vergleich von Waldgrenzen der nördlichen, inneren und südlichen Schweizeralpen. *Mitt. Nat. forsch. Ges. Bern* 52:97–120.

Estrada, J., V. Pedrocchi, L. Brotons, and S. Herrando. 2004. *Atles dels ocells nidificants de Catalunya 1999–2002*. Barcelona: Institut Català d'Ornitologia (ICO)/Lynx Edition.

Evans, K.L., P.H. Warren, and K.J. Gaston. 2005. Species–energy relationships at the macroecological scale: A review of the mechanisms. *Biol. Rev.* 80:1–25.

Gibbons, D.W., P.F. Donald, H.-G. Bauer, L. Fornasari, and I.K. Dawson. 2007. Mapping avian distributions: The evolution of atlas studies. *Bird Stud.* 54:324–34.

Hawkins, B.A., E.E. Porter, and J.A.F. Diniz-Filho. 2003. Productivity and history as predictors of the latitudinal diversity gradient of terrestrial birds. *Ecology* 84:1608–23.

Huntley, B., R.E. Green, Y.C. Collingham, and S.G. Willis. 2007. *A Climatic Atlas of European Breeding Birds*. Barcelona: Durham University, The RSPB and Lynx Edicions.

Kerr, J.T., and L. Packer. 1997. Habitat heterogeneity as a determinant of mammal species richness in high-energy regions. *Nature (London)* 385:252–54.

Kéry, M., J.A. Royle, and H. Schmid. 2008. Importance of sampling design and analysis in animal population studies: A comment on Sergio et al. *J. Appl. Ecol.* 45:981–986.

Kéry, M., and H. Schmid. 2004. Monitoring programs need to take into account imperfect species detectability. *Basic Appl. Ecol.* 5:65–73.

Kéry, M., and H. Schmid. 2006. Estimating species richness. Calibrating a large avian monitoring programme. *J. Appl. Ecol.* 43:101–10.

Körner, C. 1998. A re-assessment of high elevation treeline positions and their explanation. *Oecologia* 115:445–59.

Körner, C. 2000. Why are there global gradients in species richness? Mountains might hold the answer. *Trend Ecol. Evol.* 15:513–14.

Körner, C. 2004. Mountain biodiversity, its causes and function. *Ambio Spec. Rept.* 13:11–17.

Körner, C., and J. Paulsen. 2004. A world-wide study of high altitude treeline temperatures. *J. Biogeogr.* 31:713–32.

Lennon, J.J., J.J.D. Greenwood, and J.R.G Turner. 2000. Bird diversity and environmental gradients in Britain: A test of the species–energy hypothesis. *J. Anim. Ecol.* 69:581–98.

MacArthur, R.H. 1972. *Geographical Ecology: Patterns in the Distribution of Species*. New York: Harper & Row.

Navarro S., A.G. 1992. Altitudinal distribution of birds in the Sierra Madre del Sur, Guerrero, Mexico. *Condor* 94:29–39.

Ninyerola, M., X. Pons, and J.M. Roure. 2000. A methodological approach of climatological modelling of air temperature and precipitation through GIS techniques. *Int. J. Climatol.* 20:1823–41.

Purvis, A., and A. Hector. 2000. Getting the measure of biodiversity. *Nature* 405:212–19.

Podulka, S., R.W. Rohrbaugh Jr., and R. Bonney. 2004. Handbook of Bird Biology. Ithaca, N.Y.: Cornell Laboratory of Ornithology.

Rahbek, C. 1995. The elevational gradient of species richness: A uniform pattern? *Ecography* 18:200–205.

Romdal, R.S., J.-A. and Grytnes. 2007. An indirect area effect on elevational species richness pattern. *Ecography* 30:440–48.

Schmid, H., R. Luder, B. Naef-Daenzer, R. Graf, and N. Zbinden. 1998. *Schweizer Brutvogelatlas. Verbreitung der Brutvögel in der Schweiz und im Fürstentum Liechtenstein 1993-1996/Atlas des oiseaux nicheurs de Suisse. Distribution des oiseaux nicheurs en Suisse et au Liechtenstein en 1993–1996*. Sempach: Schweizerische Vogelwarte/Station ornithologique suisse.

Terborgh, J. 1971. Distribution on environmental gradients: Theory and a preliminary interpretation of distributional patterns in the avifauna of the Cordillera Vilcabamba, Peru. *Ecology* 52:23–40.

Terborgh, J. 1977. Bird species diversity on an Andean elevational gradient. *Ecology* 58:1007–19.

Thiollay, J.-M. 1980. L'évolution des peuplements d'oiseaux le long d'un gradient altitudinal dans l'Himalaya central. *Rev. Ecol. (Terre Vie)* 34:199–269.

Turner, J.R.G., J.J. Lennon, and J.A. Lawrenson. 1988. British bird species distributions and the energy theory. *Nature (London)* 335:539–41.

Viñas, O., and X. Baulies. 1995. 1:250 000 land-use map of Catalonia (32 000 km^2) using multitemporal Landsat-TM data. *Int. J. Remote Sens.* 16:129–46.

Wartmann, B., and R.K. Furrer. 1977. Zur Struktur der Avifauna eines Alpentales entlang des Höhengradienten. I. Veränderungen zur Brutzeit. *Ornithol. Beob.* 74:137–60.

Wiens, J.A. 1989. *The Ecology of Bird Communities*. Cambridge: Cambridge University Press.

Wink, M., and C. Wink. 1986. Diversität und Abundanz der Vogelgesellschaften von Buchen-Eichen-Hochwäldern in Relation zu Exposition, Vegetation und Höhenlage. *Ökol. Vögel* 8:179–88.

Wood, S.N. 2006. *Generalized Additive Models. An Introduction with R*. Boca Raton, Fla.: Chapman & Hall.

Wright, D.H. 1983. Species–energy theory: An extension of species--area theory. *Oikos* 41:496–506.

Zamora-Rodríguez, R. 1987. Variaciones altitudinales en la composiciòn de las comunidades nidificantes de aves de Sierra Nevada (Sur de España). *Doñana, Acta Vertebrata* 14:83–106.

10 Diverse Elevational Diversity Gradients in Great Smoky Mountains National Park, U.S.A.

Nathan J. Sanders, Robert R. Dunn, Matthew C. Fitzpatrick, Christopher E. Carlton, Michael R. Pogue, Charles R. Parker, and Theodore R. Simons

CONTENTS

Introduction 75
Methods 76
Study Area 76
Patterns of Diversity 76
Underlying Causes of Diversity Gradients 78
Results 78
Patterns of Diversity 78
Underlying Causes of Diversity Gradients 79
Discussion 83
Patterns of Diversity 83
Underlying Causes of Diversity Gradients 84
Summary 85
Acknowledgments 85
References 85

INTRODUCTION

To do science is to search for repeated patterns. . . .

—Robert MacArthur, 1972

Why does the number of species vary geographically? The earliest naturalists puzzled over this question (von Humboldt, 1808), as do many biogeographers and macroecologists today (Gaston, 2000; Hawkins et al., 2003; Currie et al., 2004). Over the last 200-plus years, the most striking geographic pattern in species richness—the decline in species richness with increasing latitude—has received the most attention (e.g., Hildebrand, 2004). Thanks to many recent theoretical developments (Colwell et al., 2004), coupled with global-scale databases (e.g., Kreft and Jetz, 2007; Jetz et al., 2007; http://www.gbif.org) and satellite technology, the number of candidate mechanisms that shape the latitudinal diversity gradient has been whittled down to a manageable number (Hillebrand, 2004).

Less well studied, however, are the factors that shape elevational diversity gradients. Because many climatic factors vary systematically along elevational gradients, as they might along latitudinal gradients, elevational diversity gradients were thought to be miniature versions of latitudinal gradients (Körner, 2000). For example, Brown (1988) wrote, "Just as change of physical conditions with altitude resembles in many respects the variation with latitude, so the decreasing diversity of most organisms with increasing elevation mirrors in most respects the latitudinal gradient of species richness." Stevens (1992) noted that, "Biologists have long recognized that elevational and latitudinal species-richness gradients mirror each other." Although the most common relationship between latitude and richness is a decline in diversity with increasing latitude, this is not the most common pattern along elevational gradients. Rahbek's (1995, 2005) thorough reviews of published studies on elevational gradients showed that mid-elevation peaks in diversity are the norm. This suggests that elevational gradients do not mirror latitudinal gradients.

Rahbek's (1995, 2005) approach to assessing how diversity varies with elevation was to count the number of published studies that showed monotonic decreasing, hump-shaped, flat-horizontal, then decreasing, increasing, or some other relationship between richness and elevation. The studies that he compiled were from various mountain ranges, and on various taxa. One reason that different patterns of elevational diversity might occur in different systems is because the scale and extent of the elevational gradients varied among studies (Rahbek, 2005) or because different mountain ranges are embedded in different regional climatic areas with different evolutionary histories.

Another approach to understand how diversity varies with elevation is to analyze the patterns of diversity for several taxa along the same elevational gradient (Pausas, 1994; Pharo et al., 1999; Kessler, 2000; Grytnes et al., 2006). Such analyses are relatively rare in the literature, perhaps because they require synthesizing multiple electronic databases. Clearly, understanding whether many taxa respond to elevation in the same ways will help uncover the underlying mechanisms. Of course, diversity does not respond to latitude or elevation per se; latitude and elevation are only surrogates for a variety of factors that shape diversity gradients (Körner, 2007). For example, both climate and area affect diversity (e.g., Currie et al., 2004; Rosenzweig et al., 1995) and vary along elevational gradients (Rahbek, 2005; Romdal and Grytnes, 2007). The strengths of examining diversity gradients for several taxa along the same elevational gradient are that (1) one can control for different environmental histories and regional factors that often exist among different mountain ranges, and (2) climatic data are often easier to obtain along a single gradient than along many gradients dispersed throughout the world. Moreover, because the factors that lead to variation in species richness may differ among taxa, comparing elevational diversity gradients across taxa may provide useful insights about the factors that shape diversity gradients more generally.

In this study, we examine elevational gradients in diversity for several taxa along a single elevational diversity gradient in Great Smoky Mountains National Park (GSMNP) in the southern Appalachians of the southeastern United States. This is a unique montane ecosystem for several reasons. First, there is a long and storied history of ecological and biodiversity research in GSMNP, going back more than seventy years (e.g., Whittaker, 1952, 1956). Second, GSMNP is one of the most well-surveyed national parks in the United States. And third, since the mid-1990s, GSMNP has hosted an All-Taxa Biodiversity Inventory (ATBI) that aims to catalog the diversity of all life in the park (Sharkey, 2001; http://www.dlia.org), and much of the data from the ATBI are freely available online for investigators to mine. Therefore, it is possible to assess diversity gradients, and their potential underlying causes, for a variety of taxa. Here, we focus on elevational diversity gradients in ants, noctuid moths, breeding birds, and beetles. In addition, eleven sites were installed for a "pilot study" to understand better how to systematically sample biodiversity. From those eleven sites, we compare diversity gradients of spiders, beetles, flies, bugs, hymenopterans, and orthopterans. Specifically, we asked three questions: (1) Do different taxa exhibit different relationships between elevation and diversity in GSMNP? (2) Does area, climatic factors, or habitat diversity account for most of the variation in species richness? and (3) Does their relative effect (importance?) vary among taxa?

METHODS

Study Area

GSMNP (area = 2,111 km^2) is located in the southern Appalachian Mountains on the border of Tennessee and North Carolina, U.S.A. Elevation ranges from 270 m to 2,025 m. Approximately 95% of GSMNP is at least partially forested, and because of the extensive elevational gradient, many different forest types are found in the park, including some of the largest tracts of primary forest in the eastern U.S.A. The high elevation forests are not entirely evergreen—in some areas Northern red oak, buckeye-yellow birch, and beech associations are common.

Patterns of Diversity

The data we analyze here were collected in a variety of ways. Below, we summarize the sampling techniques for each taxon and for the "pilot study." Importantly, we did not interpolate species richness by assuming that species were present in all elevational zones between the highest and lowest elevations at which they were collected (e.g., Grytnes and Vetaas, 2002; Grau et al., 2007), because this can inflate the signal of a mid-elevation peak in diversity when it might not actually exist.

Ants: Sanders et al. have sampled ants at twenty-nine forested sites ranging in elevation from 379 m to 1,828 m (Sanders et al., 2007; Lessard et al., 2007; Geraghty et al., 2007). The sites were all in mixed hardwood forests and located in areas away from roads, heavily visited trails, or other recent human disturbances. At each site, a 50 m x 50 m plot was placed, and we sampled ants in sixteen 1 m^2 quadrats within each site. Within each 1 m^2 quadrat, we collected the leaf litter and sifted it through a coarse mesh screen of 1 cm grid size to remove the largest fragments and concentrate the fine litter. The litter fragments that did not fit through the mesh, twigs, and sticks in each 1 m^2 quadrat were inspected for colonies. The concentrated fine litter from each of the sixteen 1 m^2 quadrats was then suspended in mini-Winkler sacks for two days in the laboratory. All worker ants that were extracted from the 1 m^2 quadrats were identified, enumerated, and stored in N. Sanders's ant collection at the University of Tennessee. A species list is available from N. Sanders. For more details of the sampling design, see Sanders et al., 2007, and Lessard et al., 2007.

Noctuid moths: Pogue et al. have sampled noctuid moths during 202 sampling bouts at 121 sites ranging in elevation from 305 m to 2,024 m in elevation. A variety of habitat types were sampled, including cove hardwood forests and old fields at low elevations, pine–oak and northern hardwood forests at mid-elevations, heath balds at mid- and high elevations, and spruce–fir forests at the highest elevations. At each site, a 15 w UV bulb attached to a box-type trap, and various types of UV or mercury vapor light, either in bucket-type traps or against a white sheet, was used to sample moths (Pogue, 2005, 2006). Sites were sampled at different times of year from 2001–2005 to capture phenological shifts in the moth fauna. We divided the elevational gradient into sixteen 100 m bands and combined the samples from all of the sites within each elevational band, as is common in many elevational diversity gradient studies (Rahbek, 2005). Observed species richness is the number of species collected within 100 m elevational bands.

This study follows the recent classification of the Noctuidae, in which the former families Pantheidae, Lymantriidae, Nolidae, and Arctiidae, (Kitching and Rawlins, 1999) are treated as subfamilies, making the Noctuidae easily defined and monophyletic (Lafontaine and Fibiger, 2006). Specimens were identified to species, and voucher specimens are stored at the U.S. National Museum, Smithsonian Institution, Washington, D.C.

Beetles: Species analyzed for this study were a subset of approximately 2,300 species currently recorded from GSMNP based on modern and historical records (Carlton and Bayless, 2007; complete checklist posted at http://entomology.lsu.edu/lsam/smokybeetles.htm). The beetle fraction analyzed here was obtained from the eleven pilot study sites, mainly from Malaise trap samples, and from collections using a diversity of methods by the team from the Louisiana State Arthropod Museum and cooperators within the beetle twig of the ATBI during 2001–2006. A large proportion of specimens were derived from forest litter sampling from seventy-two localities across the entire range of GSMNP elevations and forest types. Those samples typically consisted of 2 to 5kg of litter that had been sifted through 0.8 cm mesh wire screen. Specimens were extracted using standard Berlese funnel techniques (e.g., Carlton and Robison, 1998). Additional methods employed during the same time frame included flight intercept trapping, pitfalls, light traps, and hand collecting from vegetation, dead wood, under rocks, and fungi. Specimens were identified to species or genus, and sorted to morphospecies (for taxonomically intractable taxa). Vouchers are divided between the Louisiana State Arthropod Museum and the GSMNP collection.

Breeding birds: Simons et al. conducted 7,535 variable circular plot point transects (Reynolds et al., 1980) at 4,157 point locations from mid-May to the end of June in GSMNP (Shriner, 2001; Shriner et al., 2002; Simons et al., 2006). Points were established ~250 m from one another, mostly along low-use hiking trails. However, some points were located along roads with little traffic and along off-trail transects. At each point location, all of the birds seen or heard within 10 minutes were recorded. Observed species richness is the number of species detected within 100 m elevational bands. Birds present unique problems, which many of the other taxa in this study do not. Namely, detection probability can differ drastically among species, nocturnal birds are generally not sampled, and altitudinal migrants pose other problems as well. However, we note that many such studies of patterns of bird diversity, across spatial scales, suffer from the same shortcomings.

The "pilot study": Between January 1999 and January 2002, C. Parker organized the structured sampling of arthropod biodiversity at eleven sites ranging in elevation from 521 m to 1,944 m. At each site, ten pitfall traps were placed ~3 m apart along an approximately 30 m long transect. The pitfall traps were 6 cm diameter cups buried flush with the soil surface and were collected every two weeks. At the same sites, two Malaise traps were placed on the ground 75 to 100 m from one another, and the contents of the traps were collected every two weeks from January 1999 to January 2002. For the pilot study, observed species richness is the number of species collected at each site from 1999–2002.

Underlying Causes of Diversity Gradients

We asked how three factors might influence patterns of diversity across taxa in this system: actual evapotranspiration (AET), area, and habitat diversity. To estimate AET, we obtained temperature and precipitation data for GSMNP from the WorldClim 1.4 database (Hijmans et al., 2005). Though many environmental data are available, we limited the number of variables used here to AET because we wanted to minimize the number of collinear variables in subsequent analyses and because AET is an important correlate of diversity at broad spatial scales (Currie et al., 2004). We estimated AET based on Turc's formula (Turc, 1954; Kluge et al., 2006), where $AET = P/[0.9 + (P/L)^2]^{1/2}$ with $L = 300 + 25T + 0.05T^3$, P = annual mean precipitation, and T = mean annual temperature. AET is strongly related to net primary productivity ($r^2 = 0.93$), but the relationship is nonlinear (Kaspari et al., 2000).

To estimate area in each elevational band, we used a 30 m resolution digital elevation model of GSMNP to estimate the area of each elevational band from 400 m to > 1,900 m.

We estimated habitat diversity as the number of different habitat types in each 100 m elevational band by combining a 30 m resolution digital elevation model with a vegetation map of the GSMNP. The map was created in 2004 and is based upon 1:12,000 color-IR photography. The minimum mapping unit is 0.5 ha. The accuracy assessment indicated that the classifications were 80% accurate (Michael Jenkins and Ed Laurent, *personal communication*).

Statistical analyses. In addition to examining patterns in observed species richness, *S*, with elevation, we also calculated Fisher's α, a widely used estimate of diversity that is independent of sample size (Evans et al., 2005). Fisher's α also removes the sampling effect (i.e., the fact that diversity might be high at a site simply because there are more individuals at that site). We first related *S* and Fisher's α at each site (for the ants and for the species from the "pilot study") or in each elevational band by regressing both *S* and Fisher's α against elevation. For each richness-elevation plot, we asked whether a linear or polynomial regression best captured the relationship between richness and elevation. For the "pilot study" data, we tested for taxonomic covariance using pair-wise correlations among taxa.

To examine some potential factors which may shape variation in richness of ants, noctuid moths, beetles, and breeding birds along the elevational gradient, we used forward stepwise regression (P to enter < 0.10) to test whether AET, AET^2, area, or habitat diversity within the elevational band accounted for most of the variation in *S* and Fisher's α for each set of taxa. AET^2 accounts for potential curvilinear relationships between diversity and AET. For each taxon, we used AIC scores to determine the best model for each elevational diversity gradient. Owing to limited sample sizes (n = 11), we did not explore the potential underlying causes of the diversity from the pilot study.

RESULTS

Patterns of Diversity

Ants: We collected forty-one species of ants in forested ecosystems in GSMNP. Ant species richness declined linearly with elevation ($r^2 = 0.59$, $P < 0.0001$; Figure 10.1a). Fisher's α also declined linearly with elevation ($r^2 = 0.63$, $P < 0.0001$; Figure 10.2a).

Noctuid moths: In total, we collected 11,322 individuals and 517 species of noctuid moths. Observed species richness declined with elevation (quadratic regression: $r^2 = 0.51$, $P = 0.01$; Figure 10.1b), but leveled off at the highest elevations (above 1,500 meters). Fisher's α declined linearly with elevation ($r^2 = 0.60$, $P = 0.0004$; Figure 10.2b).

Beetles: We collected 847 species from a data set of 21,308 individuals. We note that hundreds of specimens have yet to be identified. Neither observed species richness nor Fisher's α varied systematically with elevation (Figure 10.1c, Figure 10.2c).

Breeding birds: In total, 65,489 individuals and 111 species were detected. Bird species richness exhibited a hump-shaped relationship with elevation, with richness peaking at 700 m. The diversity gradient was best described by a quadratic regression ($y = -0.0004x^2 + 0.0644x + 43.067$, $r^2 = 0.52$, $P = 0.005$; Figure. 10.1d). This relationship was driven by two low-elevation, low diversity elevational bands. These two points were relatively undersampled and clustered around a few particular locations. If they are removed from the analysis, then richness decreases monotonically with increasing elevation ($r^2 = 0.78$, $P < 0.0001$). The relationship between Fisher's α and elevation also was best explained by a polynomial regression ($= -0.000002x^2 + 0.00007x + 13.34$, $r^2 = 0.77$, $P < 0.0001$; Figure 10.2d), with diversity

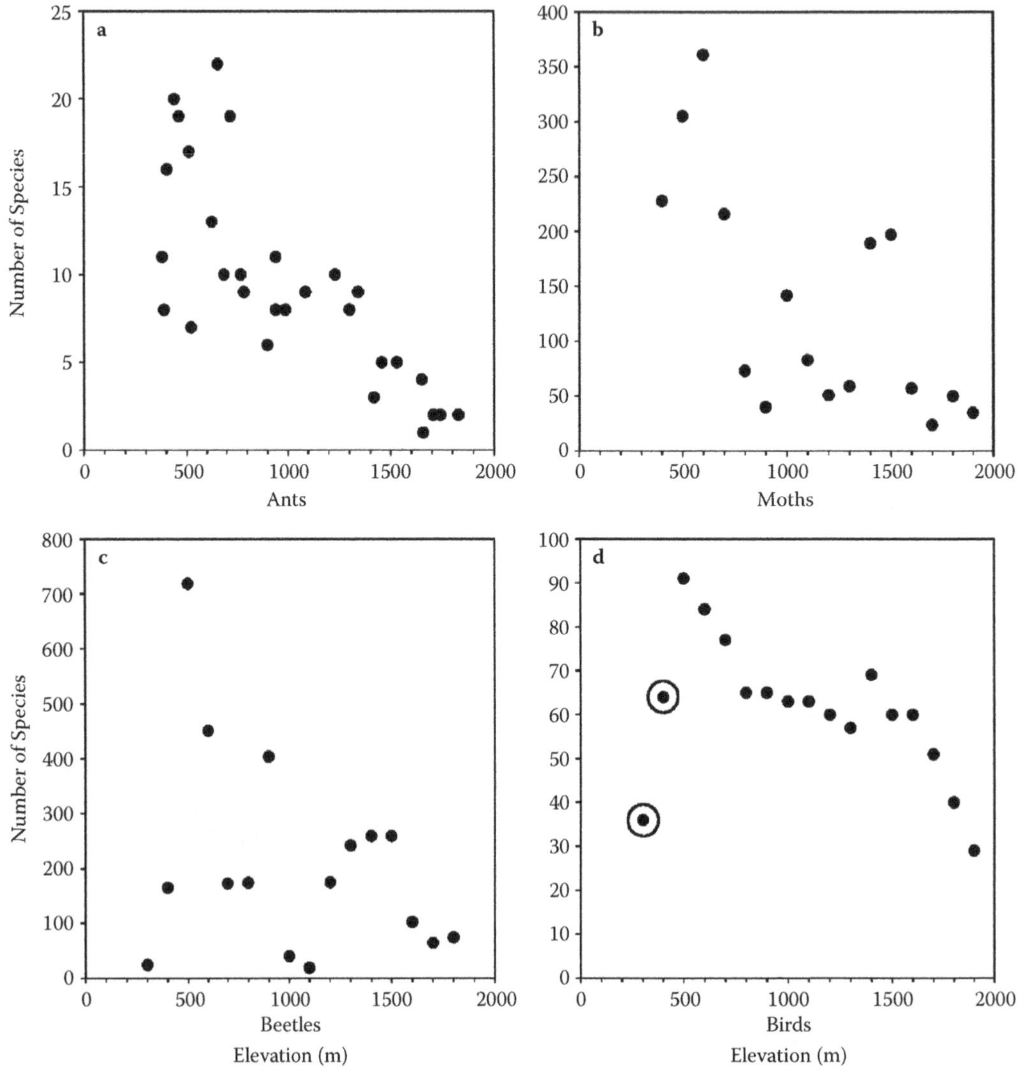

FIGURE 10.1 The relationship between species richness and elevation for ants at twenty-nine sites (a), noctuid moths in 100 m elevational bands (b), beetles in 100 m elevational bands (c), and birds in 100 m elevational bands (d) in Great Smoky Mountains National Park. In panel (d) the two circled data points indicate two low-elevation sites that were excluded from some analyses because they were not sampled as intensively as the other sites at higher elevations.

tending to be flat then declining at approximately 700 m. Again, however, we caution that the sampling was not designed to sample bird diversity, per se, and there are many confounding factors (e.g., variation in detection probability among species, altitudinal migrants, etc.) that we have glossed over in this manuscript.

The "pilot study": Richness declined with elevation for the spiders, coleopterans, and orthopterans, but there was no relationship between richness and elevation for the dipterans, hemipterans, or hymenopterans (Figure 10.3). Fisher's α, which corrects for sampling effects, did not vary systematically for any of the taxa except the orthopterans. Fisher's α of the orthopterans declined linearly with elevation ($r^2 = 0.65$, $P < 0.003$). Though the six arthropod taxa examined here exhibited different elevational diversity gradients, richness of several taxa covaried among sites (Table 10.1).

Underlying Causes of Diversity Gradients

AET declined monotonically with elevation ($r^2 = 0.93$, $P < 0.0001$), but the relationship between AET^2 and

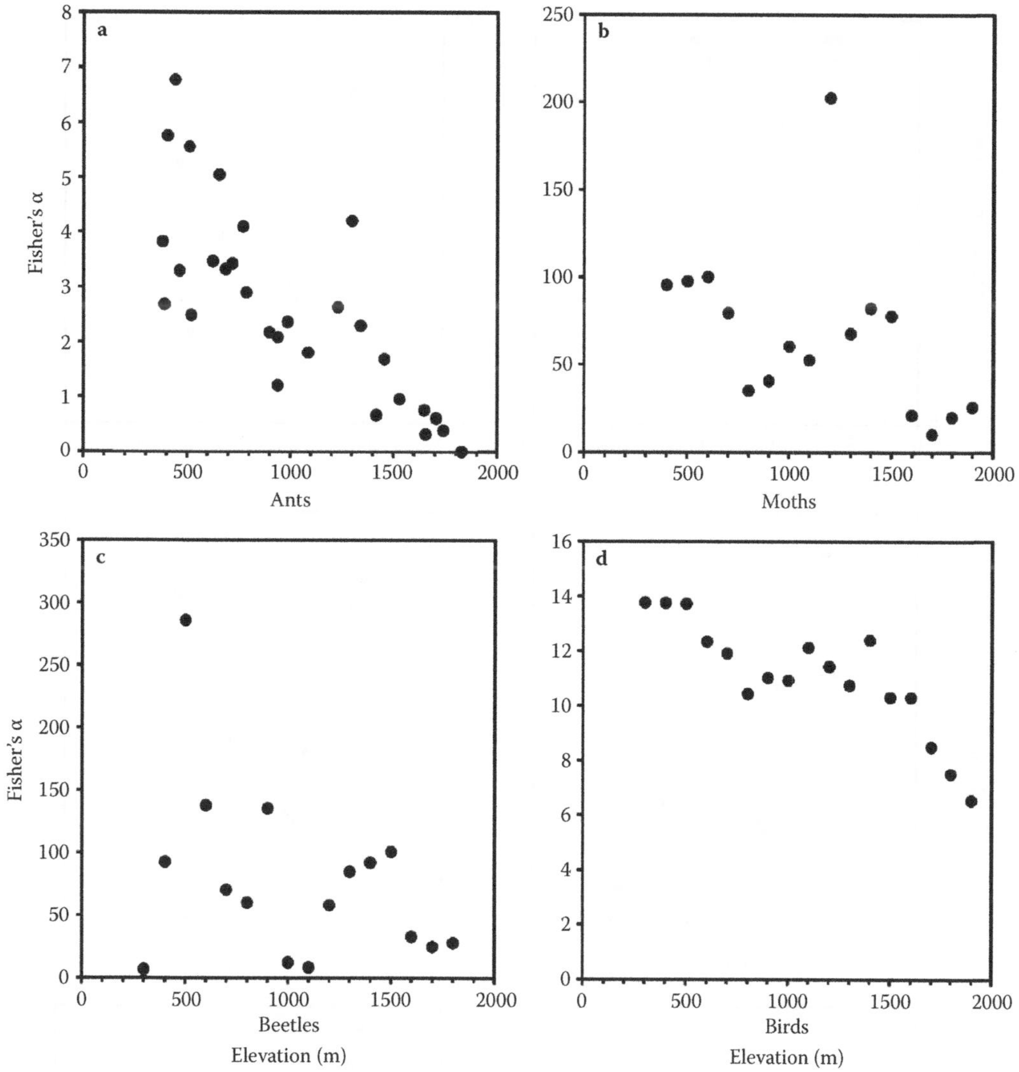

FIGURE 10.2 The relationship between Fisher's α diversity and elevation for ants at twenty-nine sites (a), noctuid moths in 100 m elevational bands (b), beetles in 100 m elevational bands (c), and birds in 100 m elevational bands (d) in Great Smoky Mountains National Park.

elevation was best described by a polynomial regression ($r^2 = 0.99$, $P < 0.0001$). The relationship between log-area of each elevational band and elevation was largely flat, then declined above 1,500 m ($r^2 = 0.94$, $P < 0.0001$). Habitat diversity was highest at low elevations, flat at mid-elevations, and then declined linearly above approximately 1,500 m. The relationship between habitat diversity and elevation was best described by a third order polynomial regression ($r^2 = 0.88$, $P < 0.0001$).

Ants: AET accounted for 60% of the variation in ant species richness (Table 10.2), and no other factor entered the stepwise regression model. Similarly, AET alone accounted for most of the variation in Fisher's α diversity of ants (Table 10.3).

Noctuid moths: Only AET^2 accounted for variation in noctuid moth richness, suggesting that the relationship between the richness of noctuid moths and AET is hump-shaped (Table 10.2). However, when we corrected for variation in moth abundance by using Fisher's α, only the number of habitats accounted for variation in moth diversity (Table 10.3).

Beetles: AET alone accounted for 27% of the variation in the number of beetle species (Table 10.2). However, correcting for variation in the number of individuals by

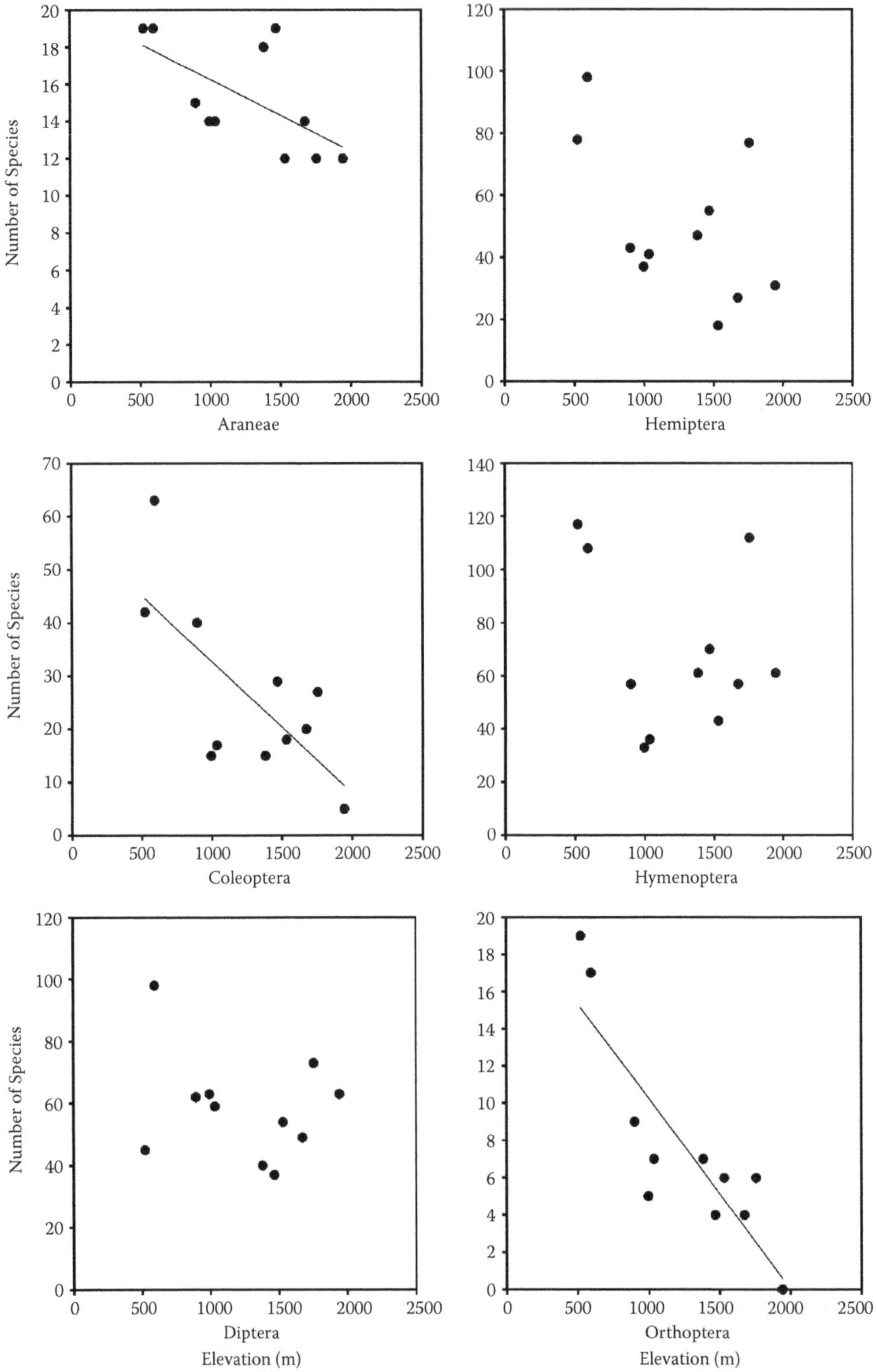

FIGURE 10.3 Results from the ATBI "pilot study." In each panel, the number of species collected at eleven sites over two years is displayed for that particular taxon. The line (where displayed) is the best-fit linear regression line. If no line is present, then the relationship was not statistically significant.

TABLE 10.1
Correlation Matrix for Five Arthropod Taxa Collected at Eleven Sites as Part of the All Taxa Biodiversity Inventory "Pilot Study" in Great Smoky Mountains National Park

	Coleoptera	Diptera	Hemiptera	Hymenoptera	Orthoptera
Araneae	**0.62**	–0.15	**0.60**	0.43	**0.64**
Coleoptera		0.50	**0.80**	**0.67**	**0.84**
Diptera			0.49	0.31	0.29
Hemiptera				**0.87**	**0.74**
Hymenoptera					**0.64**

Note: Values in the matrix are Pearson correlation coefficients. Significant ($P < 0.05$) values are indicated in bold.

TABLE 10.2
Results from Stepwise Regression Analysis of Log-Area, AET, AET², and Habitat Diversity on the Number of Species of Ants, Noctuid Moths, Beetles, and Birds in Great Smoky Mountains National Park

Taxon	Factor	Parameter	Partial r^2	P	Whole Model r^2	AIC
Ants	AET	0.036	0.60	< 0.0001	0.60	78.78
Moths	AET^2	0.0007	0.54	0.001	0.54	139.67
Beetles	AET	0.78	0.27	0.948	0.27	154.96
Birds	Log-Area	42.12	0.68	0.0001	0.94	50.89
	AET^2	0.0005	0.13	0.0009		
	No. of habitats	–1.20	0.09	0.005		
	AET	–0.39	0.03	0.03		

TABLE 10.3
Results from Stepwise Regression Analysis of Log-Area, AET, AET², and Habitat Diversity on Fisher's α Diversity of Ants, Noctuid Moths, Beetles, and Birds in Great Smoky Mountains National Park

Taxon	Factor	Parameter	Partial r^2	P	Whole Model r^2	AIC
Ants	AET	0.011	0.63	< 0.0001	0.63	6.64
Moths	No. of habitats	3.39	0.36	0.01	0.36	118.83
Beetles	AET	0.0004	0.33	0.02	0.33	124.3
Birds	No. of habitats	0.15	0.75	< 0.0001	0.83	–1.26
	AET^2	0.0006	0.83	0.02		

using Fisher's α indicates that the relationship between beetle diversity and AET is hump-shaped ($r^2 = 0.33$) because only AET^2 entered the model (Table 10.3).

Breeding birds: AET (partial $r^2 = 0.63$) and area (partial $r^2 = 0.11$) accounted for most of the variation in the number of bird species along the gradient (Table 10.2). If we remove the two low elevation, low diversity sites from the analysis, AET (partial $r^2 = 0.78$) and area (partial $r^2 = 0.08$) account for most of the variation in the number of bird species along the gradient. However, habitat diversity and AET^2 together accounted for 83% of the variation in Fisher's α diversity of birds (Table 10.3).

DISCUSSION

Our results show that patterns of species richness of a variety of taxa differ along the same elevational gradient, though AET seems to be an important correlate of diversity in this system, because AET and the focal organisms in this study respond to similar climatic drivers. Only because multiple investigators provided data to a single database was it possible to test whether the drivers of diversity were the same across taxa in this montane ecosystem. Other databases for montane ecosystems likely exist, and could be easily mined by others as a test of the generality of the patterns and their underlying causes we describe here.

Here, we first discuss the patterns of richness and taxonomic covariance along this elevational gradient. Then we move on to consider the underlying causes of the patterns.

Patterns of Diversity

A common pattern in studies of elevational diversity gradients is that diversity either declines with elevation or peaks at mid-elevations (Rahbek, 2005). We document similar patterns here: richness of ants and moths tended to decline with increasing elevation, but birds exhibited a mid-elevation peak in diversity, though with a peak near the base of the gradient. Beetles, however, exhibited no strong pattern in diversity along the elevational gradient. Other studies that have examined multiple taxa along a common elevational gradient also have found variation in diversity patterns among taxa (Pausas, 1994; Pharo et al., 1999; Kessler, 2000; Bhattarai and Vetaas, 2003; Grau et al., 2007). Interestingly, most of these studies focused solely on plants. To our knowledge, ours is one of the first to explore patterns of diversity for multiple animal taxa along the same elevational gradient.

Not surprisingly, species richness of ants declines with elevation. Most studies on elevational gradients in ant diversity have found that ant species decrease with elevation (Sanders et al., 2007; Lessard et al., 2007) or, less frequently, peak at mid-elevations (Olson, 1991; Fisher, 1996, 1998; Samson et al., 1997; Brühl et al., 1999; Sanders et al., 2003). Of those studies that found a peak in ant species richness, the peak occurred at elevations below the maximum elevation used in this study (Samson et al., 1997; Fisher, 1998).

The diversity of noctuid moths declined with elevation in this study, as in an earlier analysis of the same data (Sanders et al., in review). Of the other studies that have examined patterns in moth diversity along elevational gradients, nearly every conceivable pattern has been documented. For instance, the diversity of geometrid moths does not vary with elevation in the Andes (Brehm et al., 2003), or in tropical forests in Borneo (Schulze, 2000), or along Mt. Kilimanjaro (Axmacher et al., 2004). Similarly, the diversity of arctiid moths also is moderately constant along an elevational gradient in Borneo (Holloway, 1987). In contrast, diversity of geometrid moths along the Barva Transect in Costa Rica peaks at mid-elevations (Brehm et al., 2007). But our results here are that diversity declines nearly linearly with elevation. Of course, it could be the case that the patterns differ, but the underlying mechanisms are the same among studies, a subject to which we will return.

It is somewhat surprising that beetle diversity showed no pattern along the gradient. Most studies to date on elevational diversity gradients in beetles have focused on dung beetles (e.g., Escobar et al., 2005, and references therein), and the general pattern is that beetle diversity declines with elevation. So why is there no elevational diversity gradient for beetles in GSMNP? We can think of two reasons. First, we are ignoring a tremendous amount of trophic diversity by lumping all 847 species. It could be the case that the scarabs or the weevils or the chrysomelids alone decline with elevation. However, such an analysis is beyond the scope of this chapter. Second, though there is no strong elevational pattern, beetles at least weakly track variation in AET, suggesting that climatic variation along the elevational gradient, rather than elevation, per se, influences spatial variation in beetle diversity.

Bird diversity peaked at low-elevations in GSMNP. In a series of landmark studies on elevational diversity gradients in tropical birds, Terborgh (1971, 1977, 1985), and later Rahbek (1997), documented complex patterns of bird diversity along elevational gradients in the Andes. Terborgh's observations indicated that bird species richness declined monotonically with elevation, but when he corrected for sampling effort, the pattern was markedly hump-shaped, with a peak at mid-elevations. Similarly, Rahbek (1997) found monotonic decreases in diversity with elevation, but when he factored out the effect of area within each band, the pattern became hump-shaped. Kattan and Franco (2004) also found that diversity declined monotonically along an elevational gradient in the Andes. In our study in the southern Appalachians, the elevational diversity gradient is linear when we correct for sampling deficiencies at the lowest elevations.

Even when specifics of the diversity patterns differ among taxa, those patterns can still share common features. For example, across taxa, the peaks in diversity tended to be at approximately 500 to 600 m in elevation, the bottom or near to the bottom of our elevational gradient. In addition, the data from the "pilot study" indicate some degree of taxonomic covariance (e.g., Lamoreux et al., 2005). Gaston (2000) noted that the "lack of strong positive covariance in the species richness of higher taxa is significant in that it constrains the extent to which observed patterns in biodiversity can be extrapolated from one group to another, and from exemplar groups to biodiversity at large." Here, however, there is some suggestion that diversities are correlated among taxa. For example, coleopterans, orthopterans, hymenopterans, and hemipterans tend to covary with one another at the eleven sites sampled by Parker et al. during the pilot study. Practically speaking, this suggests that sampling one taxon, for example, the coleopterans, might be representative of the diversity of other taxa.

In this study, taxa from the pilot study covary; peaks in diversity along the elevational gradient for ants, birds, beetles, and noctuid moths are mostly congruent, but the elevational diversity patterns of the ants, birds, beetles, and moths do not necessarily mirror one another. For example, beetles and other taxa were correlated in richness in the pilot samples but not in our broader analyses of birds, ants, moths and beetles. How could this be? There are two possibilities. First, taxonomic covariance at small spatial scales tends to be more common than taxonomic covariance at large spatial scales (Gaston, 2000; but see Lamoreux et al., 2005). Particular sites within the eleven sites for the pilot study might be amenable to diverse taxa. But once scale increases to encompass 100 m elevational bands, then more habitat heterogeneity is introduced, as is variation in sampling completeness for the taxa of interest. Second, though the elevational diversity patterns at large scales might not be entirely congruent with one another, the underlying mechanisms shaping the patterns could be. This is the topic we turn to now.

Underlying Causes of Diversity Gradients

For the ants, beetles, moths, and birds, actual evapotranspiration accounted for at least a portion of the variation in species richness along the elevational gradient. However, the relationship between AET and richness varied among taxa. For example, the relationship between moth diversity and AET was hump-shaped, but ant, beetle, and bird diversity were all monotonically related to AET. There is a substantial body of literature about why energy availability and diversity should be correlated (e.g., see Clarke and Gaston, 2006, and Currie et al., 2004 for a review). Further, because AET and temperature are highly correlated in GSMNP (because rainfall is relatively invariant), we also cannot rule out mechanisms that involve temperature. AET is a strong correlate of energy availability and integrates over other factors that might influence diversity, such as season length, ambient humidity, and soil moisture. However, obtaining data on those factors, and how the focal organisms respond to them, proved beyond the scope of this study.

Along elevational gradients, the most plausible explanation is probably a "more individuals" mechanism (Kaspari et al., 2000), which posits that areas with more energy support larger populations and because populations are larger, extinction probabilities are reduced. As a result, areas with more energy have more species, simply because more individuals are supported. In the southern Appalachians, areas with higher temperature, and hence higher AET, do tend to have more individuals (of ants; Sanders et al., 2007), and sites with more individuals tend to have more species. However, other mechanisms above and beyond a "more individuals" mechanism must operate in this system because the relationship between bird and moth diversity and AET is hump-shaped. Moreover, bird diversity is best accounted for by area of each elevational band, which is common in other studies of bird diversity as well (Rahbek, 1997). Finally, area, the number of habitats, and AET all affect diversity above and beyond their effects on the number of individuals. When we correct for variation in the number of individuals by using Fisher's α, AET still accounts for the largest portion of the variation in ant and beetle diversity, but not moth and bird diversity.

Taken together, the results from our study suggest that the factors that influence elevational diversity gradients, at least for ants, moths, beetles, and birds in this system, need not be the same among taxa or affect diverse taxa in similar ways. A substantial portion of biogeographical and macroecological research has sought to elucidate the single (or handful) of mechanisms that shape broad-scale diversity patterns (e.g., Allen et al., 2002, but see Hawkins et al., 2007). But our results suggest that even along the same elevational gradient, neither patterns nor mechanisms are entirely congruent. MacArthur (1972) famously noted that, "To do science is to search for repeated patterns," but, at least in this system and in others (e.g., Grytnes et al., 2003), repeated patterns might be the exception rather than the rule. Only by mining the numerous data sets on montane biodiversity that are rapidly becoming available might ecologists come closer to understanding the general patterns (if such patterns exist) and underlying causes of elevational diversity gradients.

SUMMARY

To understand how diversity varies with elevation, a good approach is to analyze the patterns of diversity for several taxa along the same elevational gradient. Such analyses are rare, as they require synthesizing multiple electronic databases. In our study, we analyze patterns of diversity along the same elevational gradient in Great Smoky Mountains National Park (GSMNP) in the southern Appalachians (U.S.A.) for many different taxa (ants, noctuid moths, breeding birds, and beetles). GSMNP is one of the most well-surveyed national parks, with a long history and effort to catalogue the diversity of all life in the park (All-Taxa Biodiversity Inventory). Peaks in diversity along the elevational gradient are mostly congruent between the four groups of taxa studied, but elevational diversity patterns differ, and each taxon group responded to environmental variation in different ways. Particularly, actual evapotranspiration (AET) accounted for a large portion of variation in all four groups of taxa, and area and number of habitats accounted for another portion of variation. Our results suggest that even along the same elevational gradient, neither patterns nor mechanisms are entirely congruent, so at least in this system, repeated patterns are the exception rather than the rule.

ACKNOWLEDGMENTS

We all wish to thank the many students involved in the project, the wonderful staff at Great Smoky Mountains National Park, and the Friends of Great Smoky Mountains National Park, as well as Discover Life in America for partially funding the ATBI.

REFERENCES

Allen, A.P., J.H. Brown, and J.F. Gillooly. 2002. Global biodiversity, biochemical kinetics, and energetic-equivalence rule. *Science* 297:1545–48.

Axmacher, J.C., G. Holtmann, L. Scheuermann, G. Brehm, K. Müller-Hohenstein, and K. Fiedler. 2004. Diversity of geometrid moths (Lepidoptera: Geometridae) along an Afrotropical elevational rainforest transect. *Diversity and Distributions* 10:293–302.

Bhattarai, K.R., O.R. and Vetaas. 2003. Variation in plant species richness of different lifeforms along a subtropical elevation gradient in the Himalayas, east Nepal. *Global Ecology and Biogeography* 12:327–40.

Brehm G., R.K. Colwell, and J. Kluge. 2007. The role of environment and mid-domain effect on moth species richness along a tropical elevational gradient. *Global Ecology and Biogeography* 16:205–19.

Brehm, G., D. Süssenbach, and K. Fiedler. 2003. Unique elevational diversity patterns of geometrid moths in an Andean montane rainforest. *Ecography* 26:456–66.

Brown, J.H. 1988. Species diversity. In *Analytical Biogeography—An Integrated Approach to the Study of Animal and Plant Distribution*, pp. 57–89. Myers, A.A., and P.S. Giller, eds. New York: Chapman & Hall.

Brühl, C.A., M. Mohamed, and K.E. Linsenmair. 1999. Altitudinal distribution of leaf litter ants along a transect in primary forest on Mount Kinabalu, Sabah, Malaysia. *Journal of Tropical Ecology* 15:265–67.

Carlton, C.E., and V.M. Bayless. 2007. Documenting beetle diversity in Great Smoky Mountains National Park: Beyond the halfway point! (Arthropoda: Insecta: Coleoptera). Southeastern *Naturalist Special Issue* 1:183–92.

Carlton, C.E., and H.W. Robison. 1998. Diversity of litter-dwelling beetles in deciduous forests of the Ouachita highlands of Arkansas (Insecta: Coleoptera). *Biodiversity and Conservation* 7:1589–1605.

Clarke, A., and K.J. Gaston. 2006. Climate, energy and diversity. *Proceedings of the Royal Society (B)* 273:2257–66.

Currie, D.J., et al. 2004. Predictions and tests of climate-based hypotheses of broad-scale variation in taxonomic richness. *Ecology Letters* 7:1121–34.

Escobar, F., J.M. Lobo, and G. Halffter. 2005. Altitudinal variation of dung beetle (Scarabaeidae: Scarabaeinae) assemblages in the Colombian Andes. *Global Ecology and Biogeography* 14:327–37.

Evans, K.L., P.H. Warren, and K.J. Gaston. 2005. Species-energy relationships at the macroecological scale: A review of the mechanisms. *Biological Reviews* 80:1–25.

Fisher, B.L. 1996. Ant diversity patterns along an elevational gradient in the Reserve Naturelle Integrale D'Andringitra, Madagascar. *Fieldiana Zoology* 85:93–108.

Fisher, B.L. 1998. Ant diversity patterns along an elevational gradient in the Reserve Special d'AnjanaharibeSud and on the western Masoala Peninsula, Madagascar. *Fieldiana Zoology* 90:39–67.

Gaston, K.J. 2000. Global patterns in biodiversity. *Nature* 405:220–27.

Geraghty M.J., R.R. Dunn, and N.J. Sanders. 2007. Bergmann's rule in ants: Are patterns along latitudinal and elevational gradients congruent? *Myrmecological News* 10:51–58.

Grau, O., J.-A. Grytnes, and H.J.B. Birks. 2007. A comparison of altitudinal species richness patterns of bryophytes with other plant groups in Nepal, Central Himalaya. *Journal of Biogeography* 34:1907–15.

Grytnes, J.-A., E. Heegaard, and P.G. Ihlen. 2006. Species richness of vascular plants, bryophytes, and lichens along an altitudinal gradient in western Norway. *Acta Oecologica* 29:241–46.

Grytnes, J.-A., and O.R. Vetaas. 2002. Species richness and altitude: A comparison between simulation models and interpolated plant species richness along the Himalayan altitude gradient, Nepal. *American Naturalist* 159:294–304.

Hawkins, B.A., et al. 2003. Energy, water, and broad-scale geographic patterns of species richness. *Ecology* 84:3105–17.

Hawkins, B.A., et al. 2007. A global evaluation of Metabolic Theory as an explanation of diversity gradients. *Ecology* 88:1877–88.

Hijmans, R.J., S.E. Cameron, J.L. Parra, P.G. Jones, and A. Jarvies. 2005. Very high resolution interpolated climate surfaces for global land areas. *International Journal of Climatology* 25:1965–78.

Hillebrand, H. 2004. On the generality of the latitudinal diversity gradient. *The American Naturalist* 163:192–211.

Holloway, J.D. 1987. Macrolepidoptera diversity in the Indo-Australian tropics: Geographic, biotopic and taxonomic variations. *Biological Journal of the Linnean Society* 30:325–41.

Jetz, W., D.S. Wilcove, and A.P. Dobson. 2007. Projected impacts of climate and land-use change on the global diversity of birds. *PLoS* 5:e157.

Kaspari, M., S. O'Donnell, and J.R. Kercher. 2000. Energy, density, and constraints to species richness: Ant assemblages along a productivity gradient. *The American Naturalist* 155:280–93.

Kattan, G.H., and P. Franco. 2004. Bird diversity along elevational gradients in the Andes of Colombia: Area and mass effects. *Global Ecology and Biogeography* 13:451–58.

Kessler, M. 2000. Elevational gradients in species richness and endemism of selected plant groups in the central Bolivian Andes. *Plant Ecology* 149:181–93.

Kitching, I.J., and J.E. Rawlins. 1999. The Noctuoidea. *In Lepidoptera, Moths and Butterflies. Volume 1: Evolution, Systematics, and Biogeography. Handbook of Zoology Volume IV Arthropoda: Insecta Part 35*, pp. 354–401. Kristensen N.P., ed. Berlin and New York: Walter de Gruyter..

Kluge, J., M. Kessler, and R.R. Dunn. 2006. What drives elevational patterns of diversity? A test of geometric constraints, climate, and species pool effects for pteridophytes on an elevational gradient in Costa Rica. *Global Ecology and Biogeography* 15:358–71.

Kreft, H., and W. Jetz. 2007. Global patterns and determinants of vascular plant diversity. *Proceedings of the National Academy of Sciences* 104:5925–30.

Lafontaine, J.D., and M. Fibiger. 2006. Revised classification of the Noctuoidea (Lepidoptera). *Canadian Entomologist* 138:610–35.

Lamoreux, J.F., J.C. Morrison, T.H. Rickets, et al. 2005. Global tests of biodiversity concordance and the importance of endemism. *Nature* 440:212–14.

Lessard, J.-P., R.R. Dunn, and N.J. Sanders. 2007. Rarity and diversity in forest ant assemblages of the Great Smoky Mountains National Park. *Southeastern Naturalist,* special issue 1:215–228.

MacArthur, R.H. 1972. *Geographical Ecology*. Princeton, N.J.: Princeton University Press.

Olson, D.M. 1991. A comparison of the efficacy of litter sifting and pitfall traps for sampling leaf litter ants (Hymenoptera: formicidae) in a tropical wet forest, Costa Rica. *Biotropica* 23:166–72.

Pausas, J.G. 1994. Species richness patterns in the understory of Pyrenean Pinus sylvestris forest. *Journal of Vegetation Science* 5:517–24.

Pharo, E.J., A.J. Beattie, and D. Binns. 1999. Vascular plant diversity as a surrogate for bryophyte and lichen diversity. *Conservation Biology* 13:282–92.

Pogue, M.G. 2005. The Plusiinae (Lepidoptera: Noctuidae) of Great Smoky Mountains National Park. *Zootaxa* 1032:1–28.

Rahbek, C. 1995. The elevational gradient of species richness: A uniform pattern? *Ecography* 18:200–205.

Rahbek, C. 1997. The relationship among area, elevation, and regional species richness in neotropical birds. *The American Naturalist* 149:875–902.

Rahbek, C. 2005. The role of spatial scale and the perception of large-scale species-richness patterns. *Ecology Letters* 8:224–39.

Reynolds, R.T., J.M. Scott, and R.A. Nussbaum. 1980. A variable circular-plot method for estimating bird numbers. *Condor* 82:309–13.

Romdal, T.S., and J.-A. Grytnes. 2007. An indirect area effect on elevational species richness. *Ecography* 30:440–48.

Rosenzweig, M.L. 1995. *Species Diversity in Space and Time*. Cambridge: Cambridge University Press.

Samson, D.A., E.A. Rickart, and P.C. Gonzales. 1997. Ant diversity and abundance along an elevational gradient in the Philippines. *Biotropica* 29:349–63.

Sanders, N.J., J.-P. Lessard, R.R. Dunn, and M.C. Fitzpatrick. 2007. Temperature, but not productivity or geometry, predicts elevational diversity gradients in ants across spatial grains. *Global Ecology and Biogeography* 16:640–49.

Sanders, N.J., J. Moss, and D. Wagner. 2003. Patterns of ant species richness along elevational gradients in an arid ecosystem. *Global Ecology and Biogeography* 12:93–102.

Schulze, C.H. 2000. Auswirkungen anthropogener Störungen auf die Diversität von Herbivoren—Analyse von Nachtfalterzönosen entlang von Habitatgradienten in Ost-Malaysia. Ph.D. Thesism University of Bayreuth, Bayreuth, Germany.

Sharkey, M.J. 2001. The all taxa biodiversity inventory of the Great Smoky Mountains National Park. *Florida Entomologist* 84:556–64.

Shriner, S.A. 2001. Distribution of breeding birds in Great Smoky Mountains National Park. Ph.D. dissertation, Department of Zoology, North Carolina State University, Raleigh, N.C.

Shriner, S.A., T.R. Simons, and G.L. Farnsworth. 2002. A GIS-based Habitat Model for Wood Thrush (Hylocichla mustelina, in Great Smoky Mountains National Park. In *Predicting Species Occurrences: Issues of Scale and Accuracy*, pp. 529–35. Scott, J.M., ed. Washington, D.C.: Island Press.

Simons, T.R., S.A. Shriner, and G.L. Farnsworth. 2006. Comparison of breeding bird and vegetation communities in primary and secondary forests of Great Smoky Mountains National Park. *Biological Conservation* 129:302–11.

Stevens, G.C. 1992. The elevational gradient in altitudinal range: An extension of Rapoport's latitudinal rule to altitude. *The American Naturalist* 140:893–911.

Terborgh, J. 1971. Distribution on environmental gradients: Theory and a preliminary interpretation of distributional patterns in the avifauna of the Cordillera Vilcabamba, Peru. *Ecology* 52:23–40

Terborgh, J. 1977. Bird species diversity on an Andean elevational gradient. *Ecology* 58:1007–29.

Terborgh, J. 1985. The role of ecotones in the distribution of Andean birds. *Ecology* 66:1237–46.

Turc, L. 1954. Le bilan d'eau des sols: Relation entre les precipitation, l'évaporation et l'ecoulement. *Annales Agronomiques* 5:491–596.

von Humboldt, A. 1845. *Kosmos: Entwurf einer physischen Weltbeschreibung*. Tübingen, Germany: Cotta, Stuttgart.

Whittaker, R.H. 1952. A study of the summer foliage insect communities in the Great Smoky Mountains. *Ecological Monographs* 22:1–44.

Whittaker, R.H. 1956. Vegetation of the Great Smoky Mountains. *Ecological Monographs* 26:1–80.

11 Integrating Data across Biodiversity Levels

The Project IntraBioDiv

Andreas Tribsch, Thorsten Englisch, Felix Gugerli, Rolf Holderegger, Harald Niklfeld, Katharina Steinmann, Conny Thiel-Egenter, Niklaus E. Zimmermann, Pierre Taberlet, and IntraBioDiv Consortium

CONTENTS

Biodiversity Research and the Integration of Biodiversity Data of Regional and National Interest 89
An Initiative for a Cross-Nation Database for Mountain Biodiversity in Europe 90
Assessing Genetic Diversity 92
Assessing Species Diversity 97
Assessing Environmental Diversity 99
Relationships of Levels of Biodiversity with Glacial Refugia 102
Potential IntraBioDiv Impact and Dissemination of Data and Results 102
Summary 103
Acknowledgments 103
References 103

BIODIVERSITY RESEARCH AND THE INTEGRATION OF BIODIVERSITY DATA OF REGIONAL AND NATIONAL INTEREST

Advancement in biodiversity theory, new methodologies, and the loss of biodiversity due to human activities have all stimulated basic and applied research aiming for explaining and understanding biodiversity patterns from local to global levels. Important data sources for biodiversity approaches are species inventories and ecosystem surveys. Such data are often commonly produced and archived nationally. Answering basic questions in biodiversity research and ecology, however, requires data independent from administrative boundaries. To overcome such limitations, it is necessary to connect local and regional activities and resulting data by international efforts. One of these important initiatives is the Global Biodiversity Information Facility, GBIF, which encourages and offers free and open access to biodiversity data. Another limitation, despite networking across boundaries, is that data are often of different qualities and far from complete (Graham et al., 2004; Isaac et al., 2004). Thus, such common platforms have to be complemented by targeted international activities that arrive at coordinated data structure and quality to allow for approaching large-scale research questions.

The European Alps and the Carpathians make up the main part of the European mountain system. Thirteen countries share these mountains: Austria, Czech Republic, France, Germany, Italy, Liechtenstein, Monaco, Poland, Romania, Slovak Republic, Slovenia, Switzerland, and Ukraine. Existing relevant biodiversity data are scattered, have various levels of quality, and, hence, are difficult to integrate.

Genetic diversity data, collected and analyzed in a comparative way, are still rather scarce and restricted to economically important plants such as trees, on the one hand (Petit et al., 2002), and to studies of some alpine taxa, on the other hand (Schönswetter et al., 2005). Molecular methods used and the geographical distribution of samples are diverse. Thus, such data representing genetic biodiversity are difficult to synthesize (but see Petit et al., 2003, and Schönswetter et al., 2005). So far, no "database of genetic diversity" of plants is available, and theory dealing with the meaning of genetic variation is under debate and development (Moritz, 2002; DeSalle and Amato, 2004). Nevertheless, being not older than 10 to 20 years, such data are usually available and accessible via the Internet, scientific databases, or supplementary material of journal publications.

Species diversity data are strongly biased by taxonomic and floristic/faunistic tradition and knowledge. The definition of taxonomic ranks, in particular of species, will always include subjective opinion and be unstable and approximate. On the one hand, there are different "cultures" in taxonomical ranking and national traditions; on the other hand, advancement in evolutionary theory, changes in taxonomical practice, and the wish to reflect this knowledge in naming taxa are among the reasons for this instability. Such uncertainty has to be accounted for in analyses based on species inventories (Isaac et al., 2004). Although standard lists are available and frequently used in biodiversity research, integrating species diversity data (from museum collections, field observations, or literature) strongly depends on an international unification of species lists and synonyms. In well-studied groups of organisms, like mammals (Wilson and Reeder, 2005), such lists are available. Great efforts are needed for other organisms such as vascular plants, in which pragmatic, but accurate, synonymization of their scientific names is important.

Ecosystem diversity comprises the diversity of habitats, communities, and ecological processes and can hardly be measured consistently over various scales. Thus, environmental parameters that allow for describing habitat diversity are used as respective surrogates. Such environmental data are often published on a national or regional basis, but more and more data are available spanning larger areas. In the Alps, environmental diversity partly explains patterns of species diversity or allows for modelling species distributions (e.g., Wohlgemuth, 1998; Guisan and Zimmermann, 2000; Moser et al., 2005).

AN INITIATIVE FOR A CROSS-NATION DATABASE FOR MOUNTAIN BIODIVERSITY IN EUROPE

The project IntraBioDiv, described in detail in Gugerli et al. (2008), aimed at linking genetic, species, and ecosystem diversity. A new facet was the special emphasis on intraspecific, i.e., molecular–genetic diversity. The project focused on high-mountain species of vascular plants in the Alps and the Carpathians. During the project period (2004–2006) and thereafter, a team of nineteen partner institutions in ten countries has been involved in data collection, analysis, and synthesis.

The primary goals of the project were (i) to estimate the intraspecific, genetic variation of selected vascular high-mountain plant taxa over the entire Alps and the Carpathians; (ii) to record the geographic distribution of the high-mountain flora across the Alps and the Carpathians, based on available data and on new field surveys; (iii) to characterize environmental variation by generating maps of potential climatic habitat diversity of both mountain ranges; and (iv) to establish a database that combines the geographic distribution of intraspecific, interspecific, and environmental diversity. The interdisciplinary consortium aimed at a synthetic analysis to interrelate these three data sets to identify environmental surrogates for intraspecific and interspecific biodiversity.

A grid containing 388 Alpine and 377 Carpathian cells (IntraBioDiv [IBD] cells) was adopted, covering area with suitable habitat for high-mountain taxa (Figure 11.1, Table 11.1). Such a grid system allowed for integrating biodiversity data in a consistent spatial framework. Data on genetic diversity were specifically produced during the project by collecting and genetically analysing samples from selected alpine plant species (Table 11.2). Data on species diversity were obtained by merging database content of many national and regional sources. A new taxonomical standard list of high-mountain taxa formed one important basis of the species database. Data on potential habitat diversity were integrated from existing sources, setting the basis for modelling potential habitat diversity. The specific approaches chosen to gather and integrate these three data sets are described in the following paragraphs and may provide some guidance for similar undertakings in other regions of the world. Exemplary results are presented here to illustrate the value of the integrated research for a better understanding of the relationships among the three levels of biodiversity.

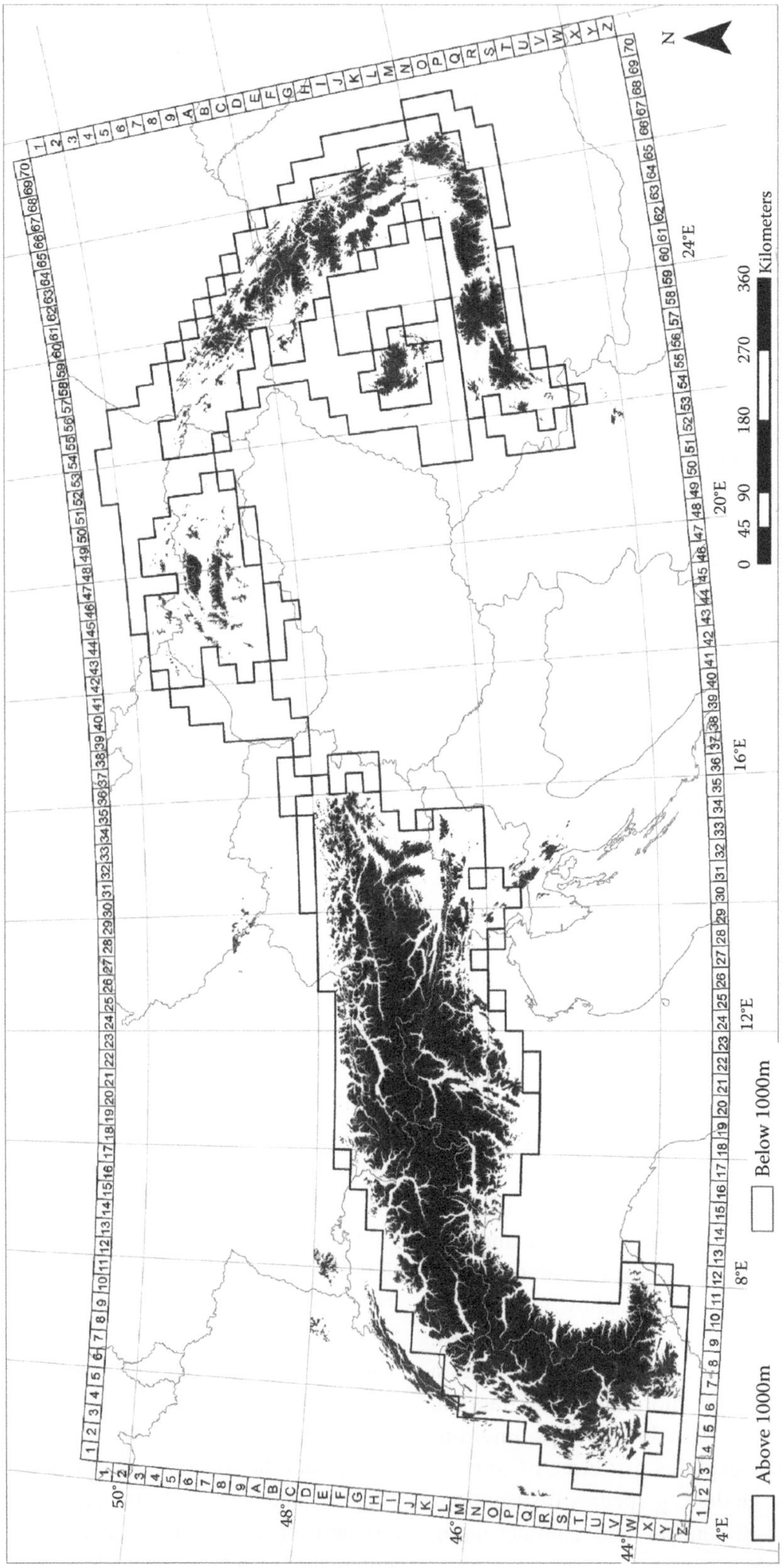

FIGURE 11.1 IntraBioDiv project range and grid system comprising the entire mountain ranges of the European Alps (west/left) and the Carpathians (east/right).

TABLE 11.1
Study Area and Number of IntraBioDiv Grid Cells for Analyzing Genetic, Species, and Habitat Diversity in the Study Ranges of the Alps and the Carpathians

		Alps		Carpathians	
	Altitudinal region	IBD Cells	Area (km²)	IBD Cells	Area (km²)
Study area for genetic sampling	Regions above 1,500 m a.s.l.	306 (79%)	174,384	104 (28%)	58,528
Selected cells for genetic sampling		149 (38%)	84,868	31 (8%)	17,380
Study area for floristic diversity	Regions above 1,000 m	353 (91%)	201,192	208 (55%)	115,843
Study area for collecting floristic distributional data	Mountainous regions below 1,000 m	35 (9%)	19,987	169 (45%)	93,876
Total study area		388	221,179	377	209,718

ASSESSING GENETIC DIVERSITY

A central aim of IntraBioDiv was to estimate the genetic variation of forty representative alpine vascular plant species of the Alps and the Carpathians (Table 11.2) sampled throughout the two mountain ranges in summer 2004. In contrast to species and ecosystem diversity, a feasible approach was to take samples only from cells comprising areas above 1,500 m a.s.l., and from every second IBD cell only in the Alps. The sampling success, i.e., the percentage of all potentially available samples from all taxa, was 80.3%, but slightly unevenly distributed across the study range and species (Figure 11.2). The genotyping success rate averaged 94.2% of all collected samples. Genetic variation was assessed in six laboratories using DNA fingerprinting methods (Amplified Fragment Length Polymorphisms, AFLPs, Vos et al., 2005; see Gugerli et al., 2008, for details). This efficient technique was chosen as it can be applied to almost any plant species with only few limitations. We avoided polyploid species with high chromosome numbers and high genome sizes, as the AFLP method does not perform well in such cases (Table 11.2). Usually three samples per location were analyzed, while this low number of local samples was compensated by the extensive number of localities considered in the sampling and a high number of polymorphic molecular markers per species (Gugerli, et al., 2008).

Crucial for the sampling success (Figure 11.2) was the careful selection of representative study species that had to occur frequently and should also be recognizable in the field. Accordingly, a random selection of study species had to be avoided, and only widespread taxa could be considered. After a preliminary selection step, sixty taxa were tested in the molecular laboratories (DNA extraction and AFLPs) before the single sampling season. Based on these results, forty-five taxa were selected that also were sufficiently widespread in the Alps and Carpathians, mostly diploid, with different habitat preferences, from different families, and representing a range of life-history traits (Table 11.2). The samples were collected during one short alpine summer, and ten teams contributed to the sampling. Some problems were caused by inaccurate sampling: Some samples were collected outside the defined grid cells, either unintended or on purpose, and in few cases, samples were misidentified. The latter were easily detected based on their distinguished genetic fingerprints and herbarium vouchers collected.

A general trend in some of the analyzed species was that genetic variation within a population was higher in the central and northern parts of the Alps, where alpine areas are larger, whereas genetic rarity was higher in peripheral areas that are glacial refugia, and where alpine areas are usually smaller and more isolated (see example of *Ranunculus alpestris* s.l. in Figure 11.3; see Paun et al., 2008). Thus, in this case, refugia do not host genetically more variable populations, as often postulated, but they host populations with significantly more rare alleles than in formerly glaciated areas (Widmer and Lexer, 2001).

It turned out to be a difficult task to combine the information of genetic variation of all study species into one "map of genetic diversity." First, not all species occur in all grid cells (Figure 11.2), and second, it is not possible to simply sum up the genetic diversity values of each species for each grid cell. One attempt to cope with these difficulties is given in Figure 11.4, showing that there is no obvious pattern of overall genetic diversity. There are, however, areas of higher diversity in the northeastern part of the Austrian Alps and in the northern Swiss Alps, while several potential refugia at the southern margin of the Alps displayed low genetic diversity.

TABLE 11.2
List of Plant Taxa Sampled for Genetic Analyses, Indicating the Number of Occurrences in IBG Grid Cells, the Distribution in Alps and Carpathians, the Number of Sampled Populations, the Description of the Elevational Distribution, and the Chromosome Number

Taxon	Family	No. Grid Cells	Alps	Carpathians	No. Sampled Populations	Elevational Range	Chromosome Number
Androsace obtusifolia All.	Primulaceae	174	widespread	rare	46	(amo) sa–alp	36, 38?, 40
Arabis alpina L.	Brassicaceae	403	widespread	widespread	155	(smo–mo) amo–alp	16, 32?
Campanula alpina Jacq.	Campanulaceae	93	E-Alps	widespread	36	sa–alp	34
Campanula barbata L.	Campanulaceae	256	widespread	absent	111	(mo) sa–alp	30, 34
Campanula serrata (Kit.) Hendrych	Campanulaceae	106	absent	widespread	22	amo–sa	34
Carex firma Mygind	Cyperaceae	229	widespread	rare	86	(mo) sa–alp (sni)	34
Carex sempervirens Vill.	Cyperaceae	325	widespread	widespread	163	(amo) sa–alp	30, 34, 56, 58, 68
Cerastium uniflorum Clairv.	Caryophyllaceae	160	widespread	rare	49	(sa) alp–sni	36
Cirsium spinosissimum (L.) Scop.	Asteraceae	241	widespread	absent	112	sa–alp	34
Dryas octopetala L.	Rosaceae	325	widespread	widespread	141	(mo) sa–alp (sni)	18
Festuca carpathica F. Dietr.	Poaceae	36	absent	widespread	9	(mo) sa–alp	28
Festuca supina Schur	Poaceae	150	rare	widespread	29	(mo) sa–alp	28, 35
Festuca versicolor Tausch s.l.	Poaceae	62	E-Alps	widespread	22	(mo) sa–alp	14
Gentiana acaulis L.	Gentianaceae	328	widespread	widespread	116	(mo) sa–alp	36
Gentiana nivalis L.	Gentianaceae	278	widespread	widespread	92	(amo) sa–alp	14
Geum montanum L.	Rosaceae	353	widespread	widespread	141	(amo) sa–alp	28, 42
Geum reptans L.	Rosaceae	181	widespread	widespread	61	(sa) alp–sni	42
Gypsophila repens L.	Caryophyllaceae	297	widespread	rare	109	(mo) sa–alp	34, 36?
Hedysarum hedysaroides (L.) Schinz & Thell. s.l.	Fabaceae	249	widespread	widespread	90	sa–alp	14
Hornungia alpina (L.) Appel s.l.[a]	Brassicaceae	266	widespread	rare	101	(amo) sa–alp (sni)	12
Hypochaeris uniflora Vill.	Asteraceae	300	widespread	widespread	90	mo–alp	10
Juncus trifidus L.	Juncaceae	265	widespread	widespread	119	sa–alp	30
Leucanthemum waldsteinii (Sch. Bip.) Pouzar[b]	Asteraceae	110	absent	widespread	27	sa–alp	18
Ligusticum mutellinoides Vill.[c]	Apiaceae	180	widespread	rare	64	alp	22
Loiseleuria procumbens (L.) Desv.[d]	Ericaceae	253	widespread	widespread	105	sa–alp	24?, 26
Luzula alpinopilosa (Chaix) Breist.	Juncaceae	246	widespread	widespread	106	sa–alp	12
Peucedanum ostruthium (L.) Koch	Apiaceae	283	widespread	absent	124	(mo) sa (alp)	22
Phyteuma betonicifolium Vill. s.l.[e]	Campanulaceae	270	widespread	absent	112	(mo) sa (alp)	24
Phyteuma confusum A. Kern.	Campanulaceae	47	E-Alps	widespread	23	sa–alp	28
Phyteuma hemisphaericum L.	Campanulaceae	210	widespread	absent	78	sa–alp (sni)	28
Primula minima L.	Primulaceae	126	E-Alps	widespread	55	(sa) alp (sni)	66
Ranunculus alpestris L. s.l.[f]	Ranunculaceae	239	widespread	rare	91	(amo) sa–alp	16

(continued)

TABLE 11.2
List of Plant Taxa Sampled for Genetic Analyses, Indicating the Number of Occurrences in IBG Grid Cells, the Distribution in Alps and Carpathians, the Number of Sampled Populations, the Description of the Elevational Distribution, and the Chromosome Number (continued)

Taxon	Family	No. Grid Cells	Alps	Carpathians	No. Sampled Populations	Elevational Range	Chromosome Number
Ranunculus oreophilus Bieb.[g]	Ranunculaceae	219	widespread	widespread	24	(mo) sa–alp	16
Ranunculus crenatus Waldst. & Kit.	Ranunculaceae	18	rare	widespread	9	alp	16
Rhododendron ferrugineum L.	Ericaceae	287	widespread	absent	131	(mo) sa (alp)	26
Rhododendron myrtifolium Schott & Kotschy	Ericaceae	42	absent	widespread	18	mo–alp	?
Salix reticulata L.	Salicaceae	270	widespread	widespread	107	sa–alp	38
Saxifraga aizoides L.	Saxifragaceae	335	widespread	widespread	143	amo–sni	26
Saxifraga stellaris L.[h]	Saxifragaceae	281	widespread	widespread	119	mo–alp	28
Saxifraga wahlenbergii Ball	Saxifragaceae	12	absent	rare	4	sa–alp (sni)	66
Sempervivum montanum L. s.l.[i]	Crassulaceae	246	widespread	widespread	25	(mo) sa–alp	38, 42
Sesleria caerulea (L.) Ard.	Poaceae	399	widespread	rare	148	(smo–mo) sa–alp	28
Soldanella pusilla Baumg.	Primulaceae	187	E-Alps	widespread	28	sa–alp (sni)	40
Trifolium alpinum L.	Fabaceae	157	widespread	absent	66	sa–alp	16
Veronica baumgartenii Roem. & Schult.	Plantaginaceae	31	absent	widespread	13	sa–alp	14

Source: Chromosome number counts from Alps and Carpathians from Dobeš and Vitek 2000; Lauber and Wagner 2001; Marhold et al. 2007).

Note: mo = montane; amo = altomontane; sa = subalpine; alp = alpine; sni = subnival.

[a] Including *Hornungia alpina* ssp. *brevicaulis* (Sternb. ex Spreng.) Appel and *H. alpina.* ssp. *austroalpina* (Trpin) Appel.

[b] = *Chrysanthemum rotundifolium* Waldst. & Kit.; *Leucanthemum rotundifolium* (Willd.) DC.

[c] = *Pachypleurum mutellinoides* (Cr.) Holub (this is the valid name now).

[d] = *Kalmia procumbens* (L.) Gift & al. ex Galasso & al. (this is the valid name now).

[e] Including *Phyteuma persicifolium* Hoppe (= *Ph. betonicifolium* ssp. *zahlbruckneri* (Vest) Hayek).

[f] Including samples of *Ranunculus bilobus* Bertol. and *Ranunculus traunfellneri* Hoppe.

[g] = *Ranunculus breyninus* Cr. (this is the valid name).

[h] *Saxifraga stellaris* ssp. *robusta* (Engl.) Murr and *S. stellaris* ssp. *prolifera* (Sternb.) Temesy.

[i] Including *Sempervivum montanum* ssp. *carpaticum* Wettst. ex Hayek and *S. montanum* ssp. *stiriacum* Wettst. ex Hayek.

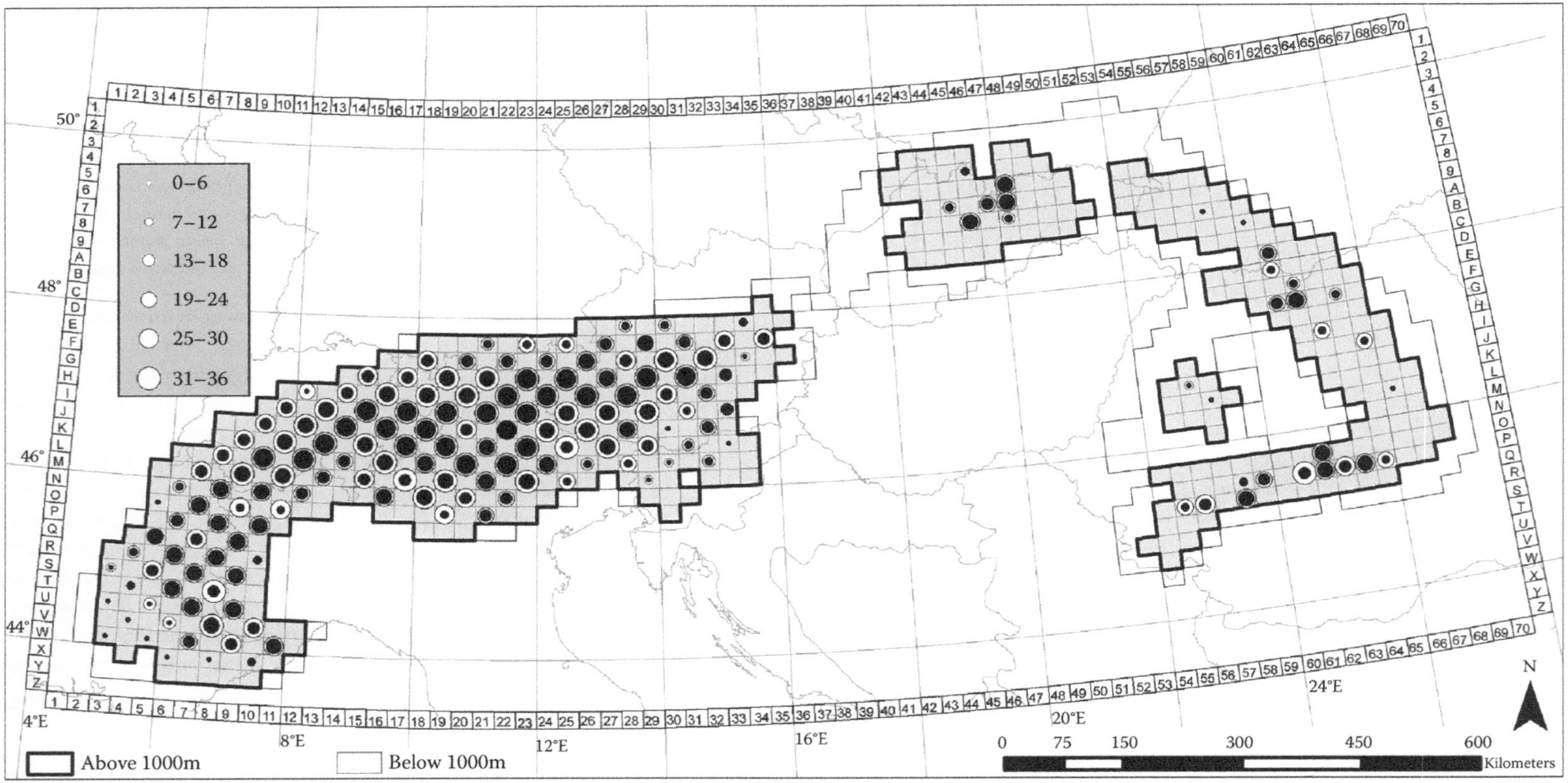

FIGURE 11.2 Sampling success of forty selected taxa analyzed for intraspecific variation in the Alps (left) and the Carpathians (right; adopted from data of Gugerli et al., 2008). Circles symbolize expected (white, data obtained from IntraBioDiv species database) versus achieved (black) sampling in each IBD grid cell.

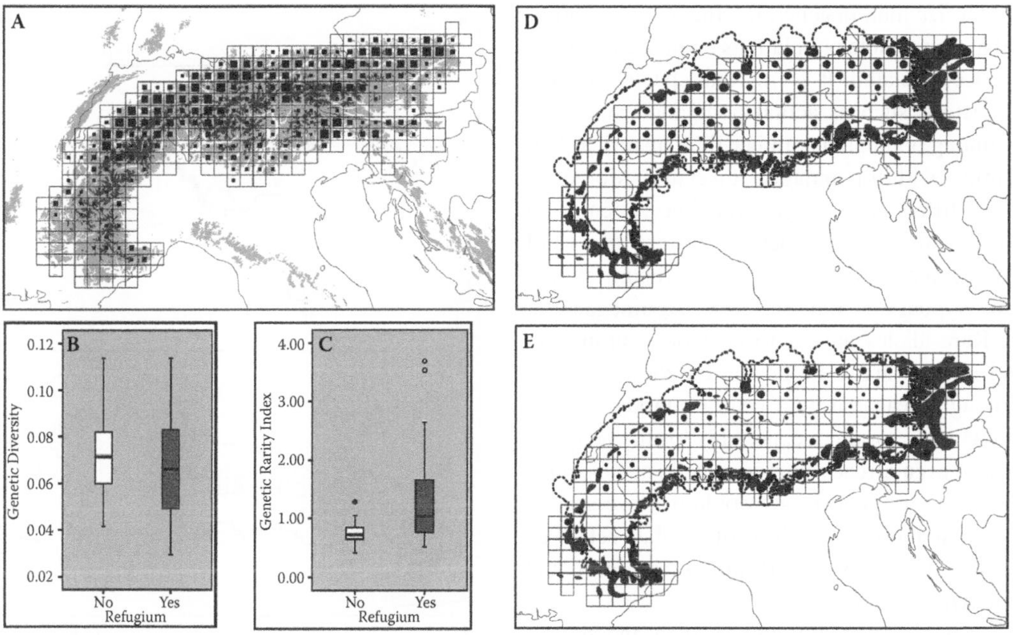

FIGURE 11.3 Sample maps from *Ranunculus alpestris* s.l. (modified from Paun et al., 2008) for the Alps. (A) Geographical distribution in the Alps, the size of quadrates symbolizes the abundance of the species in the grid cell in four categories based on the number of records per grid cell. (B, C) Genetic diversity (B) and genetic rarity (C) in and outside refugia given in boxplot diagrams (see Figure 11.9 for definitions). (D, E) Geographical pattern of genetic diversity (D) and rarity (E) within populations. Potential refugia (taken from Schönswetter et al., 2005) are given in gray. The ice extent during the Last Glacial Maximum is given as a dotted line. Larger symbols show higher genetic diversity/rarity of populations. Populations with high genetic diversity are mainly found in the northern part of the Alps. Populations that host many rare markers are mainly found in refugia in the northeastern and southern Alps.

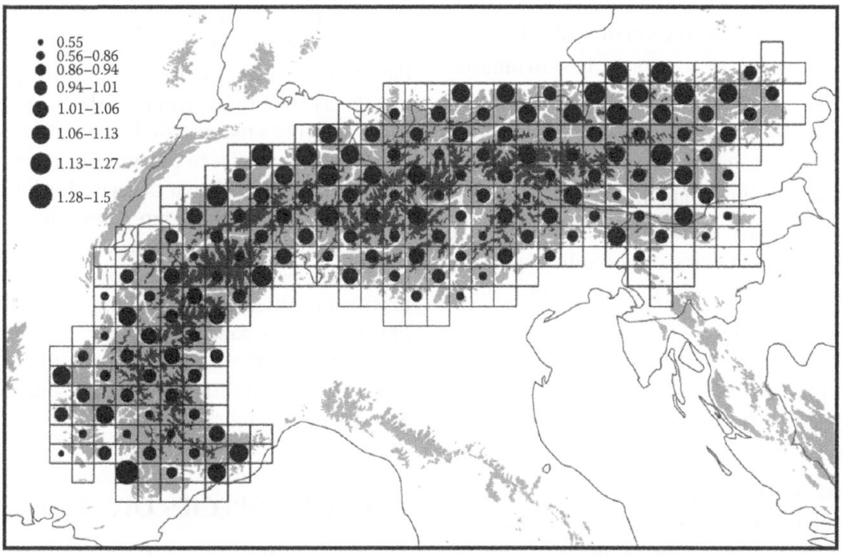

FIGURE 11.4 Total genetic diversity obtained from AFLP data sets from twenty-seven study taxa in the Alps (modified after Gugerli et al., 2008). Genetic diversity was calculated based on Nei's gene diversity for each cell, which was normalized within species to account for differences in mean diversity among species (Thiel-Egenter et al., 2008). Total cell-wise genetic diversity was calculated by averaging over all species.

We hypothesize that, in general, refugia do not harbor genetically more variable populations compared to areas that were once glaciated. It seems that species-specific ecological factors, such as actual population sizes and gene flow among populations, rather than historical factors determine within-population diversity. Moreover, each species differs in ecology, breeding system, and dispersal mechanisms, and other life-history traits, all factors shaping genetic diversity patterns (Hamrick and Godt, 1996; Nybom and Bartish, 2000; Aguinagalde et al., 2005). Rare alleles appear to be more common in refugia and isolated areas. Following the view that rare alleles, although measured by a neutral marker system, correspond with rare or unique, possibly locally adapted, phenotypes, or even already unrecognized species, refugia hosting such rare phenotypes would deserve special attention. A more thorough analysis of our data sets, e.g., an ecological outlier analysis, and similar data sets could help in answering these questions.

Most species analyzed showed a distinct and significant phylogeographic structure, i.e., alleles were not randomly distributed in the study area or simply structured through isolation by distance. Several new approaches have been applied to characterize spatial genetic patterns. One example is shown in Figure 11.5 (Thiel-Egenter et al., unpublished data), for which a new combination of methods was applied to find the locations of significant phylogeographic breaks. Other examples are published in Manel et al. (2007) and Alvarez et al. (2008, 2009). The consistent data of genetic variation of multiple alpine species also has motivated new approaches in analyzing genetic diversity. Thiel-Egenter et al. (2008) analyzed patterns of genetic diversity of twenty-two alpine species evidencing, for example, that genetic diversity was lower in the Carpathians than in the Alps in most of the species tested. Apart from *Ranunculus alpestris* (Paun et al., 2008), single-species case studies with different foci are published. The case study of *Arabis alpina* (Brassicaceae, Ehrich et al., 2007) showed a remarkable complex phylogeographic structure within the Alps and the Carpathians. In *Rhododendron ferrugineum* (Ericaceae, Manel et al., 2007), a new "moving window" method was used to evidence phylogeographic breaks and hybrid zones. In the western Alps, *Hypochaeris uniflora* (Asteraceae, Mráz et al., 2007) is one of the few taxa with higher genetic diversity and rarity in the Carpathians than in the Alps. *Campanula alpina* (Campanulaceae, Ronikier et al., 2008) is one of the taxa with the clearest and strongest phylogeographic patterns, especially within the Carpathians.

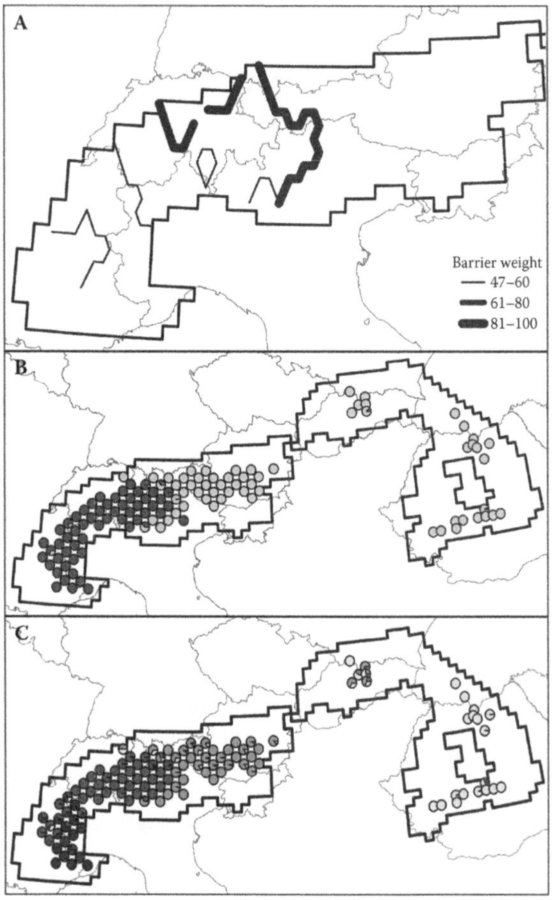

FIGURE 11.5 Examples of phylogeographic patterns revealed with different methods based on AFLP markers exemplified by the species *Juncus trifidus*, a dominating plant in alpine acidic grassland (Thiel-Egenter, unpublished). (A) Main genetic breaks by (1) applying a model-based cluster algorithm (STRUCTURE 2.1) to assign samples to genetic populations (K = 1–6 clusters); (2) calculating population pairwise Euclidean distances based on these cluster membership probabilities; and (3) geographically locating breaks between clusters based on highest distances between neighboring populations, using Monmonier's maximum difference algorithm (BARRIER 2.2). (B, C) Phylogeographic patterns of STRUCTURE results revealed. (B) Assignments at K = 2, assuming two genetic populations in the whole data set in different tones of gray. (C) Assignments at K = 4, assuming four populations.

ASSESSING SPECIES DIVERSITY

The basis for the species database was laid by a consensus taxon list of all species and subspecies that have their main altitudinal distribution at or above treeline either in the Alps or the Carpathians. This list of high-mountain taxa (HMT) of the Alps and the Carpathians was achieved by harmonizing taxonomic concepts from published floras and checklists and from regional databases (Englisch et al., in preperation), including currently recognized taxa at specific and subspecific rank. Moreover, species aggregates that comprise mostly insufficiently studied species complexes and agamospecies (e.g., *Alchemilla*, *Hieracium*, *Nigritella*, *Taraxacum*) were included. Therefore, the HMT list contains a hierarchical structure, in which one species might contain several subspecies, and species aggregates might consist of several species. This allows for calculating "taxon richness" based on genera, species, or subspecies. The HMT list comprises 2,030 high-mountain taxa (1,579 species), of which 542 taxa (406 species) are endemic to either the Alps or the Carpathians. Additional information on life forms and altitudinal distribution (montane, altomontane, subalpine, alpine, subnival) was collected based on literature sources and field observations (see Table 11.2 for examples).

To achieve the goal of obtaining distribution data across the Alps and the Carpathians, we adopted a three-step procedure: (a) critical synonymization of species names from national or regional taxon lists leading to the consensus HMT list; (b) connection of data on taxon distribution from national or regional inventories maintained by the partners involved and based on the respective synonymizing tables; and (c) careful study of taxon circumscription. The latter was done to avoid a bias of artificial distribution patterns due to different interpretation of a taxon in different countries, as well as validation of taxon occurrences throughout the range. In this way, species distribution data from sixteen national and regional floristic databases were merged. Although existing data were of appropriate quality, additional data collection was carried out by mapping species and subspecies in the field, checking herbaria and their databases, and studying the literature. Such activities were concentrated to areas which were underrepresented in databases. Finally, the species entries in the database contain accurate information on their geographical distribution. Distributions of intraspecific taxa are more problematic than the ones of species taxa, because many intraspecific taxa are not consistently recognized in all countries. Nevertheless, the database also contains distribution data of subspecies as far as available.

In total, distribution data on 1,907 HMT were included in the compiled database ("IntraBioDiv Floristic Database"; Th. Englisch et al., unpublished data). Altogether, 176,978 unique records for 674 IBD cells (Alps, 387, and Carpathians, 287; for several grid cells, especially around the Carpathians, there were no records for HMT, because these are totally missing in areas that barely reach 1,000 m.s.m.). Taxa comprising distributional data included 66 species groups ("aggregates"), 1,468 species (including 349 agamospecies), and 373 subspecies. No or incomplete distributional data could be made available on several apomictic taxa (e.g., *Hieracium*, *Taraxacum*).

Excluding apomicts, the number of taxa and species, as well as the proportion of endemic taxa/species, varied widely across countries of the Alps and the Carpathians (Table 11.3). For example, 24% of all endemic species of the Italian Alps were Alpine endemics, whereas the value for the German Alps was 6%, which has relevant implications for nature conservation strategy. In the Carpathians, a lower proportion of species was endemic. The geographical range size of taxa within the study area, measured as the number of grid occurrences, ranged from 1 to 543. Distribution maps for all HMT taxa entries will be made available in the forthcoming Distribution Atlas of the High Mountain Flora of the Alps and the Carpathians (Niklfeld et al., in preperation; examples are given in Gugerli et al., 2008). For single grid cells (IBD cells with areas above 1,000 m a.s.l. only), species richness ranged from 11 to 503 for the Alps and from 2 to 335 for the Carpathians (map in Coldea et al., 2009), the total mean was 270 for the Alps and 76 for the Carpathians (agamospecies and doubtful taxa

TABLE 11.3
Species Richness and Endemism in the High-Mountain Flora in Countries of the Alps and the Carpathians

	Species[a]	Endemism[b]
Alps	957	249 (26.0%)
France	670	105 (15.6%)
Italy	859	210 (24.4%)
Switzerland	609	59 (9.7%)
Germany	425	24 (5.6%)
Austria	709	105 (14.8%)
Slovenia	464	56 (12.1%)
Carpathians	553	65 (11.8%)
Czech Republic	39	0 (0.0%)
Slovakia	377	23 (6.1%)
Poland	336	16 (4.8%)
Ukraine	298	20 (6.7%)
Romania	458	40 (8.7%)

[a] Number of species based on the HMT list, excluding agamospecies (genera *Alchemilla*, *Hieracium*, *Taraxacum*, *Nigritella*, *Ranunculus auricomus* agg.).
[b] Number of endemic species with restricted range within the Alps/Carpathians with occurrences in (but not necessarily endemic to) the respective country.

excluded). These differences might be mainly explained by generally much smaller high-mountain areas in the Carpathians and the periphery of the Alps and a greater species pool in the centers of the Alps. However, variation in environmental factors (habitat richness) account for these differences as well.

Figure 11.6A shows the pattern of endemism of HMT in the Alps. High numbers of endemic species were mainly found in the southwestern Alps, the southern Alps, and the northeasternmost parts. Species diversity also showed an interesting pattern with rather high species richness in the southwestern Alps, where calcicole and silicicole HMT are frequent, and in some central and southern ranges with a high-mountain flora rich in mainly silicicole or calcicole species (data not shown). Neither species richness (data not shown) nor endemism was higher in refugia (Figure 11.6B). Nevertheless, there is a clear trend that endemic species accumulated in once unglaciated refugia with limited putatitve postglacial range expansions into adjacent high alpine areas. Peripheral refugia are often rather low mountains where presently few alpine species and, thus, few endemics occur. These low mountains contribute to the analysis shown in Figure 11.6B, causing that overall the number of endemics was not significantly higher in refugia than in other areas, although Figure 11.6A shows an apparent affinity of high endemism to southern and eastern refugia.

In the light of the findings outlined previously, we hypothesize that, similar to genetic diversity, species diversity in the Alps and the Carpathians is only subordinately associated with historical processes, but rather with ecological factors. Degree of endemism, in homology to genetic rarity, could be mainly affected by history and higher in or close to refugia. More detailed analyses with the IntraBioDiv data or similar data sets are necessary to reveal and understand such patterns.

ASSESSING ENVIRONMENTAL DIVERSITY

The approach used was to integrate topographical data with bioclimatic data toward modelling environmental diversity, or better termed "potential habitat diversity," in IBD grid cells. This was estimated using spatial information of three ecologically relevant climatic variables driving ecophysiological processes and, thus, considered to constrain the distribution patterns of plants in space, namely: (1) temperature in the form of annual degree-days, (2) radiation in the form of potential direct radiation, and (3) annual total precipitation. To generate maps of climate variables for the Alps and the Carpathians, we used DAYMET, a software environment that analyzes daily records of climate variables in a spatial context and allows for the interpolation of these variables in a spatially explicit manner using climate station records and a digital elevation model (DEM; Thornton et al., 1997). Each of the three climatic variables was categorized into several classes. Finally, the number of different combinations of classes ("hypercubes"; Steinmann, 2008) present in each IBD grid cell was estimated, taking into consideration only areas above a certain altitudinal threshold (e.g., 1,000 m a.s.l.).

The pattern of potential habitat diversity per grid cell revealed strong spatial gradients (Figure 11.7). Grid cells

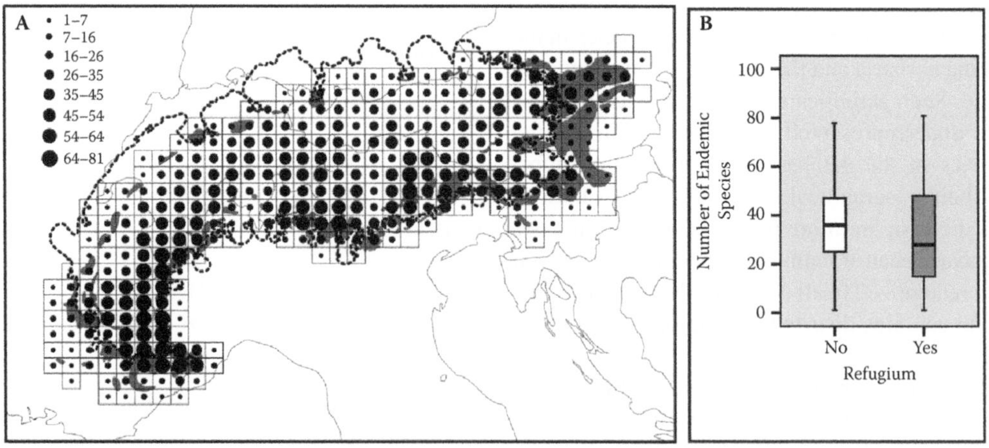

FIGURE 11.6 Number of endemic mountain species in relation to glacial refugia. (A) Data of species endemic to the Alps from the IntraBioDiv species database. Refugia (taken from Schönswetter et al., 2005) are given in gray, and the ice extent during the Last Glacial Maximum is given as a dotted line. Larger symbols show higher numbers of endemic species. (B) Boxplot diagrams of endemic species within and outside refugia (see Figure 11.9 for definitions).

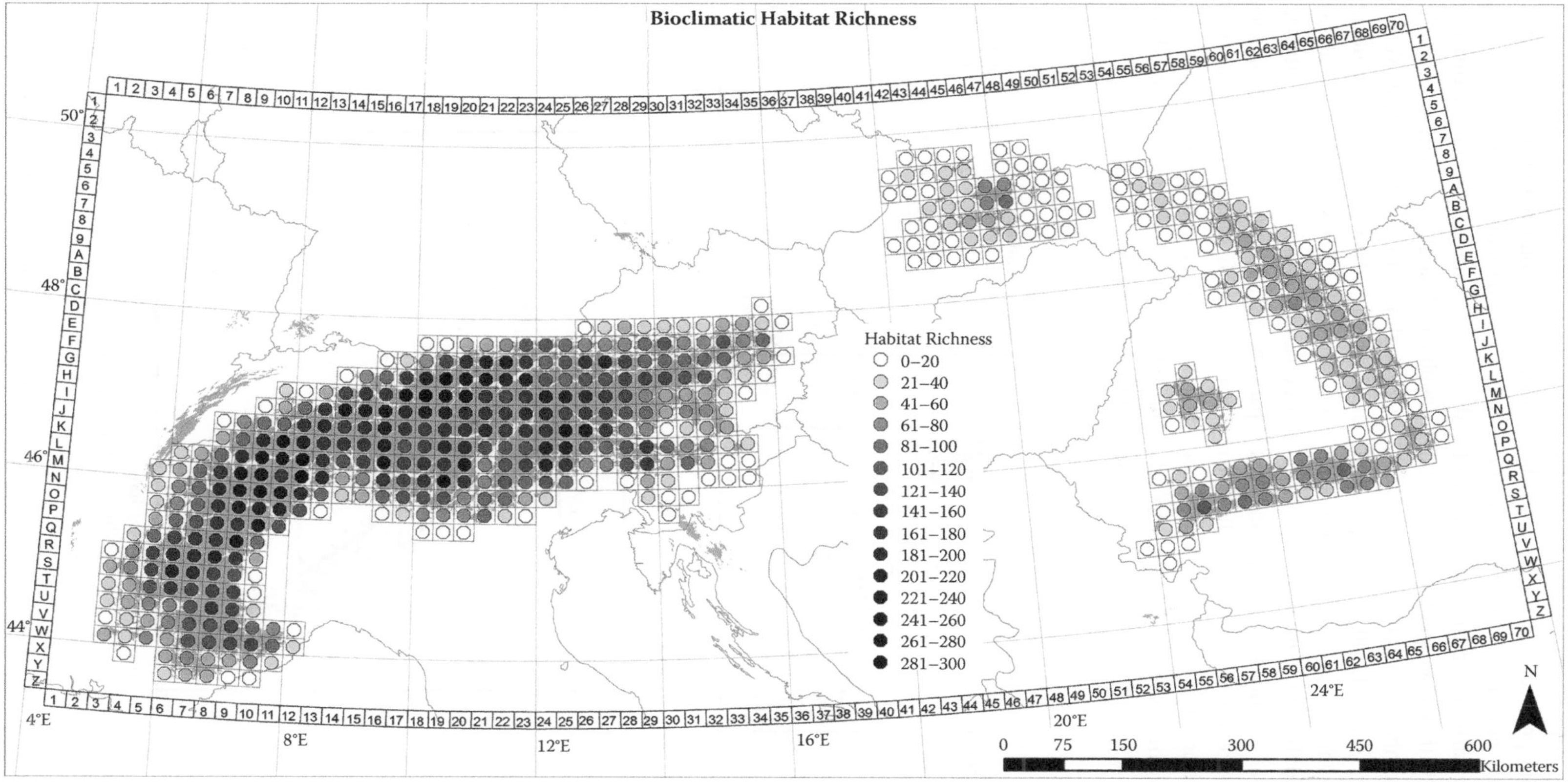

FIGURE 11.7 Map of potential habitat diversity in the Alps and the Carpathians. Darker colors symbolize higher richness of habitats in the IBD grid cells. Highest numbers are found in the central ranges in the middle and western parts of the Alps.

of the core areas of the Alps showed higher potential habitat diversity than did grid cells at the edge of the Alps. We identified two centers of potential habitat diversity in the southwestern and in the central part of the Alps that coincide with known centers of floristic richness. The highest number of potential habitat richness classes was to 287 (above 1,000 m a.s.l. per grid cell), while the lowest number was 1, thus representing a cell that barely reached the 1,000 m line.

Species diversity was highly dependent on area size (Figure 11.8A). Especially in the Alps, the area > 1,000 m a.s.l. was in clear association with the number of species. Interestingly this expected relationship was less pronounced in the Carpathians, possibly because there the timberline varies much more in elevation than in the Alps, contributing to the unexplained variance. Species richness also was clearly associated with habitat diversity (Figure 11.8B; Zimmermann and Steinmann, unpublished data), especially so in the Alps. The relation appears linear to a certain extent. Very high numbers of habitat types, however, do not result in higher species numbers. Whether such "virtual habitat types," are not occupied by species or whether other factors cause this pattern, e.g., a limited regional species pool, remains open. To explore the relationship of habitat diversity with species diversity in more detail, factors such as diversity of bedrock and timberline variation should be considered in future analyses.

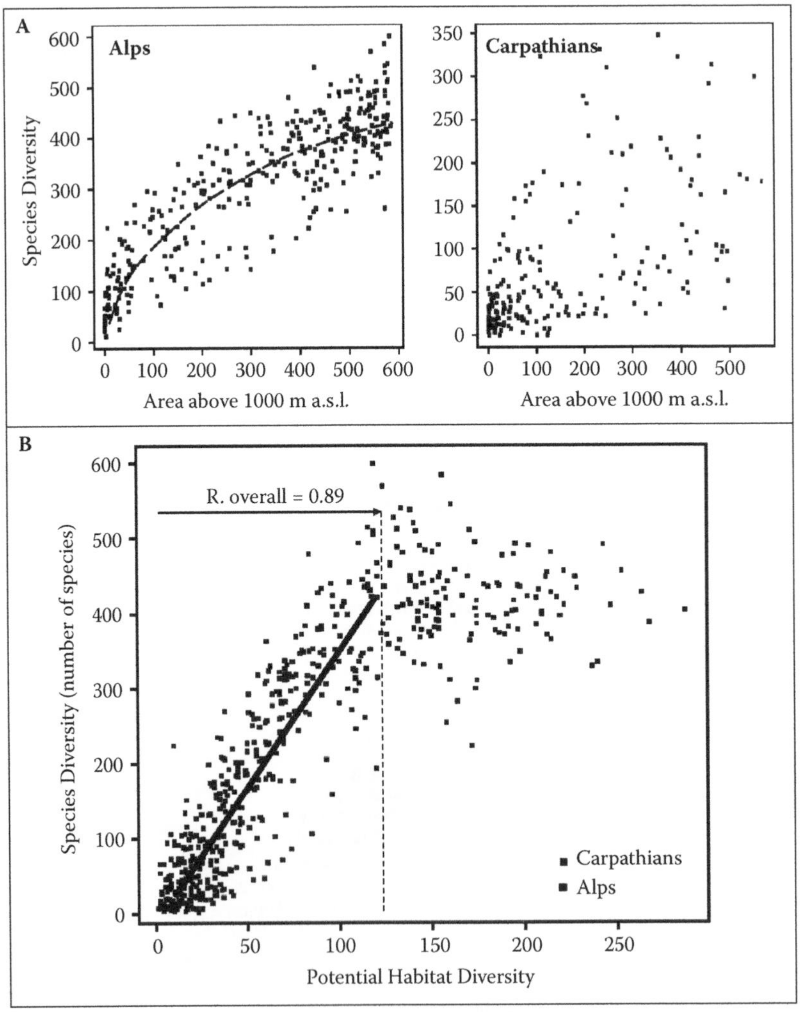

FIGURE 11.8 Scatter plots showing correlations between (A) species diversity and area in the Alps and the Carpathians, and (B) species diversity with potential habitat diversity.

RELATIONSHIPS OF LEVELS OF BIODIVERSITY WITH GLACIAL REFUGIA

Figure 11.9 shows a simple comparison of genetic, species, and habitat diversity in relation to the distribution of potential glacial refugia. Biodiversity was equal (genetic diversity) or lower (species richness and habitat diversity) in refugia compared to formerly glaciated areas. We could not identify such a trend in the Carpathians (data not shown), mainly because alpine areas there are smaller and glacial impact was much lower, which makes it difficult to define potential refugia. Thus, species richness seemed mainly driven by habitat richness and area size, whereas species rarity and endemism was higher (or at least not lower) in refugia than elsewhere. We conclude that history (survival in refugia) results in rarity and endemism, but not in diversity, which is dependent on present ecological factors. This is a trend that needs to be explored in more detail in the future.

POTENTIAL INTRABIODIV IMPACT AND DISSEMINATION OF DATA AND RESULTS

Although the present study was restricted to high-mountain vascular plant taxa, IntraBioDiv data sets have great potential to serve as an important basis for future studies. We also consider those data as highly valuable for applied conservation purposes, as the data contribute to evidencing the transnational responsibility for biodiversity. This could become especially relevant, because most Central European endemics occur in high-mountain areas, and many of these may become endangered by climate warming because the area of suitable habitats will diminish, or because environmental changes in relation to the ongoing global change may directly jeopardize population persistence. Moreover, future research with concerted efforts to tackle fundamental multidisciplinary questions in ecology and evolution will hopefully profit from the outcomes of IntraBioDiv.

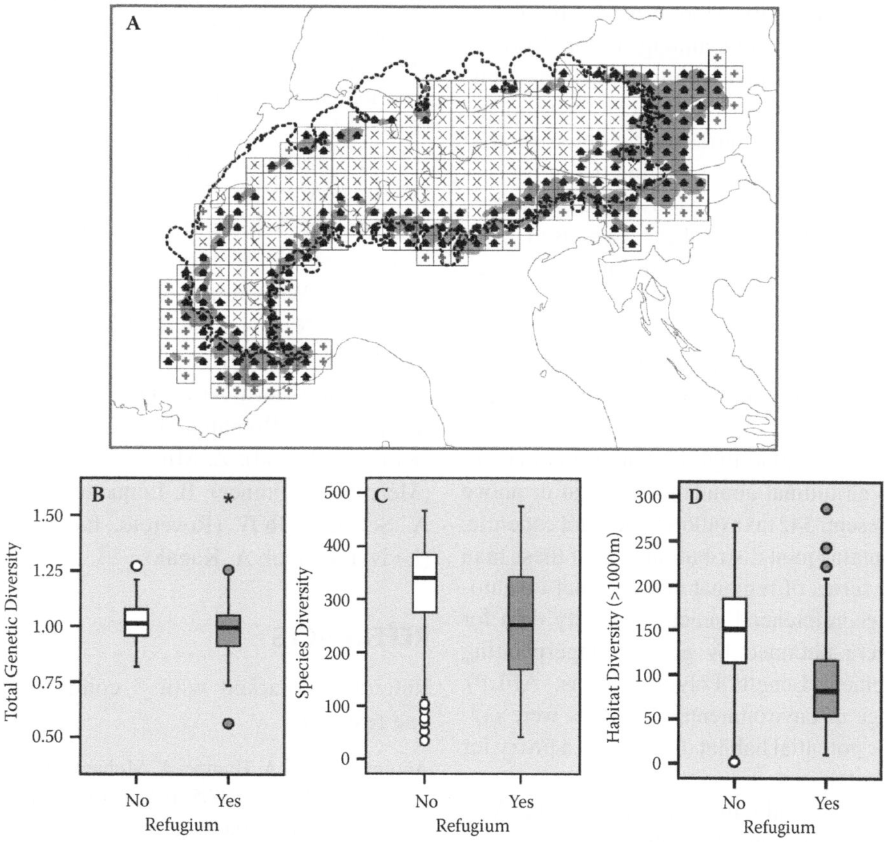

FIGURE 11.9 Biodiversity and refugia: (A) potential refugia (taken from Schönswetter et al., 2005) are given in gray. The ice extent during the Last Glacial Maximum is given as a dotted line. IntraBioDiv grid cells were assigned, containing a refugium (small black house), glaciated area (x), and peripheral border (gray diamond). (B, C, D) Boxplots of genetic diversity (B), species diversity (C), and potential habitat diversity (D) in relation to refugia. Peripheral border areas were not included in boxplots, because usually very few alpine species occur in these grid cells.

The core project funded by the European Commission was formally completed by the end of 2006. General information on biodiversity for the public is available at http://www.bio-diversity.eu. For audience interested in the IntraBioDiv project, the Web site http://www.intrabiodiv.eu is relevant. Scientific information on the project is disseminated via the latter Web site, including computing tools used such as R-scripts (http://www.intrabiodiv.eu/spip.php?article77), relevant information on the IntraBioDiv consortium, and a list of IntraBioDiv-related publications. On the Web site http://data.intrabiodiv.eu, all IntraBioDiv data will be available in due course.

SUMMARY

Biodiversity has three components—genetic, species, and ecosystem diversity—and there are complex ecological and historical factors shaping biodiversity patterns. Disentangling these factors is important for our understanding of biodiversity. Moreover, it is important to understand the relationships among the biodiversity levels. The goal of this chapter is to introduce the aims, methods, and outcomes from the project IntraBioDiv, "Tracking surrogates for intraspecific biodiversity: towards efficient selection strategies for the conservation of natural genetic resources using comparative mapping and modelling approaches." We used electronic databases and newly generated data of the rich vascular plant flora of the high-mountain regions of the Alps and the Carpathians, well suited as model regions for assessing general aspects of regional species diversity and genetic diversity, and link these with habitat diversity.

Database entries of 2,030 high-mountain taxa (1,579 species) with an altitudinal abundance peak at or above timberline are present; 542 taxa (406 species) are endemic. The database contains joint distribution data of these taxa obtained from a series of regional and national inventories. Moreover, georeferenced genetic diversity data for forty species were obtained by genetic fingerprinting (Amplified Fragment Length Polymorphisms, AFLP), and a broad range of environmental variables were collected to describe potential habitat diversity as a proxy for ecosystems diversity.

Biodiversity is not randomly distributed in the Alps and the Carpathians. Species diversity is clearly correlated with habitat diversity and higher in the southern and the western Alps than in other areas. Although genetic diversity in single species data sets does, overall genetic diversity does not show any clear pattern. Diversity of rare (endemic) species, however, tends to be higher in or close to the peripheral refugia than in the once heavily glaciated Central Alps, although present habitat diversity is lower in refugia than in other areas. This trend also is illustrated with genetic rarity of *Ranunculus alpestris*. In conclusion, genetic and species diversity might be more correlated with present ecological factors, such as population density and habitat diversity, whereas genetic and species rarity (endemism) is rather a result of historical factors.

ACKNOWLEDGMENTS

We thank the European Commission (GOCE-CT-2003-505376; the content of the publication does not represent necessarily the views of the EC or its services) and national funding agencies (Swiss State Secretariat for Education and Research, grant numbers 03.0116-1 and 03.0116-2; Erwin Schrödinger grant of the Austrian "Fonds zur Förderung der wissenschaftlichen Forschung" to A. Tribsch, J2303-BO) for their financial support. We further express our thanks to all the enthusiastic botanists that collected and deposited plants in herbaria, contributed records to floristic databases, and published floristic data.

The IntraBioDiv Floristic Group consists of: UW (Vienna, Austria): Th. Englisch, W. Gutermann, H. Niklfeld, L. Schratt-Ehrendorfer; WSL (Birmensdorf, Switzerland): T. Wohlgemuth; UREGBIO (Regensburg, Germany): W. Ahlmer, M. Scheuerer; BTF (Ljubljana, Slovenia): N. Jogan; CBNA (Gap-Charance, France): J.-P. Dalmas, L. Garraud; UT (Trieste, Italy): F. Martini; IBSAS (Bratislava, Slovakia): R. Letz, K. Marhold; IBRC (Cluj-Napoca, Romania): Gh. Coldea, M. Puşcaş; IBPAS (Kraków, Poland): Z. Mirek, H. Piekos-Mirkowa; PNM (Mercantour, France): B. Lequette; IPLA (Torino, Italy): A. Selvaggi; ROV (Rovereto, Italy): F. Prosser; LVIV (L'viv, Ukraine): A. Kagalo.

REFERENCES

References marked with * contain IntraBioDiv data and research.

Aguinagalde, I., A. Hampe, A. Mohanty, J.P. Martín, J. Duminil, and R.J. Petit. 2005. Effects of life-history traits and species distribution on genetic structure at maternally inherited markers in European trees and shrubs. *Journal of Biogeography* 32:329–39.

*Alvarez, N., N. Arrigo, and IntraBioDiv Consortium. 2008. An R (CRAN) scripts collection for computing genetic structure similarities based on STRUCTURE 2 outputs. *Molecular Ecology Notes* 8:757–62.

*Alvarez, N., C. Thiel-Egenter, A. Tribsch, R. Holderegger, S. Manel, P. Schönswetter, P. Taberlet, S. Brodbeck, M. Gaudeul, L. Gielly, P. Küpfer, G. Mansion, R. Negrini, O. Paun, M. Pellecchia, D. Rioux, P. Schönswetter, F. Schüpfer, M. Van Loo, M. Winkler, F. Gugerli, and IntraBioDiv Consortium. 2009. History or ecology? Substrate type as a major driver of spatial genetic structure in Alpine plants. *Ecology Letters* 12(doi:10.11/j.1461-0248.2005.01312.x).

*Coldea, G., I.A. Stoica, M. Puşcaş, T. Ursu, A. Oprea, and The IntraBioDiv Consortium. 2009. Alpine–subalpine species richness of the Romanian Carpathians and the current conservation status of rare species. *Biodiversity and Conservation* 18:1441–1458.

DeSalle, R., and G. Amato. 2004. The expansion of conservation genetics. *Nature Reviews Genetics* 5:702–12.

Dobeš, C., and E. Vitek. 2000. *Documented Chromosome Number Checklist of Austrian Vascular Plants*. Vienna: Museum of Natural History.

*Ehrich, D., M. Gaudeul, A. Assefa, M.A., Koch, K. Mummenhoff, S. Nemomissa, IntraBioDiv Consortium, and Ch. Brochmann. 2007. Genetic consequences of Pleistocene range shifts: Contrast between the Arctic, the Alps and the East African mountains. *Molecular Ecology* 16:2542–59.

*Englisch, Th., W. Ahlmer, G. Coldea, J.-P. Dalmas, L. Garraud, W. Gutermann, N. Jogan, A. Kagalo, B. Lequette, R. Letz, G. Mansion, K. Marhold, F. Martini, Z. Mirek, H. Piekos-Mirkowa, F. Prosser, M. Puşcaş, M. Scheuerer, L. Schratt-Ehrendorfer, A. Selvaggi, T. Wohlgemuth, H. Niklfeld, and IntraBioDiv Consortium. In preparation. *Species List of the High-Mountain Flora of the Alps and Carpathians*.

*Englisch, Th., and IntraBioDiv Consortium—Floristic Group. Unpublished. The IntraBioDiv Floristic Database, version 4.0 June 2007. Mapping the Distribution of High-Mountain taxa of the Alps and Carpathians. Maintained by Th. Englisch at University of Vienna, Department of Biogeography.

Graham, C.H., S. Ferrier, F. Huettmand, C. Moritz, and A.T. Petersone. 2004. New developments in museum-based informatics and applications in biodiversity analysis. *Trends in Ecology & Evolution* 19:497–503.

*Gugerli, F., Th. Englisch, H. Niklfeld, A. Tribsch, Z. Mirek, A. Ronikier, N.E. Zimmermann, R. Holderegger, P. Taberlet, and IntraBioDiv Consortium. 2008. Relationships among levels of biodiversity and the relevance of intraspecific diversity in conservation—a project synopsis. *Perspectives in Plant Ecology, Evolution and Systematics* 10:259–81.

Guisan, A., and N.E. Zimmermann. 2000. Predictive habitat distribution models in ecology. *Ecological Modelling* 135:147–86.

Hamrick, J.L., and M.J.W. Godt. 1996. Effects of life history traits on genetic diversity in plant species. *Philosophical Transactions of the Royal Society of London, Series B* 351:1291–98.

Isaac, N.J.B., J. Mallet, and G.M. Mace. 2004. Taxonomic inflation: Its influence on macroecology and conservation. *Trends in Ecology & Evolution* 19:464–69.

Lauber, K., and G. Wagner. 2001. *Flora Helvetica*. 3rd ed. Bern: Haupt.

*Manel, S., F. Berthoud, E. Bellemain, M. Gaudeul, J. Luikart, J.E. Swenson, P. Waits, P. Taberlet, and IntraBioDiv Consortium. 2007. A new individual-based spatial approach for identifying genetic discontinuities in natural populations. *Molecular Ecology* 16:2031–43.

Marhold, K., P. Mártonfi, P. Mereďa jun., P. Mráz, I. Hodálová, M. Kolník, J. Kučera, J. Lihová, V. Mrázová, M. Perný, and I. Valko. 2007. Karyological database of ferns and flowering plants of Slovakia. Vs. 1.0. http://147.213.100.144/webapp/index.php?lang=en; Slovac Academy of Sciences, Bratislava.

Moritz, C. 2002. Strategies to protect biological diversity and the evolutionary processes that sustain it. *Systematic Biology* 51:238–54.

Moser, D., S. Dullinger, Th. Englisch, H. Niklfeld, Ch. Plutzar, N. Sauberer, H.G. Zechmeister, G. and Grabherr. 2005. Environmental determinants of vascular plant species richness in the Austrian Alps. *Journal of Biogeography* 32:1117–27.

*Mráz, P., M. Gaudeul, D. Rioux, L. Gielly, Ph. Choler, P. Taberlet, and IntraBioDiv Consortium. 2008. Genetic structure of *Hypochaeris uniflora* (Asteraceae) suggests vicariance in the Carpathians and rapid post-glacial colonization of the Alps from an eastern Alpine refugium. *Journal of Biogeography* 34:2100–14.

Nybom, H., and I.V. Bartish. 2000. Effects of life history traits and sampling strategies on genetic diversity estimates obtained with RAPD markers in plants. *Perspectives in Plant Ecology, Evolution and Systematics* 3:93–114.

*Paun, O., P. Schönswetter, M. Winkler, A. Tribsch, and IntraBioDiv-Consortium. 2008. Evolutionary history of the *Ranunculus alpestris* group (Ranunculaceae) in the European Alps and the Carpathians. *Molecular Ecology* 17:4263–75.

Petit, R.J., U.M. Csaikl, S. Bordács, K. Burg, E. Coart, J. Cottrell, B. van Dam, J.D. Deans, S. Dumolin-Lapègue, S. Fineschi, R. Finkeldey, A. Gillies, I. Glaz, P.G. Goicoechea, J.S. Jensen, A.O. König, A.J. Lowe, S.F. Madsen, G. Mátyás, R.C. Munro, M. Olalde, M.-H. Pemonge, F. Popescu, D. Slade, H. Tabbener, D. Taurichini, S.G.M. de Vries, B. Ziegenhagen, and A. Kremer. 2002. Chloroplast DNA variation in European white oaks: Phylogeography and patterns of diversity based on data from over 2600 populations. *Forest Ecology and Management* 156: 5–26.

Petit, R.J., I. Aguinagalde, J.-L. de Beaulieu, C. Bittkau, S. Brewer, R. Cheddadi, R. Ennos, S. Fineschi, D. Grivet, M. Lascoux, A. Mohanty, G. Müller-Starck, B. Demesure-Musch, A. Palmé, J.P. Martín, S. Rendell, and G.G. Vendramin. 2003. Glacial refugia: Hotspots but not melting pots of genetic diversity. *Science* 300:1563–65.

*Ronikier, M., E. Cieślak, and G. Korbecka. 2008. High genetic differentiation in the alpine plant *Campanula alpina* Jacq. (Campanulaceae): Evidence for survival in several Carpathian regions. *Molecular Ecology* 17:1763–75.

Schönswetter, P., I. Stehlik, R. Holderegger, and A. Tribsch. 2005. Molecular evidence for glacial refugia of mountain plants in the European Alps. *Molecular Ecology* 14:3547–55.

*Steinmann, K. 2008. Testing basic assumptions of species richness hypotheses using plant species distribution data. Ph.D. thesis, University of Zürich, Zürich.

*Thiel-Egenter, C., F. Gugerli, N. Alvarez, S. Brodbeck, E. Cieślak, L. Colli, Th. Englisch, M. Gaudeul, L. Gielly, G. Korbecka, R. Negrini, M. Patrini, O. Paun, M. Pellecchia, D. Rioux, M. Ronikier, P. Schönswetter, F. Schüpfer, P. Taberlet, A. Tribsch, M. van Loo, M. Winkler, R. Holderegger, and IntraBioDiv Consortium. 2009. Effects of species traits on the genetic diversity of high-mountain plants: A multi-species study across the Alps and the Carpathians. *Global Ecology and Biogeography* 18:78–87.

Thornton, P.E., S.W. Running, and M.A. White. 1997. Generating surfaces of daily meteorological variables over large regions of complex terrain. *Journal of Hydrology* 190:214–51.

Vos, P., R. Hogers, M. Bleeker, M. Reijans, T. van de Lee, M. Hornes, A. Frijters, J. Pot, J. Peleman, M. Kuiper, and M. Zabeau, 1995. AFLP—a new technique for DNA-fingerprinting. *Nucleic Acids Research* 23:4407–14.

Widmer, A., and C. Lexer. 2001. Glacial refugia: Sanctuaries for allelic richness, but not for gene diversity. *Trends in Ecology & Evolution* 16:267–68.

Wilson, D.E., and D.A.A. Reeder, eds. 2005. Mammal species of the world. A taxonomic and geographic reference, 3rd ed. Baltimore: Johns Hopkins University Press.

Wohlgemuth, T. 1998. Modelling floristic species richness on a regional scale: A case study in Switzerland. *Biodiversity and Conservation* 7:159–77.

12 A Plant Functional Traits Database for the Alps—Application to the Understanding of Functional Effects of Changed Grassland Management

Sandra Lavorel, Sophie Gachet, Amandine Sahl, Marie-Pascale Colace, Stéphanie Gaucherand, Mélanie Burylo, and Richard Bonet

CONTENTS

Introduction 107
Methods 108
The Alpine Functional Traits Database 108
Study Sites 110
Vegetation and Plant Functional Trait Measurements 111
Ecosystem Measurements 111
Statistical Analyses 111
Results 112
Species-Level Variation in Species Trait Values 112
Functional Community Responses to Management of Subalpine Grasslands 113
Effects of Functional Traits on Ecosystem Properties 114
Discussion 115
Georeferenced Functional Databases as an Asset for Ecological Research 115
Question 1: Variability in Trait Values at the Species Level 116
Question 2: Functional Responses at Community Level 117
Question 3: Effects of Functional Traits on Ecosystem Properties 117
Conclusions 118
Summary 119
Acknowledgments 119
References 119

INTRODUCTION

Plant functional traits are powerful tools to identify generic responses of biodiversity and ecosystem function to environmental change and to understand their underlying mechanisms (Chapin et al., 2000; review by Lavorel et al., 2007). Considerable research has gone into the identification of traits relevant to the response of species and various community types to factors such as climate (Reich et al., 1999; Wright et al., 2005), atmospheric CO_2 concentration (Poorter and Pérez-Soba, 2002), grazing (Díaz et al., 2007), or fertilization (Suding et al., 2005). More recently, studies have begun to focus on traits that affect ecosystem functions, such as primary productivity (Reich et al., 1992; Garnier et al., 2004; Hodgson et al., 2005; Vile et al., 2006), litter decomposition (Cornelissen et al., 1999; Garnier et al., 2004, 2007), or nitrogen cycling (van der Krift and Berendse, 2001; Eviner, 2004; Díaz et al., 2007). In mountain regions, a few studies have examined functional trait responses to altitude and associated climate gradients (Montalvo et al., 1991; Körner, 1999; Pavón et al., 2000; Austrheim et al., 2005; de Bello et al., 2005; He et al., 2006; Klimeš, 2003), to topography (Choler, 2005, and references therein), or to land use (Austrheim et al., 1999; Austrheim and Eriksson, 2003; Maurer et al., 2003; Reiné et al., 2004; de Bello et al., 2005; Gaucherand and Lavorel, 2007; Quétier et al., 2007), and recently, analyses have been extended to include linkages from environmental responses to ecosystem effects (Díaz et al., 2007; Pérez-Harguindeguy et al., 2007; Quétier et al., 2007).

There is thus a growing amount of plant functional traits data collected at various locations around the world, along with a growing demand for such data in order to address fundamental questions about plant functioning (Reich, 2001; Wright et al., 2004; Enquist et al., 2003), as well as to analyze global or regional patterns of biodiversity distribution (Díaz et al., 2004; Thuiller et al., 2004), to predict changes in ecosystem functioning (Garnier et al., 2007; Cornelissen et al., 2007; Enquist et al., 2007), and to model vegetation dynamics over large scales (reviewed by Prentice et al., 2007). In response to both this increasing "offer" of trait data, and the pressing "demand" for its use, several groups around the world have put substantial effort into the development of large trait databases (Fitter and Peat, 1994; Grime et al., 1997; Klimeš and Klimešová, 1999; Klotz et al., 2002; Knevel et al., 2003; Díaz et al., 2004; Wright et al., 2004; Gachet et al., 2005; Moles et al., 2007). In Europe however, none of the existing databases cover the Alps, and mountain regions are only partially covered as part of broader databases. Likewise, data compilations in other parts of the world have not dedicated specific efforts to the coverage or specific analysis of alpine ecosystems (but see He et al., 2006). There is therefore a need to build plant trait databases for alpine ecosystems in order to identify generic patterns of trait and functional diversity distribution along altitude gradients or to understand the functional implications of mountain biodiversity and their impacts for society (Körner et al., 2007).

The alpine functional trait database (Sahl, 2007) has been designed to collect under a single database georeferenced information for vegetation composition, environmental variables, and plant traits of alpine ecosystems. This database makes it possible to gather under a flexible structure data collected by naturalists, conservation managers, and scientists. As a first illustration for the applicability of this and similar databases in mountain regions to address fundamental and applied questions about the biodiversity and functioning of alpine ecosystems, we asked the following questions:

1. How do trait values vary across sites at the species level? This question was addressed by comparing trait values for a few common grassland species along a climatic gradient and across grassland management states.
2. Are there generic functional responses at community level that can be detected using traits? This question is addressed by comparing responses of community-level traits to grassland management gradients within three different massifs.
3. Can traits be used to quantify the effects of environmental change on ecosystem services? This question is addressed by linking functional diversity to ecosystem functions associated with services identified by local stakeholders from the French Alps.

METHODS

THE ALPINE FUNCTIONAL TRAITS DATABASE

The information that is required to address questions about patterns of trait and functional diversity distribution along altitude gradients, or about functional implications of mountain biodiversity, belongs to four domains:

1. Environment, including climate and data about local abiotic conditions, especially soil and disturbances.
2. Functional traits of species or populations.
3. Species relative abundances within communities.
4. Taxonomic information.

All this information needs to be integrated, including through georeferencing. The alpine vegetation and functional trait database (ALTA) was designed in order to meet basic requirements for each of these domains and to provide linkages among them. These linkages are made through both the species and the geographic location (site henceforth).

Data can originate from very diverse sources including field surveys, literature including experimental studies and floras, and in some cases expert opinion. The database structure was therefore designed to be flexible and suitable for future modification. This structure is based around two components: first, the definition of the attributes of data sets (traits, data collection protocols), and second, the inclusion of data from observations. Information about data collection methods is essential for data traceability, including data on experimental protocols (how the measurement was obtained), authors (who obtained the measurement), and literature references. All taxa are standardized for taxonomy (Kerguelen, 1998; http://www2.dijon.inra.fr/flore-france/index.htm).

The database structure is presented in Figure 12.1. It includes five data packages: traits, relevés, sites, taxa, and resources. The two main packages, traits and relevés, share relationships with the taxons, sites, and resources packages. The latter three make it possible to link trait data, data on the geographic distribution of taxa, and environmental data. The database structure was developed

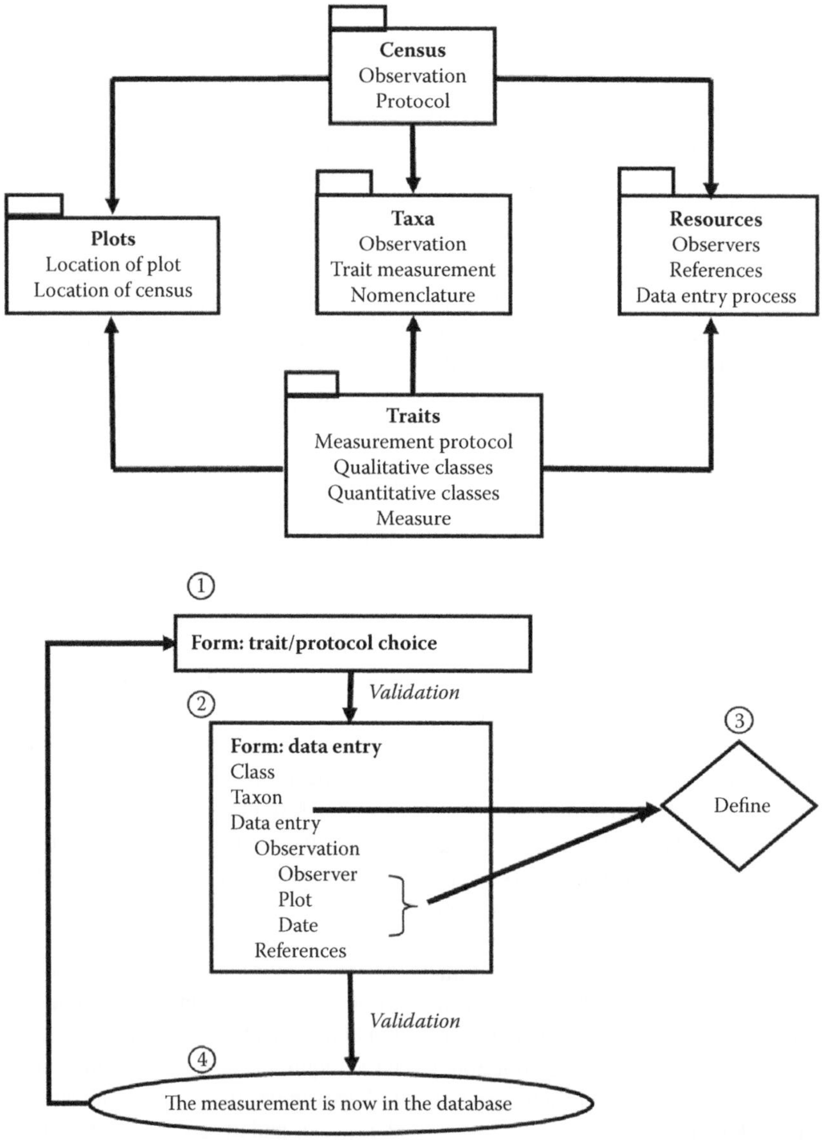

FIGURE 12.1 Simplified representation of the structure of the ALTA database. In the top panel, the boxes (folder symbol following the UML2 norm) represent the packages which structure the database. A package is a group of entities (usually called tables), listed under the package title. The arrows represent the dependency link between packages; for example, the package Census can only exist if the package Taxa exists at the same time. The bottom panel specifies the trait data entry structure. The data entry forms were created using the PHP language. The figure presents the successive (1–4) steps for the entry of a trait measurement (validation is a safety step at the beginning and at the end of the process). Boxes are the different data entry forms; arrows represent triggers.

using PostgreSGL and PostGIS (Berkeley Software Distribution license). All data are georeferenced following the OpenGIS standard, thereby making it possible to apply geographic relations to the data.

The database is informed for floristic and trait data collected initially over georeferenced locations in the entire French Alps, and ultimately over the entire European Alps. At the time of the analyses presented here, full data was available for 350 species that are dominants or subordinates in subalpine grasslands and alpine meadows at trial study sites—including those three for which data was analyzed in this paper. A total of 3,200 species are present in the 16,000 botanical inventories conducted by the French National Alpine Botanical Conservatory (CBNA), stored in a separate georeferenced database. We expect over the next two years to be able to collect and integrate full linked trait and environmental data for at least 200 more species over a range of climatically and edaphically different sites from this large database (north–south and east–west gradients).

Study Sites

In this study we analyzed available vegetation and trait data for three subalpine grassland sites in the French Alps, located along a north–south, and continentality gradient. Site characteristics are summarized in Table 12.1. Within each site floristic composition and functional traits were measured within a set of grassland plots representing different management intensities. All plots had a similar aspect (south to southwest) and moderate slope (< 25°).

At the Beaufortain site (BEAU), we studied eighteen pastures representative of different levels of pastoral management intensity (manure fertilization and grazing). We studied five vegetation classes among the ones described in the northern French Alps typology (Bornard and Dubost, 1992): fertile, fertile intermediate, moderately fertile, infertile, and moor. The five classes correspond to stages of succession developing when pastoral practices are reduced. The intensity of pastoral use defines a fertility gradient, and the succession can be reversed by reusing fertilization (Brau-Nogué, 1996). A detailed description of these pastures in terms of floristic and functional composition can be found in Gaucherand and Lavorel (2007).

At the Bauges site (BAUG), we studied eight pastures representative of fertility variations according to local abiotic conditions (slope, soil depth) and past land use. One pasture (B5) was a fertile pasture dominated by *Dactylis glomerata*. It is similar to the fertile class of the northern French Alps typology, though the abundance of *Deschampsia cespitosa*, a large tussock grass, decreases its pastoral value. A second pasture (B4) was dominated by *Deschampsia cespitosa* and *Carex ferruginea*. These species are usually reported to indicate soil moisture (Bornard and Dubost, 1992; Dorée, 1995). B4 is floristically quite different from Beaufortain pastures, and its pastoral value is slightly below that of the moderately fertile class (Bornard and Dubost, 1992). A third pasture was dominated by *Sesleria caerulea*. This class is also different from the five classes considered for Beaufortain because none of them include *Sesleria* in their floristic composition. The pastoral value of pastures dominated by *Sesleria* is somewhat lower than that of the infertile class (Bornard and Bassignana, 2001). Finally, two pastures (B1 and B2) were dominated by *Nardus stricta*, which suggests low fertility and poor pastoral value. However, one of them (B2) was a cattle camp.

At the Lautaret site (LAUT), we selected fifteen grassland plots representative of the five dominant land use trajectories (Quétier et al., 2007). Much of the hillside was converted to terraces between 1600 and 1900. This process involved plowing, which brought underlying stone to the surface. The practice has been increasingly

TABLE 12.1
Site Characteristics

Site	Beaufortain (BEAU)	Bauges (BAUG)	Lautaret (LAUT)
Location	Les Saisies	Armenaz	Villar d'Arène
Latitude, longitude	45.75° N, 6.58° E	45.57° N, 6.15° E	45.04°N, 6.34°
Altitude (m)	1,800–2,000	1,600–1,800	1,600–2,000
Tmin (°C)	–7.4	–5.6	–4.6
Tmax (°C)	11.3	13	11
Total annual precipitation (mm)	2000	1500	950
Bedrock	Calschist	Limestone	Schist

abandoned since the 1950s. Today, the majority of terraces provide fodder for livestock; traditionally through haymaking in early August, in addition to light grazing in early summer and autumn. The most-used terraced fields are fertilized with manure during May (*c* 8 kg N $ha^{-1}year^{1}$; Picart and Fleury, 1999). Sampling was performed on fifteen fields, i.e., three replicates for each of five distinct management regimes: terraces that are fertilized and mown for hay (1); mown but not fertilized (2); or neither mown nor fertilized, but lightly grazed (3); and unterraced fields that are mown (4) or unmown (5). Detailed data on floristic composition and responses of functional composition and diversity to past and present management are presented by Quétier et al. (2007), Gross et al. (2008), and Lavorel et al. (2008).

Vegetation and Plant Functional Trait Measurements

Vegetation sampling and trait measurements were conducted in 100 m^2 plots that were fenced off for the duration of measurements. We used the same protocol at the three sites. Community composition was sampled using the point quadrat method. A selected list of plant functional traits relating to morphology (e.g., vegetative and reproductive heights, lateral spread) and leaf structure known to be relevant to light capture and nutrient economy (e.g., specific leaf area, leaf dry matter, carbon, nitrogen, and phosphorus contents) were measured at each site using the standardized protocols by Cornelissen et al. (2003). Plant functional traits can vary across sites as a result of genetic variability or phenotypic response to environmental conditions. Phenotypic variation can be particularly strong for traits known as variable (e.g., height, and leaf structural and chemical traits; Garnier et al., 2007). For these, traits measurements were repeated within each land use treatment at each site.

Community weighted means (CWM) for different traits were calculated for each plot as the mean of species trait values in the relevant management treatment weighted by the relative abundance of each species (Garnier et al., 2004). CWM traits, or community functional properties (Violle et al., 2007), represent the expected trait value for a random vegetation sample within each plot.

At the Lautaret site, we also calculated functional divergence (FDvg) using the modified Rao index (Lepš et al., 2006). This index is an abundance weighted mean of all pairwise distances between trait values of species, and represent the expected difference in trait values between two randomly sampled individuals within a communities. Details on the calculation and resulting data can be found in Lavorel et al. (2008).

Ecosystem Measurements

Fertility was quantified in each pasture using the nitrogen nutrition diagnostic (nitrogen index, NI; Salette and Huché, 1989; Lemaire and Gastal, 1997; Duru et al., 2000). This method was developed specifically to characterize the mineral status of grasslands. It quantifies the dilution of nitrogen in grassland canopies during regrowth as a result of self-shading. Variations in the nitrogen index reflect any insufficient absorption of nitrogen. First developed for single-species grass stands, the model of nitrogen dilution was tested and validated for mixed-specific grasslands (Duru et al., 1997), and then for mountain pastures (1,500 to 2,000 m) of the northern French Alps (Brau-Nogué et al., 1994). We collected the aboveground biomass in each pasture at its maximum (mid-July) by harvesting aboveground biomass for four to ten subsamples representing a total sample of 1 m^2 in order to sample within pasture spatial heterogeneity. The samples were oven dried (60°C) to a constant weight, weighed, then sent for N and P analysis. The nitrogen nutrition index was then calculated as: $NNI = (\%N \times 100)/4.8\ (DM) - 0.32$, with DM the production of dry matter of the pasture expressed in tons/ha, and %N, the N content expressed as % of dry matter (Balent et al., 1997). Thus, the nitrogen nutrition index (NNI) represents the percent tissue nitrogen in the herbaceous cover compared to nonlimiting conditions. Following the same formalism, a P nutrition index can be defined to identify P limitation for vegetation growth (PNI) from the percent tissue phosphorus in the herbaceous cover compared to nonlimiting conditions (Jouany et al., 2004).

At the Lautaret site, we quantified a series of ecosystem properties within each of the fifteen plots, including biomass production, litter accumulation and decomposability, nitrogen pools and fluxes, and soil moisture across the growing season. Detailed methods for these measurements are provided by Quétier et al. (2007), Robson et al. (2007), Fortunel et al. (2009), and Gross et al. (2008).

Statistical Analyses

Floristic, trait, environmental, and ecosystem data for the three sites were entered manually into the database, although work in progress will allow import of existing data sheets, as well as databases. Relationships between species or community mean traits with environmental variables were analyzed with regression. A step-wise analysis was applied to identify relationships between ecosystem properties and components of functional diversity, as described by Díaz et al. (2007). This analysis first used four steps to identify the effects of: (1) abiotic variables (climate and soil), (2) CWM traits, (3) functional divergence of traits, and (4) abundance of dominant grass species on each ecosystem property using General Linear Modelling. A fifth step combined all significant variables from steps 1–4 and retained only the significant variables in the combined general linear model. Analyses were performed with Genstat 8.0.

RESULTS

Species-Level Variation in Species Trait Values

A query of trait data by site and species was used to identify species with multiple trait measurement records within and across sites. Grass species with records over at least two sites and a total of at least five records were selected for analysis. Here we present results for leaf dry matter content (LDMC), which is a good marker of herbaceous species responses to management and of community effects on ecosystem functioning (Garnier et al., 2004; Fortunel et al., 2008). In general, and in our database in particular, LDMC is well-correlated with specific leaf area (SLA) and leaf nitrogen content (LNC), thereby providing an easy to measure indicator of important plant functions. There was no significant effect of site, and in particular of any of the climatic variables, on variation of trait values across pastures for a given species (data not shown). On the other hand, traits varied across populations in response to site fertility, and more specifically phosphorus nutrition. For example, for *Dactylis glomerata*, there was no significant difference in LDMC across sites, but LDMC significantly decreased with the Phosphorus Nutrition Index (Figure 12.2, inset). The same relationship was marginally significant when considering species characteristic of more fertile pastures pooled together (*Agrostis capillaris, Dactylis glomerata, Phleum alpinum*, $R^2 = 0.31$, $p = 0.12$) (Figure 12.2), but there was no relationship for some other common and rather ubiquitous grasses (*Festuca rubra*). The negative trend across all common grass species was not significant ($R^2 = 0.11$, $N = 38$) (Figure 12.2), possibly due to an insufficient number of observations. Overall, species characteristic of more fertile pastures had a lower mean LDMC (mean PNI = 72%, mean LDMC = 299 mg/g, LDMC, range = 255–369 mg/g, CV = 9%) than species

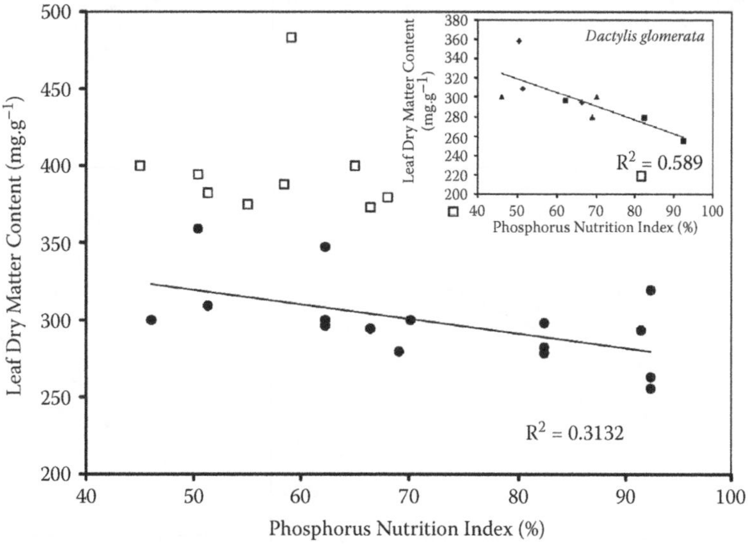

FIGURE 12.2 Variation in leaf dry matter content (LDMC, ratio of dry to fresh hydrated mass) for populations of common dominant grasses across the three study sites in response to phosphorus availability as quantified by the Phosphorus Nutrition Index (PNI, derived by an empirical relationship for stochiometry with nitrogen and dilution of nitrogen in a closed grassland stand—Duru et al., 2000; Jouany et al., 2004). Closed diamond symbols: grasses with an exploitative strategy (*Agrostis capillaris, Dactylis glomerata, Phleum alpinum*). Open square symbols: grasses with a conservative strategy (*Bromus erectus, Nardus stricta, Sesleria caerulea*). The inset figure presents variation in LDMC for populations of *Dactylis glomerata* across the three study sites in response to phosphorus availability as quantified by PNI. Square symbols: Beaufortain. Triangle symbols: Bauges. Diamond symbols: Lautaret. The significant regression coefficient is indicated for the exploitative grasses group.

characteristic of less fertile pastures (*Bromus erectus*, *Sesleria caerulea*, *Nardus stricta*) (mean PNI = 62%, mean LDMC = 393 mg/g, range = 370–484 mg/g, CV = 8%) (t-test, $p < 0.01$).

Functional Community Responses to Management of Subalpine Grasslands

The database was queried in order to obtain a table presenting for each plot within each site (taken as statistical unit) the list of community weighted mean trait values, environmental and ecosystem variables, including fertility indices. Nitrogen and phosphorus nutrition indices were not correlated, giving two independent axes in nutriment availability across the different plots. There was a distinct and generic response of community weighted mean traits to grassland management, and specifically to fertility. CWM plant vegetative height increased exponentially in response to nitrogen availability (Figure 12.3a), but showed a weak and nonsignificant response to phosphorus availability (data not shown). Both CWM LDMC and CWM SLA showed a strong response to the Phosphorus Nutrition Index, with decreasing CWM LDMC and increasing CWM SLA associated with increasing fertility (Figure 12.3b, Figure 12.3c). There was, however, no response of these leaf structural traits to nitrogen fertility (Nitrogen Nutrition Index). CWM leaf phosphorus content (LPC), which was available only for the Bauges and Lautaret sites, increased significantly in response to phosphorus availability (Figure 12.3d), whereas CWM LNC showed no response at all to variations in nitrogen availability (data not shown). Accordingly, CWM leaf N:P showed a nonsignificant decrease in response to phosphorus availability ($R^2 = 0.13$, $p > 0.10$). Overall, higher phosphorus fertility favored plants with less fibrous leaves that had a higher phosphorus content. Higher nitrogen fertility favored taller plants but had no measurable effect on leaf traits.

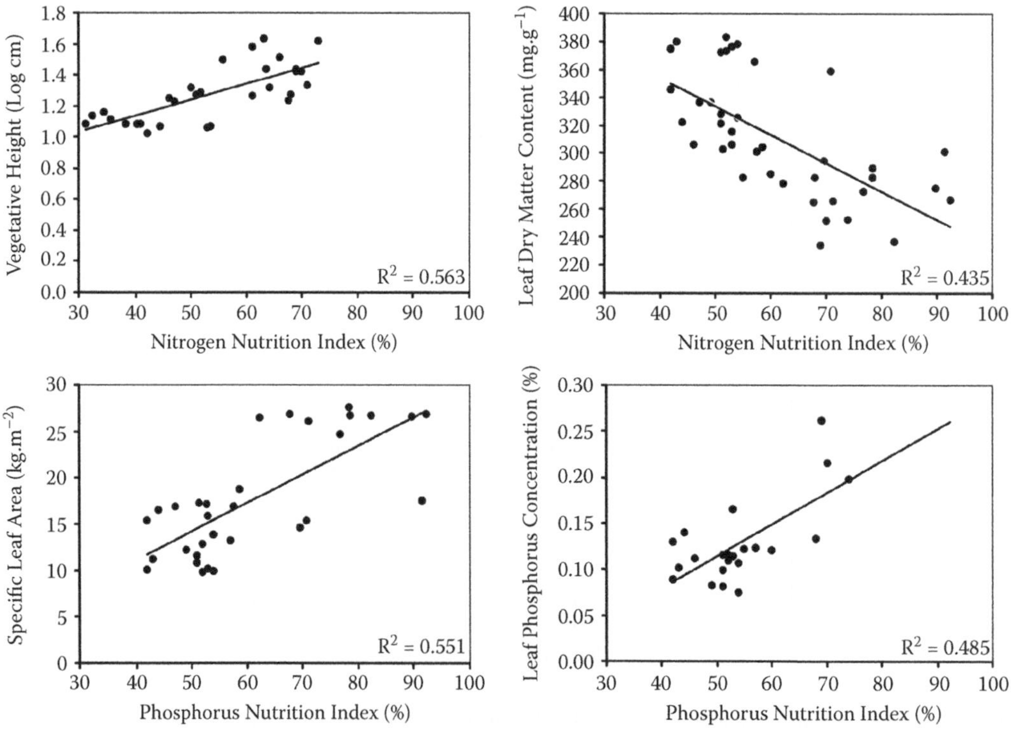

FIGURE 12.3 Response of community weighted mean traits to fertility indices. (a) Top left panel: Vegetative height and nitrogen nutrition index (NNI, derived by an empirical relationship for dilution of nitrogen in a closed grassland stand—Duru et al., 2000); (b) Top right panel: Leaf dry matter content (LDMC, ratio of dry to fresh hydrated mass) and phosphorus nutrition index (PNI, derived by an empirical relationship for stochiometry with nitrogen and dilution of nitrogen in a closed grassland stand—Duru et al., 2000; Jouany et al., 2004); (c) Bottom left panel: Specific leaf area (SLA, ratio of leaf area to dry mass) and PNI; (d) Bottom right panel: Leaf phosphorus content (LPC) and PNI. Regression coefficients are indicated on each graph.

Effects of Functional Traits on Ecosystem Properties

The plots x community-weighted mean traits, environmental variables, and ecosystem measurements table was further analyzed in order to quantify the effects of functional traits and derived functional diversity metrics (CWM and FDvg) on ecosystem properties. All these data were collected by different projects and operators with individual objectives regarding the study of community dynamics or specific ecosystem processes, and the database was essential in order to combine them and carry out such a complex integrative analysis.

Results of the significant General Linear Models for relationships between ecosystem properties and components of functional diversity at Lautaret are presented in Table 12.2. We found that ecosystem properties related to primary production, total aboveground biomass, ANPP, and SANPP were best explained by the NNI. Although ANPP was significantly related to the variety of ways in which plants acquire light, reflected by the distribution within communities of vegetative height (FDvg_VH), and of LNC (FDvg_LNC; marginally significant), these terms were not retained in the most parsimonious final model. Overall, models explained less than 50% of the total variance and were not improved by the inclusion of idiosyncratic effects, such as the abundance of *Festuca paniculata*. Hence, primary production was largely independent of functional diversity.

TABLE 12.2
Results of the General Linear Models of Ecosystem Properties Following Díaz et al. (2007)

	Individual Effects of Abiotic and Biotic Components						Combining Significant Effects	
	Step 1: Abiotic Variables		Step 2: CWM Traits		Step 3: Trait Distribution		Step 5: Final Model	
Ecosystem property	Variable	p	Variable	p	Variable	p	Model	% var.
AGB	—	—	—	—	FDvg_LHS	0.089	No suitable model	< 15
ANPP	NNI	0.045	—	—	FDvg_VH	0.040	ANPP = - 0.028 + 0.0014 NNI	44
					FDvg_LNC	0.091		
SANPP	NNI	0.045	—	—	—	—	SANPP = - 0.012 + 0.00085 NNI	22
Litt_Acc	—	—	CWM_VH	0.002	FDvg_VH	0.10	Litt = 9.4 + 0.35 CWM_VH - 7.54	75
			CWM_LNC	0.012			CWM_LNC - 0.00063	
			CWM_LTS	0.046			CWM_LTS	
Decomp	—	—	CWM_LNC	< .001	—	—	Decomp = 13.5 + 3.02 CWM_LNC	91
			CWM_Lig:N	< .001			- 0.39 CWM_Lig:N	
NO3	—	—	CWM_LNC	< .001	—	—	NO3 = -25.8 + 21.2 CWM_LNC +	76
			CWM_RL	0.044			0.0085 CWM_RL	
Den_P	—	—	CWM_LNC	0.001	FDvg_LHS	0.026	Den_P = -59.5 + 80.6 CWM_LNC	67
			CWM_RL	0.034			+ 0.042 CWM_RL	
SWC	WHC	0.003	CWM_LA	0.043	—	—	SWC = 2.07 + 1.55 WHC – 1.58	85
	AGB	< .001	CWM_RL	0.001			AGB + 0.00083 CWM_LA	
							- 0.0014 CWM_RL	

Note: Numerical results are presented for steps 1–3 and step 5. Although tested, no significant relationship was found between any of the ecosystem properties examined and the abundance of dominant grasses, hence step 4 is not presented. — : no significant effect of any variable tested within the step. Functional diversity metrics: CWM_: community weighted mean trait value (see methods); FDvg_: functional divergence based on the modified Rao index (see methods). Acronyms after CWM_ and FDvg_ refer to measured plant traits. VH: vegetative height; LNC: Leaf Nitrogen Content; LS: Leaf Tensile Strength; Lig:N: leaf lignin to nitrogen ratio; LA: Leaf Area; RL: Root Length. FDvg_LHS refers to the mean functional divergence for SLA, vegetative height and seed mass. Ecosystem properties are: AGB: total aboveground biomass; ANPP: annual net primary production; SANPP: specific annual net primary production (Garnier et al., 2004); Litt_Acc: total litter biomass at the beginning of spring; Decomp: litter decomposability assessed by Near Infrared Spectrometry; NO3: nitrate concentration measured with resin bags; Den_P: denitrification potential; SWC: mean soil water content through summer; WHC: soil water holding capacity estimated based on texture, stoniness and soil depth (see Gross et al., 2008).

In contrast, ecosystem properties associated with litter dynamics were best explained by community functional properties, including CWM leaf traits (leaf nitrogen and lignin contents; leaf tensile strength), and CWM vegetative height in the case of total accumulated litter in the spring. A marginally significant relationship was also found with the distribution of vegetative height within communities, but this term was dropped from the final model. Both for standing litter (the pool) and litter decomposability (the flux), models explained a very high proportion of the total variance. Likewise, models of the components of soil fertility involved solely community functional properties, with CWM of both leaf and root traits, and models explained a very high proportion of the total variance. Hence, processes determining litter accumulation and soil fertility were strongly related to mean functional properties of the vegetation.

Soil water content (SWC) through the summer period illustrated the case of a complex model associating both abiotic and functional diversity factors. As expected, SWC was explained by local soil conditions determining water holding capacity, and the model specified the additional contributions of community-level means for leaf area (positive effect representing shading) and root length (negative effect representing water uptake). An additional aboveground biomass term had to be included in the final model that explained a very high proportion of the total variance, and accounted for direct biomass effects on shading (water conservation) and transpiration (water demand). Hence, soil moisture through summer depended on the combination of local abiotic factors, their effects on aboveground standing biomass, and mean functional properties of the vegetation.

DISCUSSION

Georeferenced Functional Databases as an Asset for Ecological Research

The three aforementioned examples illustrate the usefulness of georeferenced functional trait databases to address important research questions on the dynamics of biodiversity and ecosystem services in alpine ecosystems. The work under way has made it possible to collect under a single database structure georeferenced information from multiple sources about plot-level species composition, environmental variables, and ecosystem processes, combined with location-specific trait data. The first asset of such a database is that it gathers under a flexible structure data collected by naturalists, conservation managers, and a range of scientists, all having different objectives at the time of collection. This data thus goes beyond the objective of a single research project, and subsets obtained through queries of the database can subsequently be used by scientists and managers alike. This is already the case in our study region, with sustained exchange of data and results between for example, the Ecrins National Park, the National Alpine Botanical Conservatory, and research teams that have or have not contributed to original data collection. Following databasing good practice, specification of methods, and observers for each record allow cross-checking or verification with contributors. It ensures later quality for the combination of data from multiple sources, while at the same time admitting into the database multiple data sources whose relevance to specific analyses is left to users to decide based on available meta-information. As with any database, the data is also stored parsimoniously and secure for the long term and allows complex cross-queries, making it possible to detect patterns in the data with the help of external data (for example, combining traits occurrence and climate). Each of the contributors benefits in the long term from the whole data set, for quality control and putting their own data in perspective, as well as for analyses they could not have conducted with their own data alone.

Second, georeferencing provides a number of key advantages for data use, and even for data collection. The first and simplest of these is the ability to identify geographical data gaps and thus to define priorities for collection. This is particularly important when aiming to link different data types and sources, for example, floristic and trait data. An essential value of georeferencing is also the ability it provides to link the database's original content with other georeferenced data contained in other databases, such as terrain (Digital Terrain Models), climate (current and future scenarios), extensive botanical surveys databases (e.g., the CBNA database for the French Alps), phylogenetic databases, taxonomic referentials and floras (for taxonomic cross-checking or for additional trait data, such as plant maximum height, phenology, and chorology), and also socioeconomic data going from detailed land use maps to demographic or economic data. Although these have not been implemented here at this stage, georeferencing is particularly interesting to develop spatially explicit queries to search most appropriate trait values for a given vegetation relevé (e.g., by geographic proximity and environmental criteria, such as bedrock and land use). Such a database structure is of great advantage to accommodate complex queries with multiple linkages across a diversity of data types.

This advantage may nevertheless also turn into a difficulty, as the complexity of the structure makes the design of queries sometimes difficult, requiring assistance from professional database specialists. For example, designing a query to calculate CWM and FDvg directly from the database has proved extremely difficult, and we have so far had to resort to offline calculation using raw data extractions. In general, the management of such large and complex databases is beyond the skills of most ecologists, and there is a need to identify *a priori* the most likely queries for a professional to program them. An interdisciplinary approach and an associated dialogue between ecology and bioinformatics are thus required. More generally, a constraint for complex databases such as ALTA is the need to plan all future data availability and potential uses before implementation. All the potential types of users (managers, scientists, politicians, public, etc.) need to specify their needs at the beginning of the work, otherwise the tool may rapidly become useless. This part of the work represented around 80% of time devoted to the database construction, a number that is not atypical for such large databases.

The flexibility of our structure nonetheless makes it possible to address a very wide range of ecological questions, of which we have illustrated three types in this article. We now outline some key scientific questions and applications for georeferenced functional trait databases relevant to ecological and evolutionary research on mountain ecosystems beyond our study area.

Question 1: Variability in Trait Values at the Species Level

Predictive studies using scenarios of land use or climate change have highlighted possible large impacts on species distributions and diversity in alpine ecosystems (Sala et al., 2000; Thuiller et al., 2005). However, individual species responses to environmental changes, including phenotypic plasticity or genetic adaptation (through selection of the fittest phenotypes) have not been taken into account by these modelling approaches, in spite of their potential to dramatically alter projected changes in biodiversity (Theurillat and Guisan, 2001; Thuiller et al., 2008). If individuals within established communities can modify their traits in response to new environmental conditions then community resistance to global change will be greater than expected. If, on the other hand, plasticity is low then there may be replacement either by new species (as predicted by current models), or also by different genotypes of the same species, either from within the community (change in dominance) or from other communities (colonization). The existence of strong differences among species in their responses to environmental change will therefore result in different trajectories of change among communities, and represents an important source of uncertainty in projections of future biodiversity. Finally, as these changes regard species traits, there also will be large uncertainties on resulting effect on ecosystem functioning (see Question 3).

Phenotypic plasticity for traits related to fitness can be advantageous either by allowing species to maintain fitness in stressful environments (jack of all trades), to increase fitness in favorable environments (master of some), or some combinations of the two previous (jack and master; Richards et al., 2006). There are some theoretical predictions on differences in the level of phenotypic plasticity between populations living at different altitudes (less plasticity and more genetic adaptation with increasing altitude; Bradshaw, 1965), or in more or less fertile conditions (greater plasticity for morphological traits in species from richer habitats—Crick and Grime, 1997; greater plasticity for physiological traits in species from poorer habitats—Grime and Mackey, 2002), between generalists and specialists (greater plasticity in generalists—van Tienderen, 1991) or between central and marginal populations within the species distribution (greater plasticity in marginal populations—Scheiner and Goodnight, 1994), but there is little empirical evidence for these hypotheses (but see Gauthier et al., 1998; Gianoli, 2004; Griffith and Sultan, 2006).

Alpine systems provide steep environmental gradients and contrasted habitat mosaics. They also are the limits to distributions of many species, and harbor varying combinations of generalist and specialist species as a function of altitude (Körner, 1999). They are therefore particularly suited to test these theoretical hypotheses. Georeferenced functional trait databases are primary tools to do this by examining the nature and sources of intraspecific variation in species traits. In the example presented for trait variation along management gradients within three massifs from the French Alps, we illustrated how intraspecific trait variation can differ across species, showing a response of leaf traits (illustrated for LDMC) to pasture fertility for common species of more fertile habitats with an exploitative strategy (lower mean LDMC) (*sensu* Grime, 1977), but no response for common species of less fertile habitats with a conservative strategy (higher mean LDMC). Tests of altitudinal effects on trait intraspecific variation, and of contrasts between species at the centre versus the edge of their range, are in progress.

For the same reasons, georeferenced functional trait databases from alpine ecosystems are primary tools to examine trait variations across species along environmental gradients and test for global patterns of trait response to climate and other environmental factors (Reich et al., 1999; Díaz et al., 2004; Wright et al., 2004; Han et al., 2005; Moles et al., 2007). Comparisons across different mountain regions in different continents that have phylogenetically distinct species pools but convergent functional characteristics will give strong power to draw conclusions from such analyses.

Such analyses, at intraspecific and interspecific levels, have potentially essential implications for conservation of functional diversity beyond the conservation of species diversity. Georeferenced functional trait databases should support the identification of functional variations within species that represent local adaptation and warrant specific conservation measures.

Finally, trait variation within species has strong implications for databasing and data use. Analyses such as the one presented here are needed to assess to what extent trait data collected under specific conditions in a given massif are applicable to other alpine areas. They should be used to design collection efforts depending on the driving variables of trait variation within and across mountain regions.

Question 2: Functional Responses at Community Level

Our example analysis on available data for three French massifs illustrated how relationships can scale from the population to species and the community. LDMC varied across populations within species (*Dactylis glomerata* and other exploitative grass species), across species (exploitative versus conservative grass species), and at the community level with a detectable response to phosphorus availability. Trait scaling relationships from individuals to communities remain a frontier in functional ecology (Enquist et al., 2007).

There is a growing literature on responses of community-level trait values (or community functional parameters, CFP; Violle et al., 2007) to environmental factors (see e.g., Lavorel et al., 2007, for a review). Fundamental questions on variations of CFP to different environmental factors require that both community composition (species relative abundances) and trait data be available. Georeferencing is also required in order to analyze the effects of climatic or topographic factors. For example, there still is little published data on how species-level trait responses to climatic gradients translate to community level (see Reich et al., 2003; Enquist et al., 2007; Pakeman et al., 2008). As for species-level responses, alpine ecosystems offer unique opportunities to describe or test such relationships, thanks to the presence of steep environmental gradients, mosaics of different habitats in relation to mesotopography (e.g., Choler, 2005), and land use (Quétier et al., 2007), and to the resulting high functional diversity. They also offer a unique opportunity to test the convergence of responses across regions and continents, and interactions of responses to factors such as soil fertility or pH with climate or latitude. Finally, relationships between community-level trait values and environmental factors can be used as a means to project functional responses to changes in environmental factors or land use (e.g., Quétier et al., 2007), and thereby forecast global change effects on alpine biodiversity.

Questions on trait responses at the community level could be extended to consider variations in other dimensions of functional diversity than just community mean, and to examine how the spread of trait values within communities varies along environmental gradients (Mason et al., 2005). Important questions on the structure and dynamics of communities regard whether communities are organized by trait convergence or trait divergence (Grime, 2006; Wilson, 2007). Constraining conditions, such as those experienced in alpine ecosystems, are expected to lead to trait convergence, and this hypothesis could be examined across altitudinal gradients, or mosaics of habitats with differing levels of severity (e.g., different lengths of snow persistence; Choler 2005; Wipf et al., 2006). Variations in functional diversity, including functional divergence (the spread of trait values within a community) are relevant to mechanisms of species coexistence, such as niche separation (i.e., divergence), and examination of their changes along gradients of environmental severity is a means to address hypotheses, such as the effects of environmental severity on the importance of competition (Callaway et al., 2002).

Question 3: Effects of Functional Traits on Ecosystem Properties

There is increasing evidence for relationships between plant traits and ecosystem functioning (Lavorel and Garnier, 2002; Eviner and Chapin, 2003; Díaz et al., 2007; de Bello et al., submitted). The Lautaret case study illustrated the variety of ways in which ecosystem properties can be related to abiotic factors and components of functional diversity. There was support for the biomass ratio hypothesis (Grime, 1998), as community-level trait means, which are determined by the trait values of dominant species, rather than the distribution of traits (FDvg) ultimately explained ecosystem properties. In addition, FDvg–LHS was never significant (marginal significant relationship with peak standing green biomass AGB, explaining only 15% of the total variance). Given that FDvg_LHS was correlated with Simpson species diversity (Lavorel et al., unpublished) this indicated an absence of effects of species diversity per se. The biomass ratio hypothesis can be seen as a scaling relationship from species and community-level traits to ecosystem functioning (Garnier et al., 2004). Effects of the diversification of traits within communities were not shown in our analyses, but these are usually expected for stability functions (Díaz and Cabido, 2001; Hooper et al., 2005), which were not examined here.

Functional trait databases can therefore be used to address research questions on the related dynamics of biodiversity and the functioning of alpine ecosystems (Körner, 1993; Körner et al., 2007). These regard in particular the assessment of overlaps between those traits that determine response to environmental factors (e.g., climate, soil fertility, disturbance through management regimes) and traits that determine effects on ecosystem functioning (Lavorel and Garnier, 2002). This overlap was illustrated by a previous study at the Lautaret site, which showed that community-level means for plant height and some leaf traits (LNC, LDMC) are involved in both the response of community composition to grassland management and effects on ecosystem properties, such as litter dynamics, primary production from mown fields, or fodder digestibility (Quétier et al., 2007).

Our analysis expanded on this previous analysis to examine whether relationships can be identified not only for community-level trait values, but also the distribution of trait values within the community, thereby testing the hypothesis that functional divergence may enhance ecosystem functioning through mechanisms of functional complementarity (Hooper et al., 2005; Wright et al., 2006). We were not able to find strong supporting evidence for this hypothesis in our study system, but trait databases should allow such tests over wider sets of localities where measurements of floristic composition and ecosystem functioning also are available. Furthermore, such analyses using the framework for decomposition of the variance in ecosystem properties (Díaz et al., 2007) would enable the exploration of how increasing abiotic constraints interact with biotic trait-based effects, a question that has hardly been assessed with quantitative data thus far.

Quantitative models of ecosystem properties with abiotic and trait-based variables are an essential tool to project changes in ecosystem service delivery in response to global change (Quétier et al., 2007; Quétier et al., 2009). In our case study, we can predict that under scenarios of grassland management (1) changes in fodder production would be directly driven by changes in fertilization practices rather than by effects on biodiversity; (2) changes in ecosystem properties associated with litter dynamics and nutrient cycling would be affected not only by direct and immediate effects of management (e.g., cessation of mowing, changes in grazing intensity, fertilization), but also in the long term by all their effects on plant functional diversity; and (3) because soil water holding capacity, which is a legacy of plowing over the historical period (Bakker et al., 2008), will not change, changes in soil water content through summer will be the result of management effects on plant functional diversity.

CONCLUSIONS

A georeferenced database of plant functional traits for alpine regions makes it possible to address key questions in fundamental ecology and specific questions relevant to the conservation and management of biodiversity in alpine ecosystems. In this paper, we illustrated some of the domains of applicability of such a database, thereby demonstrating its usefulness to test ecological theory and enhance the knowledge of alpine ecosystems by collecting under a single flexible and evolving structure information gathered by scientists, naturalists, and managers, and making it available to all to evaluate their own data and address their specific questions with a large collective data set. Trait databases are also particularly powerful when they can be coupled with georeferenced floristic databases, as well as climatic, geographic (e.g., terrain models, land use variables), or taxonomic and phylogenetic databases, allowing complex queries across a diversity of data types to address state-of-the-art research questions, as well as management objectives. Such data systems should thus be developed to support research on alpine ecosystems and provide a basis for their management in the context of global change. The flexibility of the structure will make it possible to integrate new questions as required by future studies and users, although flexibility and the capacity for the structure to evolve mean that the database should at any point in time be considered a perfectible work in progress. Such flexibility and complexity come at a major cost in terms of specialized skills, meaning that assistance by databasing and geomatics experts cannot be dispensed with, and that a constant dialogue between these specialists and users is required, a short-term cost, but a long-term benefit to both groups. Finally, trait databases such as the one demonstrated here for plants need to be developed for other groups of organisms than plants (e.g., birds, insects, soil biota), where equally relevant questions may be addressed. For this, the building of an interdisciplinary research community, among ecologists and also with bioinformatics specialists, as well as with managers, appears as a necessary pathway.

SUMMARY

Plant functional traits are powerful tools to identify generic responses of biodiversity and ecosystem function to environmental change and to understand their underlying mechanisms. A collaborative initiative was launched to collect under a single, georeferenced database information for vegetation composition, environmental variables, and plant traits of alpine ecosystems. This database makes it possible to gather under a flexible structure data collected by naturalists, conservation managers, and scientists, and to use this data to address both fundamental research and management questions. As a first test for the applicability of this database, we asked the following questions:

1. What are the patterns of trait variability at species level, according to environmental variation? This question was addressed by comparing trait values for a few common grassland species along a climatic gradient and across grassland management states.
2. Are there generic functional responses that can be detected using traits? This question was addressed by comparing responses of community-level traits to grassland management gradients within three different massifs.
3. Can traits be used to quantify the effects of environmental change on ecosystem services that are relevant to local stakeholders? This question was addressed by linking functional diversity to ecosystem functions associated with services identified by local stakeholders.

The key strengths of the database, through its flexible structure, its ability to accommodate data from a diversity of sources, and the capability to link it to other georeferenced databases (e.g., terrain, climate, botanical, social) are discussed. We also emphasize the associated complexities, which come at the short-term cost but long-term benefit of requiring a sustained interdisciplinary dialogue with bioinformatics specialists.

ACKNOWLEDGMENTS

The construction of the ALTA database was funded by the Ecrins National Park. Data used for this article was obtained through the VISTA project (EVK2-2002-CT00386), the Office National de la Chasse et de la Faune Sauvage, and CEMAGREF. We acknowledge numerous field helpers and students who have participated in data collection in the field and in the lab, and in data management (with special thanks to Aurélie Thébault for the first set of database entries).

REFERENCES

Austrheim, G., and O. Eriksson. 2003. Colonization and life history traits of sparse plant species in subalpine grasslands. *Canadian Journal of Botany* 81:171–82.

Austrheim, G., M. Evju, and A. Mysterud. 2005. Herb abundance and life-history traits in two contrasting alpine habitats in southern Norway. *Plant Ecology* 179:217–29.

Austrheim, G., E.G.A. Olsson, and E. Grontvedt. 1999. Land-use impact on plant communities in semi-natural sub-alpine grassland of Budalen, central Norway. *Biological Conservation* 87:369–79.

Bakker, M.M., G. Govers, A. Van Doorn, F. Quétier, D. Chouvardas, and M.D.A. Rounsevell. 2008. The response of soil erosion and sediment export to land-use change in four areas of Europe: The importance of landscape pattern. *Geomorphology* 98:213–26.

Balent, G., M. Duru, A. Gibon, D. Magda, and J.P. Theau. 1997. Prairies permanentes de milieu océanique et de montagne humide. Outil de diagnostic agro-écologique et guide pour leur utilisation. INRA, Toulouse, France.

Bornard, A., and M. Bassignana. 2001. Typologie agro-écologique des végétations d'alpages en zone intra-alpine des Alpes du Nord Occidentales Cemagref, Grenoble.

Bornard, A., and M. Dubost, M. 1992. Diagnostique agro-écologique de la végétation des alpages laitiers des Alpes du Nord humides: établissement et utilisation d'une typologie simplifiée. *Agronomie* 12:581–99.

Bradshaw, A.D. 1965. Evolutionary significance of phenotypic plasticity in plants. *Advances in Genetics* 13:115–55.

Brau-Nogué, C. 1996. Dynamique des pelouses d'alpages laitiers des Alpes du Nord Externes. Ph.D., Université Joseph Fournier, Grenoble 1, France, Grenoble.

Brau-Nogué, C., M. Bassignana, and A. Bornard, A. 1994. Diagnostic de nutrition minérale de l'herbe par analyse du végétal: Application aux pelouses d'alpages. *Fourrages* 137:43–59.

Callaway, R.M., R.W. Brooker, P. Choler, Z. Kikvidze, C.J. Lortie, R. Michalet, L. Paolini, F.I. Pugnaire, B. Newingham, B.J., Cook, and E.T. Aschehoug. 2002. Positive interactions among alpine plants increase with stress. *Nature* 417:812–20.

Chapin, F.I., E.S. Zavaleta, V.T. Eviner, R.L. Naylor, P.M. Vitousek, H.L. Reynolds, D.U. Hooper, S. Lavorel, O.E. Sala, S.E., Hobbie, M.C. Mack, and S. Diaz. 2000. Consequences of changing biotic diversity. *Nature* 405:234–42.

Choler, P. 2005. Consistent shifts in alpine plant traits along a mesotopographical gradient. *Arctic, Antarctic, and Alpine Research* 37:444–53.

Cornelissen, J.H.C., S. Lavorel, E. Garnier, S. Díaz, N. Buchmann, D.E. Gurvich, P.B. Reich, H. ter Steege, H.D. Morgan, M.G.A. van der Heijden, J.G. Pausas, and H. Poorter. 2003. Handbook of protocols for standardised and easy measurement of plant functional traits worldwide. *Australian Journal of Botany* 51:335–80.

Cornelissen, J.H.C., N. Pérez-Harguindeguy, S. Díaz, J.P. Grime, B. Marzano, M. Cabido, F. Vendramini, and B. Cerabolini. 1999. Leaf structure and defence control litter decomposition rate across species and life forms in regional flora on two continents. *New Phytologist* 143:191–200.

Cornelissen, J.H.C., P.M. van Bodegom, R. Aerts, T.V. Callaghan, R.S.P. van Logtestijn, J. Alatalo, F.S.I. Chapin, R. Gerdol, J. Gudmundsson, D. Gwynn-Jones, A.E. Hartley, D.S. Hik, A. Hofgaard, I.S. Jonsdottir, S. Karlsson, J.A. Klein, J. Laundre, B. Magnusson, A. Michelsen, U. Molau, V.G. Onipchenko, H.M. Quested, S.M., Sandvik, I.K. Schimidt, G.R. Shaver, B. Solheim, N.A. Soudzilovskaia, A., Stenström, A. Tovlanen, O. Totland, N. Wada, J.M. Welker, W. Zhao, and M.O.L. Team. 2007. Global negative vegetation feedback to climate warming responses of leaf litter decomposition rates in cold biomes. *Ecology Letters* 10:619–27.

Crick, J.C., and J.P. Grime. 1987. Morphological plasticity and mineral nutrient capture in two herbaceous species of contrasted ecology. *New Phytologist* 107:403–14.

De Bello, F., S. Lavorel, S. Díaz, R. Harrington, R. Bardgett, P. Cipriotti, H. Cornelissen, C. Feld, D. Hering, P. Martins da Silva, S. Potts, L. Sandin, J.P. Sousa, J. Storkey, and D. Wardle. Submitted. Functional traits underlie the delivery of ecosystem services across different trophic levels. *Frontiers in Ecology and the Environment.*

de Bello, F., J. Leps, and M.-T. Sebastià. 2005. Predictive value of plant traits to grazing along a climatic gradient in the Mediterranean. *Journal of Applied Ecology* 42:824–33.

Diaz, S., and M. Cabido. 2001. Vive la différence: Plant functional diversity matters to ecosystem processes. *Trends in Ecology & Evolution* 16:646–55.

Diaz, S., J.G. Hodgson, K. Thompson, M. Cabido, J.H.C. Cornelissen, A. Jalili, G. Montserrat-Martí, J.P. Grime, F. Zarrinkamar, Y. Astri, S.R. Band, S. Basconcelo, P. Castro-Díez, G. Funes, B. Hamzehee, M. Koshnevi, N. Pérez-Harguindeguy, M.C. Pérez-Rontomé, F.A. Shirvany, F. Vendramini, S. Yazdani, R. Abbas-Azimi, A. Bogaard, S. Boustani, M. Charles, M. Dehghan, L. de Torres-Espuny, V. Falczuk, J. Guerrero-Campo, A. Hynd, G. Jones, E. Kowsary, F. Kazemi-Saeed, M. Maestro-Martinez, A. Romo-Diez, S. Shaw, B. Siavash, P. Villar-Salvador, and M. Zak. 2004. The plant traits that drive ecosystems: Evidence from three continents. *Journal of Vegetation Science* 15:295–304.

Diaz, S., S. Lavorel, F. De Bello, F. Quétier, K. Grigulis, and T.M. Robson. 2007. Incorporating plant functional diversity effects in ecosystem service assessments. *Proceedings of the National Academy of Sciences* 104:20684–89.

Díaz, S., S. Lavorel, S McIntyre, V. Falczuk, F. Casanoves, D. Milchunas, C. Skarpe, G. Rusch, M. Sternberg, I. Noy-Meir, J. Landsberg, W. Zangh, H. Clark, and B.D. Campbell. 2007. Grazing and plant traits—a global synthesis. *Global Change Biology* 13:313–41.

Dorée, A. 1995. *Flore pastorale de montagne.* Paris: Cemagref.

Duru, M., G. Lemaire, and P. Cruz. 1997. Grasslands. In *Diagnosis of the Nitrogen Status in Crops*, pp. 59–72. Lemaire, G., ed. Berlin: Springer-Verlag.

Duru, M., J.-P. Théau, P. Cruz, and C. Jouany. 2000. Intérêt, pour le conseil, du diagnostic de nutrition azotée de prairies de graminées par analyse de plantes. *Fourrages* 164:381–95.

Enquist, B.J., E.P. Economo, T.E. Huxman, A.P. Allen, D.D. Ignace, and J.F. Gillooly. 2003. Scaling metabolism from organisms to ecosystems. *Nature* 423:639–42.

Enquist, B.J., A.J. Kerkhoff, S.C. Stark, N.G. Swenson, M.C. McCarthy, and C.A. Price. 2007. A general integrative model for scaling plant growth, carbon flux, and functional trait spectra. *Nature* 449:218.

Eviner, V.T. 2004. Plant traits that influence ecosystem processes vary independently among species. *Ecology* 85:2215–29.

Eviner, V.T., and F.S. Chapin. 2003. Functional matrix: A conceptual framework for predicting multiple plant effects on ecosystem processes. *Annual Review of Ecology and Systematics* 34:455–85.

Fitter, A., H.J. and Peat. 1994. The ecological flora database. *Journal of Ecology* 82:415–25.

Fortunel, C., E. Garnier, R. Joffre, E. Kazakou, H. Quested, K. Grigulis, S. Lavorel, and VISTA Consortium. 2009. Plant functional traits capture the effects of land use change and climate on litter decomposability of herbaceous communities in Europe and Israel. *Ecology*. 90:598–611.

Gachet, S., E. Véla, and T. Tatoni. 2005. BASECO: A floristic and ecological database of the Mediterranean French flora. *Biodiversity and Conservation* 14:1023–34.

Garnier, E., J. Cortez, G. Billès, M.-L. Navas, C. Roumet, M. Debussche, G. Laurent, A. Blanchard, D. Aubry, A. Bellmann, C. Neill, and J.-P. Toussaint. 2004. Plant functional markers capture ecosystem properties during secondary succession. *Ecology* 85:2630–37.

Garnier, E., S. Lavorel, P. Ansquer, H. Castro, P. Cruz, J. Dolezal, O. Eriksson, C. Fortunel, H. Freitas, C. Golodets, K. Grigulis, C. Jouany, E. Kazakou, J. Kigel, M. Kleyer, V. Lehsten, J. Leps, T. Meier, R. Pakeman, M. Papadimitriou, V. Papanastasis, H. Quested, F. Quétier, T.M. Robson, C. Roumet, G. Rusch, C. Skarpe, M. Sternberg, J.P. Theau, A. Thébault, D. Vile, and M.P. Zarovali. 2007. A standardized methodology to assess the effects of land use change on plant traits, communities and ecosystem functioning in grasslands. *Annals of Botany* 99:967–85.

Gaucherand, S., and S. Lavorel. 2007. A new protocol for a quick survey of functional traits values in a plant community. *Austral Ecology*, 32:927–36.

Gauthier, P., R. Lumaret, and A. Bédécarrats. 1998. Ecotype differentiation and coexistence of two parapatric tetraploid subspecies of cocksfoot *Dactylis glomerata* in the Alps. *New Phytologist* 139:741–50.

Gianoli, E., and M. González-Teuber. 2005. Environmental heterogeneity and population differentiation in plasticity to drought in Convolvulus Chilensis (Convolvulaceae). *Evolutionary Ecology* 19:603–13.

Griffith, T.M., and S.E. Sultan. 2006. Plastic and constant developmental traits contribute to adaptive differences in co-occurring *Polygonum* species. *Oikos* 114:5–14.

Grime, J.P. 1977. Evidence for the existence of three primary strategies in plants and its relevance to ecological and evolutionary theory. *The American Naturalist* 111:1169–94.

Grime, J.P. 1998. Benefits of plant diversity to ecosystems: Immediate, filter and founder effects. *Journal of Ecology* 86:902–6.

Grime, J.P. 2006. Trait convergence and trait divergence in the plant community: Mechanisms and consequences. *Journal of Vegetation Science* 17:255–60.

Grime, J.P., and J.M.L. Mackey. 2002. The role of plasticity in resource capture by plants. *Evolutionary Ecology* 16:299–307.

Grime, J.P., K. Thompson, R. Hunt, J.G. Hodgson, J.H.C. Cornelissen, I.H. Rorison, G.A.F. Hendry, T.W. Ashenden, A.P. Askew, S.R. Band, R.E. Booth, C.C. Bossard, B.D. Campbell, J.E.L. Cooper, A.W. Davison, P.L. Gupta, W. Hall, D.W., Hand, M.A. Hannah, S.H. Hillier, D.J. Hodkinson, A. Jalili, Z. Liu, J.M.L. Mackey, N. Matthews, M.A. Mowforth, A.M. Neal, R.J. Reader, K. Reiling, W. Ross-Fraser, R.E. Spencer, F. Sutton, D.E. Tasker, P.C. Thorpe, and J. Whitehouse, J. 1997. Integrated screening validates primary axis of specialisation in plants. *Oikos* 79:259–281.

Gross, N., T.M. Robson, S. Lavorel, C. Albert, Y. Le Bagousse-Pinguet, and R. Guillemin. 2008. Plant response traits mediate the effects of subalpine grasslands on soil moisture. *New Phytologist* 180:652–62.

Han, W., J. Fang, D. Guo, and Zhang. 2005. Leaf nitrogen and phosphorus stoichiometry across 753 terrestrial plant species in China. *New Phytologist* 168:377–85.

He, J.-S., Z. Wang, X. Wang, B. Schmid, W. Zuo, M. Zhou, C. Zheng, M. Wang, and J. Fang. 2006. A test of the generality of leaf trait relationships on the Tibetan Plateau. *New Phytologist* 170:835–48.

Hodgson, J.G., G. Montserrat-Martí, J. Tallowin, K. Thompson, S. Diaz, M. Cabido, J.P. Grime, P.J. Wilson, S.R. Band, A. Bogard, R. Cabido, D. Caceres, P. Castro-Díez, C. Ferrer, M. Maestro-Martinez, M.C., Pérez-Rontomé, M. Charles, J.H.C. Cornelissen, S. Dabbert, N. Pérez-Harguindeguy, T. Krimly, F.J. Sijstma, D. Strijker, F. Vendramini, J. Guerrero-Campo, A. Hynd, G. Jones, A. Romo-Diez, L. de Torres Espuny, P. Villar-Salvador, and M.R. Zak. 2005. How much will it cost to save grassland diversity? *Biological Conservation* 122:263–273.

Hooper, D.U., J.J. Ewel, A. Hector, P. Inchausti, S. Lavorel, D. Lodge, M. Loreau, S. Naeem, B. Schmid, H. Setälä, A.J. Symstad, J. Vandermeer, and D.A. Wardle. 2005. Effects of biodiversity on ecosystem functioning: A consensus of current knowledge and needs for future research. *Ecological Monographs* 75:3–35.

Jouany, C., P. Cruz, P. Petibon, and M. Duru. 2004. Diagnosing phosphorus status of natural grassland in the presence of white clover. *European Journal of Agronomy* 21:273–85.

Kerguélen, M. 1998. Index synonymique de la Flore de France. INRA.

Klimeš, L. 2003. Life-forms and clonality of vascular plants along an altitudinal gradient in E Ladakh (NW Himalayas). *Basic and Applied Ecology* 4:317–28.

Klimeš, L., and J. Klimešová. 1999. CLO-PLA2—a database of clonal plants in central Europe. *Plant Ecology* 141:9–19.

Klotz, S., I. Kühn, and W. Durka. 2002. BIOLFLOR—Eine Datenbank mit biologisch-ökologischen Merkmalen zur Flora von Deutschland. *Schriftenreihe für Vegetationskunde* 38:1–334.

Knevel, I.C., R.M. Bekker, J.P. Bakker, and M. Kleyer. 2003. Life-history traits of the northwest European flora: The LEDA database. *Journal of Vegetation Science* 14:611–14.

Körner, C. 1993. Scaling from species to vegetation: The usefulness of functional group. In *Biodiversity and Ecosystem Function*, Vol. 99, pp. 117–140. Schulze, E.D., and H.A. Mooney, eds. Berlin: Springer-Verlag.

Körner, C. 1999. *Alpine Plant Life: Plant Ecology of High Mountain Ecosystems.* Heidelberg, Germany: Springer-Verlag.

Körner, C., M. Donoghue, T. Fabbro, C. Häuse, D. Nogués-Bravo, M.T.K. Arroyo, J. Soberon, L. Speers, E.M. Spehn, H. Sun, A. Tribsch, P. Tykarski, and N. Zbinden. 2007. Creative use of mountain biodiversity databases: The Kazbegi Research Agenda of GMBA-DIVERSITAS. *Mountain Research and Development* 27:276–81.

Lavorel, S., S. Diaz, J.H.C. Cornelissen, E. Garnier, S.P. Harrison, S. McIntyre, J. Pausas, N. Pérez-Harguindeguy, C. Roumet, and C. Urcelay. 2007. Plant functional types: Are we getting any closer to the Holy Grail? In *Terrestrial Ecosystems in a Changing World*, pp. 171–186. Canadell, J., L.F. Pitelka, and D. Pataki, eds. Berlin, Heidelberg: Springer-Verlag.

Lavorel, S., and E. Garnier. 2002. Predicting the effects of environmental changes on plant community composition and ecosystem functioning: Revisiting the Holy Grail. *Functional Ecology* 16:545–56.

Lavorel, S., K. Grigulis, S. McIntyre, D. Garden, N. Williams, J. Dorrough, S. Berman, F. Quétier, A. Thébault, and A. Bonis. 2008. Assessing functional diversity in the field—methodology matters! *Functional Ecology* 22:134–147.

Lemaire, G., and F. Gastal. 1997. N uptake and distribution in plant canopies. In *Diagnosis on the Nitrogen Status in Crops*, pp. 3–44. Lemaire, G. ed. Springer-Verlag, Heidelberg, Germany.

Leps, J., F. De Bello, S. Lavorel, and S. Berman. 2006. Quantifying and interpreting functional diversity of natural communities: Practical considerations matter. *Preslia* 78:481–501.

Mason, N.W.H., D. Mouillot, W.G. Lee, and J.B. Wilson. 2005. Functional richness, functional evenness and functional divergence: The primary components of functional diversity. *Oikos* 111:112–18.

Moles, A.T., D.D. Ackerly, J.C. Tweedle, J.B., Dickie, R. Smith, M.R., Leishman, M.M., Mayfield, A. Pitman, J.T. Wood, and M. Westoby. 2007. Global patterns in seed size. *Global Ecology and Biogeography* 16:109–16.

Montalvo, J., M.A. Casado, C. Levassor, and F.D. Pineda. 1991. Adaptation of ecological systems: Compositional patterns of species and morphological and functional traits. *Journal of Vegetation Science* 2:655–66.

Pakeman, R.J., J. Lepš, E. Garnier, S. Lavorel, P. Ansquer, H. Castro, P. Cruz, J. Doležal, O. Eriksson, C. Golodets, M. Kleyer, T. Meier, M. Papadimitriou, V.P. Papanastasis, H. Quested, F. Quétier, G. Rusch, M. Sternberg, J.P. Theau, A. Thébault, and D. Vile. 2008. Relative climatic, edaphic and management controls of plant functional trait signatures. *Journal of Vegetation Science*. In press.

Pavón, N., H. Hernández-Trejo, and V. Rico-Gray. 2000. Distribution of plant life forms along an altitude gradient in the semi-arid valley of Zapoltitlán, Mexico. *Journal of Vegetation Science* 11:39–42.

Pérez-Harguindeguy, N., S. Diaz, F. Vendramini, D.E. Gurvich, A.M. Cingolani, M.A. Giorgis, and M. Cabido. 2007. Direct and indirect effects of climate on decomposition in native ecosystems from Argentina. *Austral Ecology* 32:749–57.

Picart, E., and P. Fleury. 1999. *Valorisation de l'herbe dans le canton de La Grave—Villar d'Arène*. Chambéry, France: Groupement d'Intérêt Scientifique des Alpes du Nord.

Poorter, H., M. and Pérez-Soba. 2002. Plant growth at elevated CO2. In *The Earth System: Biological and Ecological Dimensions of Global Environmental Change, Vol. 2*, pp. 489–96. Mooney, H.A., and J. Canadell, eds. Chichester: John Wiley and Sons.

Prentice, I.C., A. Bondeau, W. Cramer, S.P. Harrison, W. Lucht, S. Sitch, B. Smith, and M. Sykes. 2007. Dynamic global vegetation models: Tools to understand the biosphere. In *Terrestrial Ecosystems in a Changing World*, pp. 175–194. Canadell, J., L.F. Pitelka, and D. Pataki, eds. Heidelberg, Germany: Springer-Verlag, Berlin.

Quétier, F., S. Lavorel, S Daigney, and J. De Chazal. 2009. Functional approaches to assess vulnerability to land use change of multiple ecosystem service delivery. *Journal of Land Use Science*, in press.

Quétier, F., A. Thébault, S. and Lavorel. 2007. Linking vegetation and ecosystem response to complex past and present land use changes using plant traits and a multiple stable state framework. *Ecological Monographs* 77:33–52.

Reich, P.B. 2001. Body size, geometry, longevity and metabolism: Do plant leaves behave like animal bodies? *Trends in Ecology & Evolution* 16:674–80.

Reich, P.B., D.S. Ellsworth, M.B. Walters, J.M. Vose, C. Gresham, J.C. Volin, and W.D. Bowman. 1999. Generality of leaf trait relationships: A test across six biomes. *Ecology* 80:1955–69.

Reich, P.B., M.B. Walters, and D.S. Ellsworth. 1992. Leaf lifespan in relation to leaf, plant and stand characteristics among diverse ecosystems. *Ecological Monographs* 62:356–92.

Reich, P.B., I.J. Wright, J. Cavender-Bares, J.M. Craine, J. Oleksyn, M. Westoby, and M.B. Walters. 2003. The evolution of plant functional variation: Traits, spectra, and strategies. *International Journal of Plant Sciences* 164:143–64.

Reiné, R., C. Chocarro, and F. Fillat. 2004. Soil seed bank and management regimes of semi-natural mountain meadow communities. *Agriculture, Ecosystems and Environment* 104:567–75.

Richards, C.L., O. Bossdorf, N.Z. Muth, J. Gurevitch, and Pigliucci, M. 2006. Jack of all trades, master of some? On the role of phenotypic plasticity in plant invasions. *Ecology Letters* 9:981–93.

Robson, T.M., S. Lavorel, J.C. Clément, and X. Le Roux. 2007. Neglect of mowing and manuring leads to slower nitrogen cycling in subalpine grasslands. *Soil Biology and Biogeochemistry* 39:930–41.

Sahl, A. 2007. Traits biologiques de la flore de l'Arc Alpin, Université Montpellier 2, Montpellier.

Sala, O.E., F.S. Chapin III, J.J. Armesto, E. Berlow, J. Bloomfield, R. Dirzo, E. Huber-Sanwald, L.F. Huenneke, R.B. Jackson, A. Kinzig, R. Leemans, D.M. Lodge, H.A. Mooney, M. Oesterheld, N. LeRoy Poff, M.T. Sykes, B.H. Walker, M. Walker, and D.H. Wall. 2000. Global biodiversity scenarios for the year 2100. *Science* 287:1770–74.

Salette, J., and L. Huché. 1991. Diagnostic de l'état de nutrition minéral d'une prairie par l'analyse du végétal: principes, mise en oeuvre, exemples. *Fourrages* 125:3–18.

Scheiner, S.M., and C.J. Goodnight. 1994. The comparison of phenotypic plasticity and genetic variation in populations of the grass Danthonia spicata. *Evolution* 38:845–55.

Suding, K.N., S.L. Collins, L. Gough, C. Clark, E.E. Cleland, K.L. Gross, D.G. Milchunas, and S. Pennings. 2005. Functional- and abundance-based mechanisms explain diversity loss due to N fertilization. *Proceedings of the National Academy of Sciences* 102:4387–92.

Theurillat, J.-P., and A. Guisan. 2001. Potential impact of climate change on vegetation in the European Alps: A review. *Climatic Change* 50:77–109.

Thuiller, W., C. Albert, M.B. Araujo, P.M. Berry, M. Cabeza, A. Guisan, T. Hickler, G. Midgley, J. Paterson, F.M. Schurr, M.T. Sykes, and N.E. Zimmermann. 2008. Predicting global change impacts on plant species distributions: Future challenges. *Perspectives in Plant Ecology, Evolution and Systematics* 9:137–52.

Thuiller, W., S. Lavorel, M.B. Araujo, M.T. Sykes, and I.C. Prentice. 2005. Climate change threats to plant diversity in Europe. *Proceedings of the National Academy of Sciences* 102:8245–50.

Thuiller, W., S. Lavorel, G.F. Midgley, and S. Lavergne. 2004. Relating plant traits and species distributions along bioclimatic gradients for 88 *Leucadendron* species. *Ecology* 85:1688–99.

van der Krift, T.A.J., and F. Berendse. 2001. The effect of plant species on soil nitrogen mineralization. *Journal of Ecology* 89:555–61.

Van Tienderen, P.H. 1991. Evolution of generalists and specialists in spatially heterogeneous environments. *Evolution* 45:1317–31.

Vile, D., B. Shipley, and E. Garnier. 2006. Ecosystem productivity relates to species' potential relative growth rate: A field test and a conceptual framework. *Ecology Letters* 9:1061–67.

Violle, C., M.L. Navas, D. Vile, E. Kazakou, C. Fortunel, I. Hummel, and E. Garnier. 2007. Let the concept of trait be functional! *Oikos* 116:882–92.

Wilson, J.B. 2007. Trait-divergence assembly rules have been demonstrated: Limiting similarity lives! A reply to Grime. *Journal of Vegetation Science* 18:451–52.

Wipf, S., C. Rixen, and C.P.H. Mulder. 2006. Advanced snowmelt causes shift towards positive neighbour interactions in a subarctic tundra community. *Global Change Biology* 12:1496–1506.

Wright, I.J., P.B. Reich, H.J.C. Cornelissen, D.S. Falster, E. Garnier, K. Hikosaka, B.B. Lamont, W. Lee, J. Oleksyn, N. Osada, H. Poorter, R. Villar, D.I. Warton, and M. Westoby. 2005. Assessing the generality of global leaf trait relationships. *New Phytologist* 166:485–96.

Wright, I.J., P.B. Reich, M. Westoby, D.D. Ackerly, Z. Baruch, F. Bongers, J. Cavender-Bares, F.S.I. Chapin, J.H.C. Cornelissen, M. Diemer, J. Flexas, E. Garnier, P.K. Groom, J. Gulias, K. Hikosaka, B.B. Lamont, T. Lee, W. Lee, C. Lusk, J.J. Midgley, M.-L. Navas, Ü. Niinemets, J. Oleksyn, N. Osada, H. Poorter, P. Poot, L. Prior, V.I. Pyankov, C. Roumet, S.C. Thomas, M.G. Tjoelker, E. Veneklaas, and R. Villar. 2004. The worldwide leaf economics spectrum. *Nature* 428:821–27.

Wright, J.P., S. Naeem, A. Hector, C.L. Lehman, P.B. Reich, B. Schmid, and D. Tilman. 2006. Conventional functional classification schemes underestimate the relationship with ecosystem functioning. *Ecology Letters* 9:111–20.

13 Using Species Occurrence Databases to Determine Niche Dynamics of Montane and Lowland Species since the Last Glacial Maximum

Robert Guralnick and Peter B. Pearman

CONTENTS

Introduction 125
Material and Methods 127
Study Species and Occurrence Data 127
Fossil Species Occurrence Data 127
Data Layers for Present and Past Climates and Landscapes 127
Modeling Approach 128
Comparisons of Hindcasted Distribution with Fossil Record Data 128
Results 129
Discussion 130
Use of Occurrence Databases 130
Possible Explanations for Differential Niche Conservatism 131
Next Steps 132
Summary 134
Acknowledgments 134
References 134

INTRODUCTION

Species distributions are shifting northwards and to higher elevations as a result of ongoing climate warming (Parmesan, 2006). These shifts appear at both long and short time scales (e.g., since 21kyBP, Guralnick, 2007; Grayson, 2005; the last one hundred years, Pounds et al., 2005; Beever et al., 2003) and are in agreement with those predicted by niche-based species distribution models (SDMs, Guisan and Zimmermann, 2000; Araújo et al., 2005). Although this pattern of shifts is likely broadly valid, individual species display idiosyncratic responses to climate change (Parmesan and Yohe, 2003). This suggests the importance of predicting the response of individual species to climate change and of understanding how and why these responses vary among species. Two factors that likely explain this variation are the differing physiological tolerances of species and whether species populations can adapt to rapidly changing climates. Species with populations that differ in dispersal capabilities, phenology, average fitness, or levels of heritable genetic variation also should differ in ability to adapt and migrate in response to environmental change (Holt et al., 2003). Therefore, determining the contributions of niche shifts and migration to species response to climate change is an important step toward improving predictions of biotic responses to global warming (Pearman et al., 2008b).

Climate tracking of species via migration and adaptation to new conditions are not mutually exclusive. Adaptation and migration likely occur simultaneously, as suggested by evidence of past population differentiation, the separation of populations into refugia that varied in environmental conditions, and the adaptation of current populations along latitudinal gradients (Davis and Shaw, 2001). The position of populations in relation to changing range boundaries also might influence the response of populations to climate change. Adaptation of populations at range margins can, in theory, be facilitated by intermediate levels of gene flow, low differences in the relative fitness of source and sink populations, and temporally autocorrelated variation in fitness in sink habitats (Holt, 1996; Antonovics et al., 2001; Holt and Gomulkiewicz, 2004). Trailing edge populations are more likely to be locally adapted, and adapt more slowly to new conditions, because of increasing isolation and decreasing gene flow from other populations (Hampe and Petit, 2005). These patterns suggest that niches are not identical and static over the range of a species. Instead, novel climate conditions and continued influx of new genotypes as material for selection can lead to new adaptations and niche shifts.

Evidence for the local adaptation of populations to environmental variation across the range of species is provided by experiments in which plants have been reciprocally transplanted across altitudinal, moisture, and latitudinal gradients (Rehfeldt et al., 1999; Davis and Shaw, 2001). Studies of invasive species suggest that niches can shift within one hundred years or less (Broennimann et al., 2007; Fitzpatrick et al., 2007). Furthermore, hindcasting SDMs (transferring models back in time) shows that plants vary in the amount of estimated niche shift they have experienced since the mid-Holocene, 6kyBP, (Pearman et al., 2008a). However, studies of ecological niche conservatism (the tendency for species niches to remain little changed over time) and niche shift in mammal species have concluded that niche shifts since Last Glacial Maximum (LGM) are uncommon.

Martinez-Meyer et al. (2004) and Martinez-Meyer and Peterson (2006) used modern climate data sets, paleoclimate reconstructions at the LGM, and fossil and modern occurrence data now available from globally distributed databases to test for niche conservatism. Their test consisted of forecasting present distributions based on niche models that were calibrated with data from LGM and hindcasting distributions at LGM based on models of species niches currently, and then examining the overlap. The authors concluded that niches were generally conserved over species ranges, with some exceptions. Their results suggest that any niche shift across a species range might be swamped by trends toward niche conservation, at least for mammals and plants. Nonetheless, the authors also note that they were only able to examine this question utilizing coarse-scale LGM paleoclimate data layers that contained few variables: cloud cover, mean annual temperature, and precipitation. Measures of climate variability over seasons, for example, were unavailable. These limitations to climate databases have largely disappeared, which warrants re-examination of the question of niche conservatism versus shift by employing new data on climate and expanded fossil and modern occurrence information.

In this chapter, we reexamine the question of niche conservation versus niche shift by using data on LGM and recent species occurrences and high-resolution bioclimatic data (in geographic information system layers) that describe recent and LGM climate. Such fossil and LGM species occurrences and paleoclimate layers have only recently been compiled and made available to the broader research community. Repositories, such as Faunmap (Graham et al., 1994) and the Global Biodiversity Information Facility (http://data.gbif.org), and modelled climate layers, such as those provided by WorldClim, provide a starting point for monitoring past, present, and future global biodiversity changes (Guralnick et al., 2007).

We compare putative niche shifts in groups of species that are either characteristic of temperate areas with little topographic relief (six species), or of montane or arctic areas (six species) in North America. We reason that montane species might track environmental changes more effectively than flatland species, because climate changes along elevational gradients occur over much shorter distances than do changes across latitudinal gradients. However, topographically and environmentally heterogeneous montane habitat also might promote opportunities for local adaptation (niche shift) within species, owing to the progressive isolation of montane populations during climate warming. Given these contradictory predictions, we consider the null hypothesis that no difference in degree of niche shift exists between montane and flatland species.

The species that we chose all have sufficient, range-wide occurrence data, which allow modelling of current ecological niches, and are available from global biodiversity occurrence repositories and local, online database resources. Enough high-quality fossil occurrence data, mammal fossils, and fossil pollen are available to describe the distributions of the species at LGM. Using these data and newly available, fine-scale estimations of LGM climate (Waltari et al., 2007; Peterson and Nyari, 2008; Hijmans et al., in preparation), we model potential present and LGM distributions for each taxon and then determine whether known fossil occurrences from LGM fall within the predicted suitable habitat. If niches are conserved, these fossil distributions should match predicted suitable habitat at LGM. If species niches have shifted, fossil occurrences should not be found in areas that we predict were suitable at LGM. Overall patterns across montane and lowland species could provide further insights into the magnitude and direction of observed niche shift. The work we report here shows the value of both modern and fossil species-occurrence data sets, and maps of past and present climate for estimating climate change effects in areas with differing topography, and especially in mountainous regions.

MATERIAL AND METHODS

Study Species and Occurrence Data

We used networked biodiversity information systems of occurrence data from natural history collections (e.g., Global Biodiversity Information Facility; see Guralnick et al., 2007, and the Mammal Networked Information System; see Stein and Wieczorek, 2004, and other DiGIR providers, the Canadian Biodiversity Information Facility and Mapstedi) to collate modern georeferenced species occurrence information (4,388 total records), to select the focal taxa, and to identify their distributions. When processing records downloaded from natural history collections, we first removed duplicate records for multiple occurrences of a species at a site. To avoid imprecise occurrence data, only records with radii of geographic uncertainty of less than 10 km (Wieczorek et al., 2004) were retained for analysis. For those records that did not contain geographic uncertainty measurements, we examined uncertainty extent with the Biogeomancer workbench (http://bg.berkeley.edu/latest), which provides a standardized measurement of radius uncertainty comparable to those from manual georeferencing methods (Guralnick et al., 2006).

Fossil Species Occurrence Data

For mammal occurrence records, we used the Faunmap database of fossil mammal species occurrences (Graham et al., 1994) as a means to compare against the models of species paleodistributions. The records of interest in Faunmap were those within the "window" of the LGM. In Faunmap, almost all occurrence records have a time range (e.g., 20.5 to 25kya) as opposed to a single date. We therefore had two criteria for selecting species occurrences at or near LGM. First, the records must have starting and ending dates that fall within a 10 to 35kya time range, and second, the range must include times in what is defined as the Full Glacial period, from 15 to 20kya. Thus, records that were from 10 to 14kya were excluded, while records from 10 to 18kya were not. This criterion represents a balance between accumulating enough records for a reasonable test while still being strict enough to avoid records coming from the early Holocene period of rapid warming. Between 15 to 11kya, the position and extent of ice sheets and climate, and likely species distributions, were all changing rapidly (Thompson et al., 1993; Webb et al., 1993).

To identify plant species occurrence records of the two plant taxa (spruce, genus *Picea* and hickory, genus *Carya*), we used the North American Pollen Database (http://www.ncdc.noaa.gov/paleo/napd.html), which includes pollen records from multiple studies throughout the region and over multiple time periods from the Pleistocene to the present. Unlike Faunmap, temporal control over known time of deposition of the pollen is typically excellent, and all pollen records denoting the presence of spruce pollen are between 18.5 and 23 kyBP; most are between 20 and 21kyBP. Table 13.1 summarizes the source and occurrence data for fossil plants and mammals.

For both the plant and mammal fossils, we are making an untested assumption that the fossils come from source populations as opposed to sinks, and that fossil were deposited *in situ* and not transported far away from their original place of death. These assumptions are unlikely to be violated very often, however. Given preservation likelihoods, the probability is higher that we are sampling well-established, large populations in suitable areas of a range rather than a rare accidental outside of a range edge. As well, most sites from Faunmap (Cannon, 2004) and from pollen databases appear to be representative of the local community at the time of deposition.

Data Layers for Present and Past Climates and Landscapes

We used current and LGM monthly climate data at 2.5′ spatial resolution. Waltari et al. (2007) discuss the process of layer development. In summary, LGM climate data were based on two general circulation model (GCM) simulations: the Community Climate System Model (CCSM) (Collins et al., 2004), and the Model for Interdisciplinary Research on Climate (MIROC, version 3.2; Hasumi and Emori, 2004; http://www.pmip2.cnrs-gif.fr). Each model was initially generated at a spatial resolution of 2.8°, and these outputs were further processed by interpolating differences between LGM and recent (preindustrial) conditions based on the WorldClim (http://www.worldclim.org/) data set to create monthly climate surfaces at 2.5′ (1×1 km) spatial resolution. Niche-based SDMs for both the present and the LGM were based on the nineteen bioclimatic variables in the WorldClim data set (Hijmans et

TABLE 13.1
List of Taxa Utilized, Sources of Data, and Number of Occurrences for Modern and Fossil Data

Taxon	Mountain or Flatland	Present Occurrence Source	No. Occurrences	Data Source Fossil Occurrences	No. Occurrences
Tamias striatus	Flatland	GBIF	144	Faunmap	16
Synaptomys cooperi	Flatland	GBIF	113	Faunmap	23
Microtus pinetorum	Flatland	GBIF	96	Faunmap	17
Microtus ochrogaster	Flatland	GBIF	308	Faunmap	13
Spermophilus richardsonii	Flatland	GBIF	106	Faunmap	12
Carya	Flatland	Whitmore et al., 2005	910	NA Pollen Database	21
Thomomys talpoides	Mountain	GBIF, Mapstedi	705	Faunmap	22
Neotoma cinerea	Mountain	GBIF, Mapstedi	509	Faunmap	19
Picea	Mountain	GBIF, CBIF	611	NA Pollen Database	31
Microtus montanus	Mountain	GBIF, Mapstedi	417	Faunmap	12
Marmota flaviventris	Mountain	GBIF, Mapstedi	228	Faunmap	19
Phenacomys intermedius	Mountain	GBIF, Mapstedi	229	Faunmap	23

al., 2005). These variables represent summaries of means and variation in temperature and precipitation, and likely characterize dimensions of climate particularly relevant in determining species distributions (Hijmans et al., 2005).

Modeling Approach

For this preliminary analysis, we used Maxent version 2.3 (Philips et al., 2006) for construction of niche-based SDMs of present and past species distributions, based on the principle of maximum entropy as constrained by the input occurrence data. Maxent generates SDMs using only presence records, contrasting them with pseudoabsence data resampled from the remainder of the study area. When running the program, we used the default convergence threshold (10–5) and maximum number of iterations (500), reserving 25% of localities for model testing. We let the program select both suitable regularization values and functions of environmental variables automatically, which it achieves based on considerations of sample size. Maxent outputs are continuous probability values, scaled by a factor of 10 and ranging from 0 to 100, and they serve as an indicator of relative suitability of habitat for the species. We then used Maxent's project function in order to reproject the current climate niche onto climate landscapes in the past, utilizing the CCSM and MIROC climate layers for LGM, discussed earlier. In order to create categorical maps showing likely suitable and unsuitable areas, as opposed to continuous measures of suitable, we selected two thresholds. The first is the lowest presence threshold (Pearson et al., 2007), which shows the smallest distribution that has no omission errors; that is, all the training points are included in the final suitable habitat map. Second, for an even more stringent threshold, we used a threshold of 50 for Maxent (M50). These thresholds were used for predictions of modern and past potential distributions (Figure 13.1 and Figure 13.2). The conservative thresholds represent areas of very high modelled suitability, past and present, but are conservative, making errors of omission more likely, while the least presence threshold includes areas with lower modelled suitability, increasing the probability of making errors of commission. We use both thresholds for the sake of comparison.

Comparisons of Hindcasted Distribution with Fossil Record Data

Given that the purpose of this study is exploratory regarding measuring niche conservation, we do not attempt a full statistical analysis of whether niches are conserved or not. Statistical approaches that can overcome potential sampling biases and spatial autocorrelation are in development. Instead, we focus on simple summaries of the data and heuristic examination of map outputs. The main data summary step is to determine the number of fossil records located in predicted suitable habitats, based on the hindcasted niche modelling results at the Least Presence Threshold, compared to the total number of fossil occurrences. If niches are conserved, the expectation is that a large proportion of fossil localities will be located in predicted suitable habitats based on the hindcasted potential distribution at LGM. In order to determine whether time averaging may have affected results for the mammal

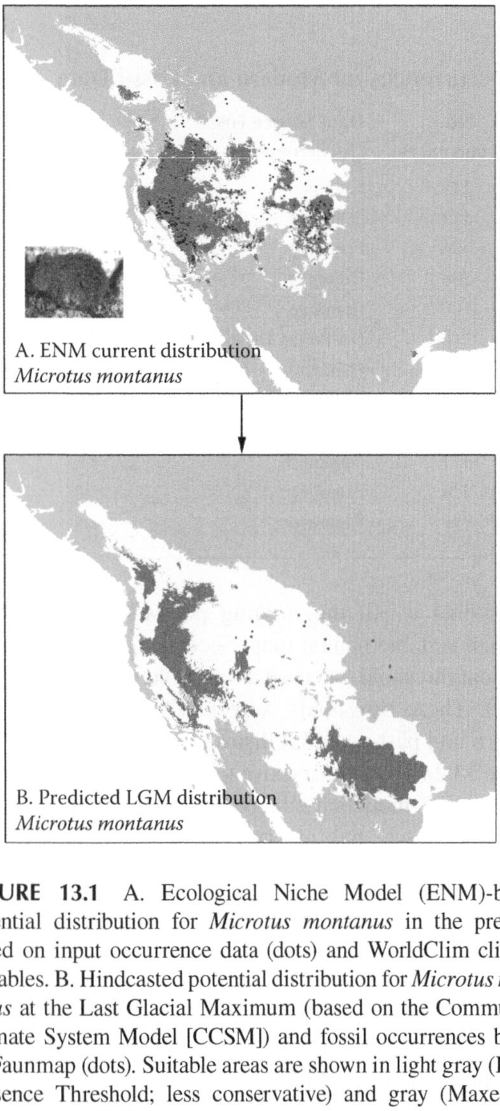

FIGURE 13.1 A. Ecological Niche Model (ENM)-based potential distribution for *Microtus montanus* in the present, based on input occurrence data (dots) and WorldClim climate variables. B. Hindcasted potential distribution for *Microtus montanus* at the Last Glacial Maximum (based on the Community Climate System Model [CCSM]) and fossil occurrences based on Faunmap (dots). Suitable areas are shown in light gray (Least Presence Threshold; less conservative) and gray (Maxent50; more conservative).

species, we also heuristically examined whether the fossil localities predicted not in suitable habitats were more likely to be ones that have either wide data ranges (e.g., 10–35kya) or whose estimated dates include a period of rapid warming (e.g., 12–16kya).

Results

Based on data from fossil and modern occurrence databases and climate data sets from the present and the LGM, we document an overall pattern of niche conservation in mountainous regions and niche differentiation in flatland areas. Before discussing this result further, we first discuss some data quality issues. We identified twelve species with sufficient occurrence data for our purposes. Ten of the twelve species are mammals, the others are trees (the conifer *Picea* and the dicot *Carya*; Table 13.1), the pollen of which could only be identified to the genus level. Both modern and fossil occurrence of spruce and hickory, therefore, includes the multiple species of those two taxa found in North America. The ten mammals are identified to species level. The species vary in the degree to which fossil data coincide with predicted LGM distribution. The latitude of the LGM distribution of fossils of *Microtus montanus*, a mountain species, largely corresponds to the distribution predicted by the SDM (Figure 13.1). Furthermore, as Figure 13.2 shows, the overlap between the latitudes of suitable habitats and fossil localities is greater for the mountain taxa *Picea* (spruce) and *Marmota flaviventris* (the yellow-bellied marmot) (Figure 13.2B, Figure 13.2C) than for two flatland species (*Tamias striatus*, the eastern chipmunk, and *Microtus pinetorum*, the woodland vole; Figure 13.2A, Figure 13.2D). These latter species have predicted LGM distributions that are offset southwards from the fossil localities, suggesting that the niches of these species have shifted. In the case of the other lowland species, the modelled LGM distributions are all located further south than most of the fossil localities.

Taxa that occur in mountainous areas have higher rates of overlap between fossil localities and predicted distributions than do flatland species (Table 13.2). It appears that taxa that are restricted to western North America mountainous areas (e.g., *Marmota flaviventris, Microtus montanus, Neotoma cinerea, Thomomys talpoides*) show little or no niche shift. The matches between niche modelling results and known fossil occurrences range from 59 to 89%. Wide-ranging taxa native to both the mountainous areas in western North America and to lowland areas in boreal eastern North America show weaker niche conservatism, which is indicated by mismatches between eastern fossil localities and predicted niche model suitable habitats. For these taxa, matches between hindcasted niche models and fossil distributions range from 21 to 75%, depending on the paleoclimate reconstructions used. In contrast, niche modelling results and fossil occurrences overlap 0 to 54% for taxa that are restricted to the lowlands. One taxon, *Spermophilus richardsonii*, occurs in lowland areas in western North America as opposed to eastern North America, and also shows relatively poor niche conservatism. It therefore appears the result is not simply due to comparing different regions.

We did not find strong indications that fossil occurrences outside the hindcasted suitable ranges were more likely to have wide temporal uncertainty or were likely deposited at

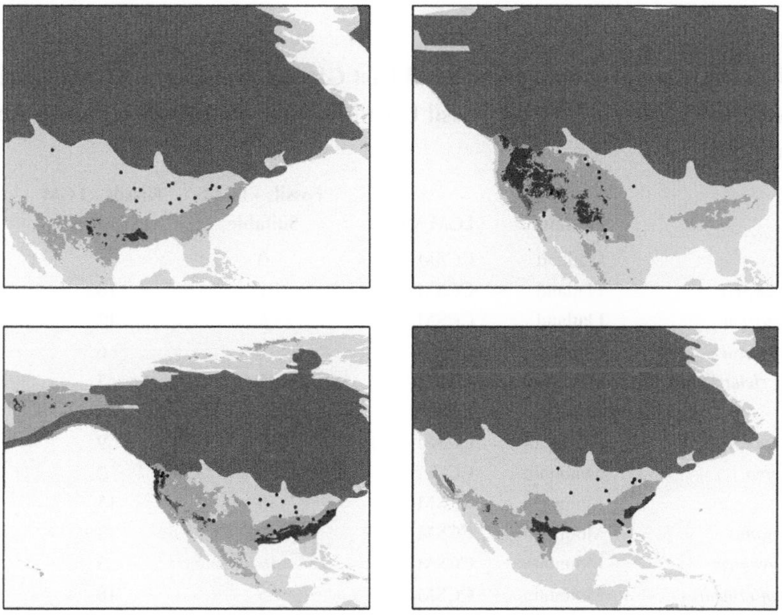

FIGURE 13.2 Predicted Last Global Maximum (LGM) distributions for four taxa under niche conservation and actual fossil occurrence data (green dots). Light gray and gray areas are suitable at the LGM; gray areas represent most suitable conditions under a conservative threshold (M50) while light gray areas are based on the less conservative Least Presence Threshold. The darkest gray area in northern North America is predicted extent of continental glacial ice sheets. (A) *Tamias striatus* distribution based on the MIROC LGM climate data along with fossil localities. (B) *Marmota flaviventris* distribution based on the Model for Interdisciplinary Research on Climate (MIROC) LGM climate data along with fossil localities. (C) *Picea* LGM distribution based on the CCSM LGM climate data along with fossil localities. (D) *Microtus pinetorum* distribution based on MIROC LGM climate data along with fossil localities.

times of rapid warming. For *Marmota flaviventris*, the two fossil localities outside the ENM-predicted suitable habitat for both paleoclimate models are indeed records with wide age uncertainty or at the edge of the Full Glacial period; one record has a date range at 10 to 35kya, the other has a date of 13 to 15kya and is just north of the predicted LGM distribution. For *Neotoma cinerea*, however, the two records that are not within hindcasted suitable habitat have dates at 22 to 35kya and 26kya, which were likely as cold as the conditions at LGM. For flatland taxa, most fossil occurrences are not in suitable habitats, and these include records unambiguously aged to the Full Glacial period. Thus, the few records in suitable habitats are not any more likely to be more constrained in either in date or geographic range than are records falling in unsuitable habitats.

DISCUSSION

Use of Occurrence Databases

One of the goals of this study is to show how databases of fossil and modern species occurrences can be used to test key questions concerning the responses of species to continued climate perturbations, in this case how responses may differ depending on affinity for mountainous or lowland habitat. We used locations of mammal fossils from Faunmap and fossil pollen data from the North American Pollen Database to infer past species distributions. Data from global repositories, such as the Global Biodiversity Information Facility, and more regional repositories, such as the Canadian Biodiversity Information Facility and the Mapstedi project, were used to accumulate species occurrences needed to model the current climatic niche of species. Niche modeling, in turn, requires not just species occurrences, but also fine scale georeferenced reconstructions of current and paleoclimates in order to model areas of suitable habitat both past and present.

High precision georeferenced occurrence data and fine scale digital climate maps of both present and past climate is crucial for modeling the distributions of taxa that live in topographically heterogeneous environments and is increasingly available. New paleoclimate data layers include measures of seasonality, coldest and warmest monthly temperatures, and monthly precipitation, in addition to mean annual temperature and precipitation (as used by Martinez-Meyer et al., 2004). These climate measurements may be particularly relevant, given that

TABLE 13.2
Results of Comparisons between Predicted Last Glacial Maximum (LGM) Based on Niche Conservation against Actual Fossil Localities from Faunmap or North American Pollen Database

Taxon	Mountain/ Flatland	LGM Model	Fossils - LGM Suitable	Fossils - LGM Unsuitable	% Overlap[a]
Tamias striatus	Flatland	CCSM	0	16	0.0
Synaptomys cooperi	Flatland	CCSM	7	16	30.4
Microtus pinetorum	Flatland	CCSM	5	12	29.4
Microtus ochrogaster	Flatland	CCSM	7	6	53.8
Spermophilus richardsonii	Flatland	CCSM	5	7	41.7
Carya	Flatland	CCSM	5	15	25.0
Thomomys talpoides	Mountain	CCSM	13	9	59.1
Neotoma cinerea	Mountain	CCSM	17	2	89.5
Picea	Mountain	CCSM	16	15	51.6
Microtus montanus	Mountain	CCSM	9	3	75.0
Marmota flaviventris	Mountain	CCSM	16	3	84.2
Phenacomys intermedius	Mountain	CCSM	5	18	21.7
Tamias striatus	Flatland	MIROC	2	14	12.5
Synaptomys cooperi	Flatland	MIROC	5	18	21.7
Microtus pinetorum	Flatland	MIROC	3	14	17.6
Microtus ochrogaster	Flatland	MIROC	7	6	53.8
Spermophilus richardsonii	Flatland	MIROC	4	8	33.3
Carya	Flatland	MIROC	12	8	60.0
Thomomys talpoides	Mountain	MIROC	14	8	63.6
Neotoma cinerea	Mountain	MIROC	17	2	89.5
Picea	Mountain	MIROC	24	7	77.4
Microtus montanus	Mountain	MIROC	9	3	75.0
Marmota flaviventris	Mountain	MIROC	14	5	73.7
Phenacomys intermedius	Mountain	MIROC	9	14	39.1

[a] Percent overlap refers to number of matches of suitable LGM habitat predicted by niche modeling to fossil localities divided by total number of localities.

seasonal temperature variation may be as important as mean temperature in determining species responses to climate change (Bradshaw and Holzapfel, 2006). These fine scale, high-quality data sets allow the global change community unprecedented opportunities to test important and unresolved questions, such as niche conservatism through time, using data mining approaches (Körner et al., 2007).

Possible Explanations for Differential Niche Conservatism

Although the work presented is preliminary, an intriguing result is that mountain taxa appear to show less niche shift compared to flatland taxa. There are multiple possible explanations for this pattern. First, the paleoclimate layers we used are potentially inaccurate. If conditions were warmer in North America at the LGM than these layers depict, then there might be better correspondence of fossil localities and predicted ranges at LGM. However, our predictions for distributions at LGM are quite good for montane taxa in western North America. Thus, the models do not fail ubiquitously across North America. Further, analyses based on two separate general circulation models generate the same pattern: niche conservation in montane regions and niche shift in flatland areas. In the same regard, our inclusion of fossil occurrences with relatively wide date uncertainty or occurrence records falling outside of the Full Glacial period are unlikely to explain the difference between mountain and flatland groups regarding magnitude of niche shifts.

A possible methodological consideration is whether there may be greater overprediction of suitable habitat areas in mountainous terrain than in lowlands. Given uncertainties in exact locations of species occurrences, and given that the spatial scale of both present and past climate layers, although much improved, still averages over local landscape variation in mountainous terrain, overprediction could potentially inflate matches between fossil localities and hindcasted areas of suitable habitat. For flatland species, minor shifts in location unlikely lead to dramatically different environments; i.e., local topography is likely less important to climate variation than are latitudinal gradients. The overprediction of the ranges of mountain species compared to lowland species unlikely explains the pattern of less niche shift in the former, because hindcasted LGM ranges of mountain taxa do not appear to greatly overpredict ranges, but instead match known distributions quite well, e.g., for *Marmota flaviventris* (Figure 13.2B). However, spatially explicit statistical approaches should be used to examine this issue further.

Different magnitudes of niche shift in regions with different topographies are unlikely an artifact of methodology, but, instead, likely have a basis in biology and geography. This is because climate changes along elevation gradients differ from those along gradients of latitude. Along latitudinal gradients, climate warming creates new climate landscapes that are almost certainly unique compared to past conditions. For example, although higher mean annual temperatures shift northwards as climate warms, other climate factors are decoupled from this warming; areas with equivalent mean annual temperatures in the north and south of North America still have different amounts of overall seasonality. Thus, lowland species are almost certainly challenged with new climate regimes, and these daily and seasonal factors could be essential in determining a species' niche. Figure 13.3 presents a map-based view of temperature and seasonality differences between the LGM and present showing that while mean annual temperature changes dramatically over time along gradients of latitude and longitude, climate seasonality shows little change with either. Therefore, species likely shift into new seasonality regimes if their ranges shift northwards. In mountainous areas, similar changes in climate landscapes do not necessarily develop; daily and seasonal variability may not differ much among areas with very different average yearly temperatures. Thus, mountain taxa, especially near the trailing edge and range core, may be able to shift upwards or downwards in elevation during climate change, finding a climate regime to which they are already adapted. Nonetheless, North America mountain taxa experience more variation in temperature at southern range edges compared to northern edges, suggesting that adaptation might occur during warm periods when mountain populations are separated from one another (Guralnick, 2006).

The fossil localities of the flatland species we examined are found further north than the hindcasted species distributions. Thus, taxa at LGM are found in colder areas than expected if niche conservatism holds true. One possible explanation for this pattern is that species currently have yet to migrate into suitable habitats farther to the north due to either dispersal limitation or other biotic factors such as competition and predation. Another potential explanation is that adaptation to warmer conditions is occurring as populations are challenged with new climate regimes, and these new regimes select the most-fit variants. If adaptation has occurred, then the direction would be toward tolerance to warmer, more seasonal conditions than experienced by ancestral populations. This is because any range shift toward the north increases the magnitude of seasonality that organisms experience. Distinguishing among these two alternatives requires more careful analysis of many factors, including changes in dispersal rates over time, the distributions of competitors and predators, and birth and growth rates across the differing conditions within a species range.

Next Steps

Models of species response to ongoing climate change are dependent on whether species conserve their niches or not, over shorter or longer time scales (Pearman et al., 2008a). If the answer is that species can adapt quickly to changing conditions, then simple distributional modelling will almost certainly overestimate possibilities for population or species extinctions. In alpine areas, where populations and species may be particularly prone to extinction due to continued climate warming, information on whether these biological units may be able to adapt to changes is particularly relevant. Empirical studies of niche dynamics are essential to understand whether there are rules governing the likelihood, taxonomic distribution, and magnitude of niche shifts. Although the present work and the work of others (e.g., Martinez-Meyer et al., 2004; Martinez-Meyer and Peterson, 2006) has examined this question empirically utilizing fossil and modern occurrence data for plants and mammals; actual statistical measurement of shifts of climate niches has just begun (Pearman et al., 2008a). Given the potential

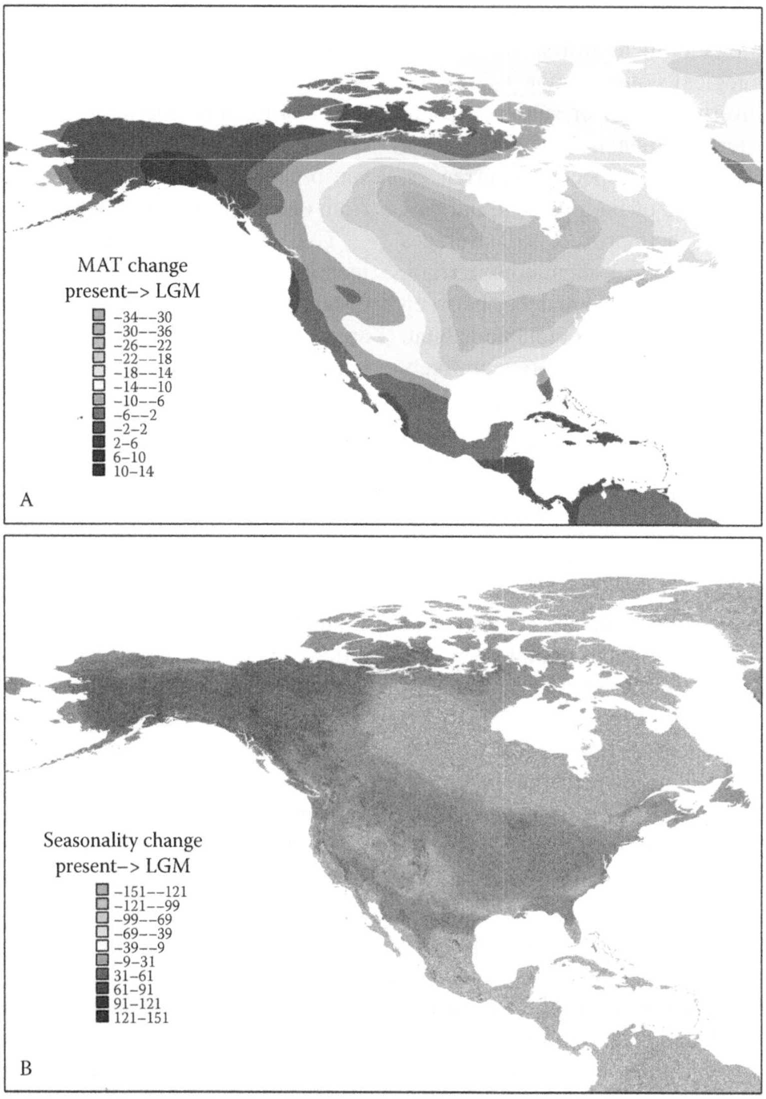

FIGURE 13.3 Differences in temperature and seasonality based on the Community Climate System Model (CCSM) paleoclimate models for the Last Glacial Maximum and WorldClim for the present, showing that the seasonality of temperature (Panel B) varies only slightly over the time period and not in relation to latitudinal or longitudinal patterns, while mean annual temperature (MAT) in Panel A varies in relation to both.

biases and spatial autocorrelation in both fossil and modern occurrence data sets, determining the rate and magnitude of niche shifts in a statistical framework remains an essential next step.

Modeling niches of species lineages or populations that were historically separated during the last ice age, as opposed to the whole species, also will be important. Estimating climatic niches of lineages in the present provides a simple test for niche shifts. Similarly, the integration of models of past and present distributions with phylogeographic and population genetic data is under way (e.g., Waltari et al., 2007; Carstens and Richards, 2007) and will provide essential tests for niche shifts over multiple time scales. Ultimately, a full synthesis of phylogenetic information in the present, along with ancient DNA from the past and high resolution climate layers over multiple periods of time, will provide the machinery for assessing the relative importance of climate and other drivers in generating the response of species to environmental change in differing regions, especially topographically heterogeneous environments such as mountains.

SUMMARY

Testing questions of niche conservation requires a multidisciplinary approach and a wide range of data types, from paleoclimate reconstructions to demographic inferences from ancient DNA. Here we utilized some of these data types (past and present climate reconstructions, modern and fossil species occurrence data) to show that mountain-dwelling species show stronger niche conservatism than those with affinities for flatlands. For the future, the integration of fossil and modern species occurrence data, climate and earth observing data, and phylogenetic information extracted from modern and ancient organisms will benefit from a stronger database and informatics approach. All such data sets need to be stored in common formats, published and registered, and made freely available. We also envision the development of community workbenches where users can access past experiments, perform their own experiments utilizing existing data sets, and upload new data sets for exploration and analyses. In order to truly falsify (or support) niche conservation in different kinds of habitat with different topographic complexities, we will need to utilize community data sharing approaches so that we can adequately judge the preponderance of evidence.

ACKNOWLEDGMENTS

We thank C. Körner and E. Spehn of the Global Mountain Biodiversity Assessment for the opportunity to present work on ecological niche modelling results at the Copenhagen workshop. We also thank the numerous museums whose data is available on networked biodiversity information systems of natural history collection data, such as Mammal Network Information System (MaNIS) and Global Biodiversity Information Facility (GBIF). All data sets used for analysis are available upon request. P.B. Pearman was supported by the European Union's FP-6 ECOCHANGE project and thanks A. Guisan for discussions on niche dynamics.

REFERENCES

Antonovics, J., T.J. Newman, and B.J. Best. 2001. Spatially explicit studies on the ecology and genetics of population margins. In *Integrating Ecology and Evolution in a Spatial Context*, pp. 97–116. Silvertown, J., and J. Antonovics, eds. London: Blackwell.

Araújo, M.B., R.G. Pearson, W. Thuiller, and M. Erhard. 2005. Validation of species–climate impact models under climate change. *Global Change Biology* 11:1504–13.

Beever, E.A., P.F. Brussard, and J. Berger. 2003. Patterns of apparent extirpation among isolated populations of pikas (*Ochotona princeps*) in the Great Basin. *Journal of Mammalogy* 84(1):37–54.

Bradshaw, W.E., and C.M. Holzapfel. 2006. Climate change: Evolutionary response to rapid climate change. *Science* 312(5779):1477–78.

Broennimann, O., U. Treier, H. Müller-Schärer, W. Thuiller, A.T. Peterson, and Guisan. 2007. Evidence of climatic niche shift during biological invasion. *Ecology Letters* 10:701–9.

Cannon, M.D. 2004. Geographic variability in North American mammal community richness during the terminal Pleistocene. *Quaternary Science Reviews* 23(9–10):1099–1123.

Carstens, B.C., and C.L. Richards. 2007. Integrating coalescent and ecological niche modeling in comparative phylogeography. *Evolution* 61:1439–54.

Collins, W.D., M. Blackmon, C. Bitz, G. Bonan, C.S. Bretherton et al. 2004. The community climate system model: CCSM3. *Journal of Climate* 19:2122–43.

Davis, M.B., and R.G. Shaw. 2001. Range shifts and adaptive responses to quaternary climate change. *Science* 292:673–79.

Fitzpatrick, M., J. Weltzin, N. Sanders, and R. Dunn. 2007. The biogeography of prediction error: Why does the introduced range of the fire ant over-predict its native range. *Global Ecology and Biogeography* 16:24–33.

Graham, R.W., E.L. Lundelius Jr., and FAUNMAP Working Group. 1994. FAUNMAP: A database documenting Late Quaternary distributions of mammal species in the United States. *Illinois State Museum Scientific Papers*, XXV, Nos. 1 & 2.

Grayson, D.K. 2005. A brief history of Great Basin pikas. *Journal of Biogeography* 32(12):2103–11.

Guisan, A., and N.E. Zimmermann. 2000. Predictive habitat distribution models in ecology. *Ecological Modelling* 135:147–86.

Guralnick, R.P. 2006. The legacy of past climate and landscape change on species' current experienced climate and elevation ranges across latitude: A multispecies study utilizing mammals in western North America. *Global Ecology and Biogeography* 15(5):505–18.

Guralnick, R., J. Wieczorek, R.J. Hijmans, R. Beaman, and the Biogeomancer Working Group. 2006. Biogeomancer: Automated georeferencing to map the world's biodiversity data. *PLoS Biology* 4(11):1908–09.

Guralnick, R.P. 2007a. Differential effects of past climate warming on mountain and flatland species' distributions: A multispecies North American mammal assessment. *Global Ecology and Biogeography* 16(1):14–23.

Guralnick, R.P., A.W. Hill, and M. Lane. 2007. Towards a collaborative, global infrastructure for biodiversity assessment. *Ecology Letters* 10(8):663–72.

Hampe, A., and R.J. Petit. 2005. Conserving biodiversity under climate change: The rear edge matters. *Ecology Letters* 8:461–67.

Hasumi, H., and S. Emori. 2004. K-1 coupled GCM (MIROC) description Tokyo: Center for Climate System Research, University of Tokyo.

Hijmans, R.J., S.E. Cameron, J.L. Parra, P.G. Jones, and A. Jarvis, A. 2005. Very high resolution interpolated climate surfaces for global land areas. *International Journal of Climatology* 25:1965–78.

Holt, R.D. 1996. Demographic constraints in evolution: Towards unifying the evolutionary theories of senescence and niche conservatism. *Evolutionary Ecology* 10:1–11.

Holt, R.D., and R. Gomulkiewicz. 2004. Conservation implications of niche conservatism and evolution in heterogeneous environments. In *Evolutionary Conservation Biology*, pp. 244–264. Ferrière, R., U. Dieckmann, and D. Couvet, eds. Cambridge: Cambridge University Press.

Holt, R.D., R. Gomulkiewicz, and M. Barfield. 2003. The phenomology of niche evolution via quantitive traits in a "black-hole" sink. *Proceedings of the Royal Society of London Series B-Biological Sciences* 270:215–24.

Körner, Ch., M. Donoghue, T. Fabbro, Ch. Häuser, D. Nogués-Bravo, M.T. Kalin Arroyo, J. Soberon, L. Speers, E.M. Spehn, H. Sun, A. Tribsch, P. Tykarski, and N. Zbinden. 2007. Creative use of mountain biodiversity databases: The Kazbegi Research Agenda of GMBA-DIVERSITAS. *Mountain Research and Development* 27(3):276–81.

Martínez-Meyer, E., A.T. Peterson, and W.W. Hargrove. 2004. Ecological niches as stable distributional constraints on mammal species, with implications for Pleistocene extinctions and climate change projections for biodiversity. *Global Ecology and Biogeography* 13:305–14.

Martínez-Meyer, E., and A.T. Peterson. 2006. Conservatism of ecological niche characteristics in North American plant species over the Pleistocene-to-Recent transition. *Journal of Biogeography* 33:1779–89.

Parmesan, C. 2006. Ecological and evolutionary responses to recent climate change. *Annual Review of Ecology, Evolution and Systematics* 37:637–90.

Parmesan, C., and G. Yohe. 2003. A globally coherent fingerprint of climate change impacts across natural systems. *Nature* 421:37–42.

Pearman, P.B., C.F. Randin, O. Broennimann, P. Vittoz, W.O. van der Knaap, R. Engler, G. Le Lay, N.E. Zimmermann, and A. Guisan, A. 2008a. Prediction of plant species distributions across six millennia. *Ecology Letters* 11:357–69.

Pearman, P.B., A. Guisan, O. Broennimann, and C.F. Randin. 2008b. Niche dynamics in space and time. *Trends in Ecology & Evolution* 23:149–58.

Pearson, R.G., W. Thuiller, M.B. Araújo et al. 2006. Model-based uncertainty in species range prediction. *Journal of Biogeography* 33:1704–11.

Peterson, A.T., and Á. Nyári. 2008. Ecological niche conservatism and Pleistocene refugia in the thrush-like mourner, Schiffornis sp., in the Neotropics. *Evolution* 62(1):173–83

Phillips, S.J., R.P. Anderson, and R.E. Schapire. 2006. Maximum entropy modeling of species geographic distributions. *Ecological Modelling* 190:231–59.

Pounds, J.A., M.P.L. Fogden, and K.L. Masters. 2005. Responses of natural communities to climate change in a highland tropical forest. In *Climate Change and Biodiversity*. Lovejoy, T., and L. Hannah, eds. New Haven, Conn.: Yale University Press.

Rehfeldt, G.E., C.C. Ying, D.L. Spittlehouse, and D.A. Hamilton. 1999. Genetic responses to climate in *Pinus contorta*: Niche breadth, climate change, and reforestation. *Ecological Monographs* 69:375–407.

Stein, B.R., and J. Wieczorek. 2004. Mammals of the world: MaNIS as an example of data integration in a distributed network environment. *Biodiversity Informatics* 1:14–22.

Thompson R.S., C. Whitlock, P.J. Bartlein, S.P. Harrison, and W.G. Spaulding. 1993. Climatic changes in the western United States since 18,000 yr B.P. In *Global Climates Since the Last Glacial Maximum*, pp. 468–513. Wright Jr., H.E., J.E. Kutzbach, T. Webb III, W.F. Ruddiman, F.A. Street-Perrott, and P.J. Bartlein, eds. Minneapolis, Minn.: University of Minnesota Press.

Waltari, E., R. Hijmans, A. Peterson, Á. Nyári, S. Perkins, and R. Guralnick. 2007. Locating Pleistocene refugia: Comparing phylogeographic and ecological niche model predictions. *PLoS ONE* 2(7):e563.

Webb III, T., P.J. Bartlein, S.P. Harrsion, and K.H. Anderson. 1993. Vegetation, lake level, and climate change in eastern North America. In *Global Climates Since the Last Glacial Maximum*, pp. 415–467. Wright Jr., H.E., J.E. Kutzbach, T. Webb III, W.F. Ruddiman, F.A. Street-Perrott, and P.J. Bartlein, eds. Minneapolis, Minn.: University of Minnesota Press.

Whitmore, J., K. Gajewski, M. Sawada, J.W. Williams, B. Shuman, P.J. Bartlein, T. Minckley, A.E. Viau, T. Webb III, S. Shafer, P. Anderson, and L. Brubaker. 2005. Modern pollen data from North American and Greenland for multi-scale paleoenvironmental applications. *Quaternary Science Reviews* 24:1828–48.

Wieczorek J., Q. Guo, and R.J. Hijmans. 2004. The point-radius method for georeferencing locality descriptions and calculating associated uncertainty. *International Journal of Geographical Information Science* 18:745–67.

14 A Georeferenced Biodiversity Databank for Evaluating the Impact of Climate Change in Southern Italy Mountains

Roberto Pizzolotto, Maria Sapia, Francesco Rotondaro, Stefano Scalercio, and Pietro Brandmayr

CONTENTS

Introduction 137
Methods 138
The Study Area 138
Data Collection 139
The Database Construction 139
Results 139
The Database 139
Quantifying Slope Shifts 142
Comparing Different Mountain Ranges in Mediterranean Europe and Elsewhere 144
A Research Protocol for the HBDB 145
Conclusions 145
Summary 145
References 146

INTRODUCTION

Temperature decreases with elevation in mountains, which is the main factor driving the adaptation of living organisms to mountain environments (Körner, 2003).

The influence of temperature (accompanied with atmospheric pressure and water saturation deficit of the air) along the elevational gradient again changes with latitude. Within the same latitude, the variation of the ecological or geographical characteristics of the area (i.e., the associated bedrock, soil chemistry and geomorphology, landscape or biome history, geological history) are additional drivers of evolution of particular biota, which can then characterize in their unique composition even a single mountain range (Table 14.1).

While it is possible to outline a global trend for all elevational gradients, i.e., the decline in temperature and atmospheric pressure with elevation on a broad scale, temperature and microclimate in addition strongly vary on a small scale due to differences in geomorphology, aspect, and slope. Therefore, differences in biotic composition of mountain organisms can be explained on a very small scale only if abiotic factors were taken into account on a small grid too, such as differences in bedrock characteristics within and among mountain ranges, to adequately analyze fine resolution species data (i.e., well-georeferenced ones). Apart from geomorphology, historical factors are crucial to be included in such a database for a deeper understanding of the ecological and evolutionary drivers of the mountain fauna. Isolation is important; in fact, it is the main reason why mountain ranges, which seem to be very similar and belong to the same geographic area (e.g., Northmediterranean Peninsulae, Italy, versus Greece, or

TABLE 14.1
Levels of Complexity within Different Geographical Scales

Global	Latitudinal	Morphological
Altitudinal gradients of temperature and atmospheric pressure	Mediterranean region (summer aridity)	Pollino Mt. vs. Parnaso Mt.
	Amazonian region (rainfalls)	Stelvio Mt. vs. Marmolada Mt.
	Tibet (temperature)	

Note: The three columns could be read also as worldwide, regional, local. The narrower the region, the more the variables to be taken into account for biodiversity analysis.

Appennines versus Alps), show partially differentiated faunas. Mountain species are affected not only by the past geographic connections of the mountain range (Holdhaus, 1954), but especially by their paleoecological history (e.g., ice age, Holdhaus and Lindroth, 1939), and they are often marked by a specific evolutionary history (Darlington, 1943; Lindroth, 1949; Brandmayr, 1991). Thus, even very similar mountain massifs may be inhabited by either taxonomically or ecologically vicariant species. A good example of how alpine ground beetle faunas and distribution vary on a small scale can be found in Casale and Vigna Taglianti (1993) for the western Alps. Ecologically vicariant species followed the same "evolutionary pathway" (in the sense of Erwin, 1979), and this gives us the opportunity to discover common ecological drivers of evolution.

In the following, we will exemplify such rules, using a database for species diversity on the local species community level. Species communities of ground beetles and their ecological characteristics are powerful indicators of change of the ecosystem, and are frequently used as environmental or ecological indicators (McGeoch, 1998). Indicator values of ground beetles were used for different purposes to analyze climate change (Ashworth, 1996), succession following glacier retreat in the Alps (Gobbi et al., 2006, 2007), habitat fragmentation (Spence et al., 1996; Lövei et al., 2006), or to assess the influence of stress factors emerging from management practices in grasslands (Blake et al., 1996; Rushton et al., 1989), in croplands (Kromp, 1999), and forest clearing in temperate (Szyszko, 1983) or boreal zones (Niemelä et al., 1993). Carabid beetles were shown to be valuable biodiversity indicators for assessing environmental conditions in Italy (Pizzolotto, 1994a, 1994b; Rainio and Niemelä, 2003), and were standardized recently by Brandmayr et al. (2005) using their biological traits (morphology, diet, life cycle, habitat preference, behavior) described in Thiele (1977) and summarized by Lövei and Sunderland (1996).

In this chapter, we propose a model procedure for georeferenced database development that we used for Pollino National Park, the largest Italian nature reserve, with a calcareous-dolomitic mountain range from 0 to 2264 m a.s.l. The animal data for this database has been collected in about 30 years of ground beetle samplings, coordinated by P. Brandmayr.

METHODS

The Study Area

Pollino National Park extends over ca. 190,000 ha and encloses the highest mountain ranges at the southern end of the calcareous and karstic Apennine, with peaks that reach 2,267 m on the tip of Serra Dolcedorme. Peaks above 1,500 m show evident traces of the last glaciations (small moraines and snow depressions or dolines). There are three major biomes (Avena and Bruno 1975):

- The sclerophyll belt from 0 to 1,200 m, mostly with evergreen *Quercus ilex* forests or shrubs (macchia), more abundant on the western (Thyrrhenian) side of the park, where a large wilderness area is found in the Orsomarso Mountains.
- The sub-Mediterranean-temperate broad leaved summer green vegetation belt from 600 to 2,100 m, with several types of *Quercus* forests (*Q. pubescens*, *Q. cerris*, *Q. petraea*) and a large *Fagus* zone (1,200 to 2,000 m, subdivided into "warmer aspects," Aquifolio-Fagetum, and cooler ones, Asyneumati-Fagetum) forming the treeline at 1,800 to 1,900 m.
- The alpine zone, with remnants of high mountain (above treeline) biomes of the Mediterranean area as represented, for instance, by *Festuca bosniaca* and *Carex kitaibeliana* natural grasslands (from 2,150 m upwards).

The sclerophyll belt has been widely transformed into pasture lands or crops and pine reforestations over large areas. Typical open landscape is represented by a mix of seminatural dry grasslands and scrubland types on calcareous substrate (Habitat 6020 of the EU Habitat Directive 92/43, Natura 2000 network). A very special type of herbaceous vegetation are *Stipa austro-italica* grasslands (an Annex II priority graminaceous species), a hot spot of species richness for many animal groups: Orthopterans, Hymenoptera, Lepidoptera, birds (Brandmayr et al.,

2002, and unpublished data). The *Quercus/Fagus* temperate belt was converted into pastures, today less overgrazed than in the past, whereas the summit regions are now all protected, with grazing pressure reduced after the 1980s. This allowed regeneration of the old *Pinus leucodermis* stands, an endemic pine species that covers rocky ridges and southern-exposed cliffs. The habitat map of the park encloses more than fifty vegetation types ("associations"), belonging to about twenty codified habitats of the EU Directive or to the corresponding Corine categories. In the years immediately after 2000, the park administration provided a georeferenced vegetation map that runs on ARC-VIEW and ARC-GIS software.

Data Collection

Ground beetles were captured using pitfall traps, plastic vessels with a top diameter of 9 cm and 11 cm depth, filled with 200 ml of a mixture of wine vinegar and 5 to 6% formalin. After 2001, the formalin was replaced by 5% concentration of ascorbic acid. The traps had a small hole near the top to prevent flooding after heavy rain, and were spaced about 15 m from each other. The traps were checked and refilled monthly, to achieve a "full year sample" covering the activity season from the last winter frosts to the first autumn frosts. In the sclerophyll belt, the activity season lasts nearly a full year. A variable number of traps was used depending on the size of the sampling site, in forests normally 6 to 8 traps were buried, in open or shrub habitats 3 to 4 traps were set up, in small biotopes (dolines, karstic sinks) often 1 to 2 only.

Species assemblages were organized in annual samples, comparing the species' abundances by individuals and traps over the standard period of ten days, and then averaged for the entire season (annual activity density, aAD. The length of the season is defined as the number of days from the first setting of the traps until their removal in autumn).

The Database Construction

The Habitat Biodiversity Data Base (HDBD in the following) for Pollino National Park has been constructed on the basis of a limited number of sampling sites, listed in Table 14.2 and covering an altitude range of approximately 1,700 m. This altitudinal span corresponds more or less to a mean difference in annual temperature from 16°C to 5 to 4.5°C, provided that the main slope exposure remains the same (e.g., S–SSW).

The HDBD is based on the interaction between a database management system (Microsoft Access) and a Geographical Information System (GIS; ESRI ArcView). The database has three major entry tables for the sites, the species, and for the abundance of the species (i.e., aAD) in the sites. Each species is described by the following entries: Italian checklist code; genus and species name; author and description year; chorotype; distribution in Italy; endemic status; photo; and notes on ecology. Each sampling site is described by the following criteria: acronym of the site; altitude; slope; aspect; geographical coordinates; vegetation; Natura 2000 habitat code and denomination; short description; and photo. The relations among the tables allow for the production of new tables bearing information about, e.g., community structure (i.e., what species and their abundance in a sample site), habitat faunistic composition (i.e., the species list, or the list and abundance of the chorotypes), and auto-ecology of the most frequent species.

The database has been linked with the GIS by means of a relation, which links every vegetation patch to the corresponding habitat code in the database. In this way, all biological information (index-card contents, species lists) can be linked to a specific vegetation patch. The vegetation map has been improved by adding new georeferenced sampling sites, especially small karstic habitats.

RESULTS

The Database

The HBDB so far includes 82 carabid species collected in the 1977–2004 survey carried out in Pollino National Park. The records of the Coleopteran fauna of this area currently amounts to 1,824 species belonging to 86 families (Angelini, 1986), and 156 species are ground beetles in the widest sense (Coleoptera: Carabidae). The logic of the HBDB is shortly summarized in Figure 14.1 showing the main search routines. An outstanding feature of HBDB is the ability to get information on animal (and plant) communities by starting directly from the habitat patch shown in the GIS map. This allows nonspecialists, e.g., national park administrators, but also experts (zoologists or botanists) to immediately get an idea of the community structures for a given habitat. Further searches allow the obtainment of an overview of the species and their ecology, their habitat choice, biological features, and their geographic distribution.

The database mainly serves two important purposes. First, it provides rapid information on the species present in an area. A park administrator can enter nearly all data collected over decades of field surveys, starting possibly from the early historical faunistic and floristic inspections, which then can be used for research and conservation management.

TABLE 14.2
Sampling Stations, their Altitude, and Vegetation Type

No.	Sampling Sites	Altitude	Vegetation Type and Biome/Belt
1	PFeBos-99	2075	*Festuca bosniaca* grass mats (high mountain)
2	PCaKi1-99	2260	*Carex kitaibeliana* stripped grassland (high mountain)
3	PCaKi2-99	2260	*Carex kitaibeliana* stripped grassland (high mountain)
4	PNaLup-99	2230	Acidophilic *Nardus stricta* pasture (treeline doline)
5	PSeBrm-04	2130	Seslerio-Brometum grasslands (upper Fagus belt)
6	PSeBr-77	2000	*Sesleria nitida* grasslands (upper Fagus belt)
7	PGhia-04	2050	"Calcareous scree, sparse herb. veg., Fagus belt"
8	PGhiMcl-04	2090	Crevices (cm 20–30) in vertical rock walls
9	PMA1-77	1540	*Meum athamanticum* mesic pastures (upper Fagus)
10	PMA1-04	1540	"Same site of PMA1-77, 27 years after"
11	PIngMA-04	1800	Water sinks in *Meum athamanticum* pastures
12	PMA2-04	1860	*Meum athamanticum* pastures (upper Fagus belt)
13	PAsF-77	1580	Asyneumati-Fagetum high trunk Fagus stand
14	PAsF-04	1580	"Same site of PAsF-77, 27 years after"
15	PAqF-77	1485	Aquifolio-Fagetum high tr. thermoph. Fagus st.
16	PAqF-04	1485	"Same site of PAqF-77, 27 years after"
17	PJun-04	1280	Dwarf *Juniperus* pastures of the lower Fagus belt
18	PSaBr-04	1200	Brushy dry grassland of the upper Quercus belt
19	PIngh1-99	2240	"Calcareous sinkhole, traps at 1,5 metres depth"
20	PIngh2-99	2230	"Calc. sinkhole within PNaLup-99, (doline at treeline)"
21	PIngh1-04	2240	Same site of PIngh1-99 after 5 years
22	PQuP2-99	725	"Young *Quercus pubescens* st., lower Quercus belt"
23	PQuP1-99	710	Sparse aged *Quercus* trees on calc. conglomerates
24	PRim-99	730	*Pinus halepensis* reforestation (lower *Quercus* belt)
25	PCult-99	700	Cereal cropland on the "Petrosa" conglomerates
26	PStA1-99	830	Stipa austro-italica garigue (lower *Quercus* belt)
27	PStA2-99	615	Stipa austro-italica garigue (sclerophyll belt?)
28	PGinSJ-99	610	Garigue with *Spartium junceum* (sclerophylls?)
29	PPyA-99	535	Garigue with *Pyrus amygdaliformis* (sclerophylls?)

Second, the HBDB allows for the planning of further surveys, to repeat samplings in the more endangered habitats or biotopes, to make predictions and study biodiversity trends over time, and, finally, to plan biodiversity assessments after extreme events or tracking the consequences of more gradual changes, such as global climatic change. In the following we will illustrate examples of assessments, based on the carabid beetles communities recorded so far.

Example 1: Single species indicators of environmental change (early warning species)

In 1999 a new carabid species was recorded in Pollino National Park, at the southern fringe of its geographical distribution: *Cychrus attenuatus latialis*. Very small populations of this Central European snail predator were known to live on the Apennine mountain tips, from Abruzzo to Basilicata (Mt. Sirino). The first Pollino populations were discovered in 1999 by Sapia and Rotondaro (2006) in a karstic sink ("Pingh1"), and we then detected a slight decrease in population density during five years (i.e., 0.3 of aAD in 1999, while 0.27 of aAD in 2004). In Central Europe, *C. attenuatus* lives in cool forests, while the Italian endemic subspecies (*C.a. latialis*) prefers cold summit habitats where a conspicuous amount of snow accumulates during the winter, or deep rock crevices that maintain a high humidity level throughout the summer. The database will permit entering and monitoring the species' fate over the next decades as a possible "early warner" of irreversible invertebrate extinction as a consequence of climate change.

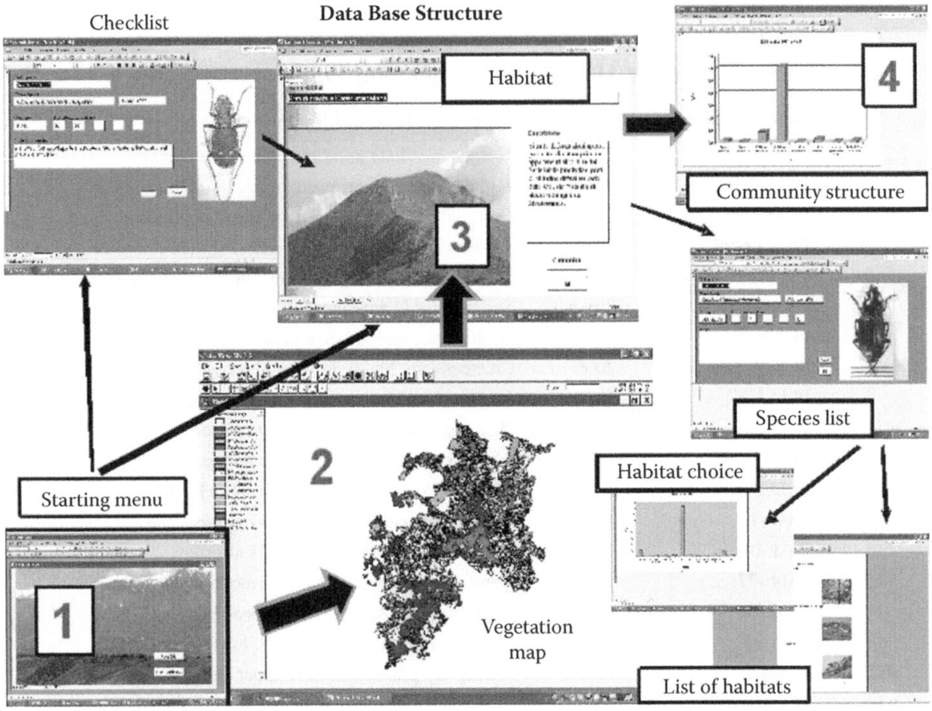

FIGURE 14.1 Schema of the habitat biodiversity database, showing the main search routines (arrows).

Example 2: Communities (i.e., species assemblages) as indicators of environmental change

HBDB offer the highest predictive value if it allows construction of time series data for annual community samplings made in the same habitat over several decades. Three sites in particular, two beech forests and one *Meum athamanticum* pasture, have been trapped by pitfalls in 1977 and in 2004 (PAsF, PAqF, PMA1 in Table 14.2). Table 14.3 presents the changes observed in the pasture: Among the nineteen species collected, five "new entries" have been recorded in 2004, which are all thermophilous species. Among them, *Calathus fuscipes* is particularly impressive, because the species was absent in 1977, while it then became the dominant species in 2004. Of the fourteen species found in 1977, two microtherm forest indicators disappeared in 2004 (*Haptoderus apenninus, Trichotichnus nitens*), together with some high montane pasture dwellers (e.g., *Carabus violaceus picenus*), while some others showed strong reductions (*Calathus melanocephalus, Zabrus costae*). Meanwhile, xerophilous (*C. sirentensis*) or euryedaphic species (*Steropus melas*) became the two other most dominant species.

TABLE 14.3
Species Abundance (Annual Activity Density) and Substitution in 1977–2004 in the Pasture PMA1

Species	1977	2004
Amara sicula	1.65	0.03
Calathus melanocephalus	1.25	0.19
Nebria kratteri	0.75	1.10
Notiophilus aestuans	0.73	0.09
Calathus fracassii	0.71	0.25
Calathus sirentensis	0.50	10.09
Steropus melas	0.40	2.54
Zabrus costae	0.34	0.03
Pterostichus bicolor	0.30	0.53
Synuchus vivalis	0.18	0.16
Trichotichnus nitens	0.14	0.00
Percus bilineatus	0.04	0.00
Haptoderus apenninus	0.42	0.00
Carabus violaceus picenus	0.36	0.00
Pseudoophonus rufipes	0.00	0.03
Notiophilus quadripunctatus	0.00	0.06
Cymindis scapularis	0.00	0.16
Cymindis etrusca	0.00	0.31
Calathus fuscipes latus	0.00	27.77

The approach based on the community structure suggests that species substitutions and losses are consistent with a strong warming effect.

Similar changes have been recorded after twenty-seven years in a tall Asyneumati-Fagetum forest (PAsF-77 versus PAsF-04, Table 14.4). Six new species, all of them thermophilous or species tied to warmer forest types (e.g., *Calathus montivagus*) appeared. However, the abundance ranking of the first ten species is hardly affected, and the first three dominant carabids remain unchanged. This is in accordance with our expectations, as the Asyneumati-Fagetum is a northern-exposed cool beech forest, unlikely to show strong signals of warming.

The Aquifolio-Fagetum "warmer" beech forest (PAqF-77 versus PAqF-04, Table 14.5), provides a contrasting signal: On one side, there are strong indications for a change in land use, namely the end of logging in the '80s, which improved soil conditions. As a consequence, the population densities of some forest dwellers increased strongly, and the annual activity density of all species increased from 21 to 109 individuals per trap per decade in 2004. We observed four thermophilous "new entries," among which the large *Carabus lefebvrei* is widespread in the lower forest belts.

TABLE 14.4
Species Abundance (Annual Activity Density) and Substitution in 1977–2004 in the Beech Wood PAsF

Species	1977	2004
Calathus fracassii	71.44	53.03
Calathus rotundicollis	14.75	42.33
Pterostichus bicolor	10.23	17.75
Trechus obtusus lucanus	6.16	1.23
Haptoderus apenninus	1.93	1.79
Trichotichnus nitens	1.47	0.18
Nebria kratteri	0.92	1.10
Cychrus italicus	0.26	1.11
Pterostichus micans	0.07	0.09
Leistus spinibarbis fiorii	0.05	0.01
Abax parallelepipedus curtulus	0.05	0.02
Leistus fulvibarbis	0.01	0.01
Clinidium canaliculatum	0.03	0.00
Calathus sirentensis	0.01	0.00
Platyderus canaliculatus	0.01	0.00
Leistus sardous	0.00	0.01
Nebria brevicollis	0.00	0.02
Carabus lefebvrei bayardi	0.00	0.02
Steropus melas	0.00	0.03
Calathus montivagus	0.00	0.04
Trechus quadristriatus	0.00	0.05

TABLE 14.5
Species Abundance (Annual Activity Density) and Substitution in 1977–2004 in the More Thermoxerophilic Aquifolio-Fagetum PAqF

Species	1977	2004
Calathus fracassii	12.95	58.05
Pterostichus micans	3.68	9.09
Nebria kratteri	1.23	11.76
Steropus melas	1.13	9.10
Calathus rotundicollis	0.76	9.91
Cychrus italicus	0.43	1.35
Abax parallelepipedus curtulus	0.39	2.77
Calathus montivagus	0.24	0.41
Percus bilineatus	0.13	0.07
Pterostichus bicolor	0.11	5.51
Trechus obtusus lucanus	0.01	0.04
Haptoderus apenninus	0.01	0.23
Synuchus vivalis	0.01	0.40
Carabus violaceus picenus	0.07	0.00
Calosoma sycophanta	0.01	0.00
Laemostenus acutangulus	0.01	0.00
Calathus fuscipes latus	0.01	0.00
Carabus coriaceus	0.00	0.01
Notiophilus substriatus	0.00	0.02
Trichotichnus nitens	0.00	0.04
Trechus quadristriatus	0.00	0.07
Carabus lefebvrei bayardi	0.00	0.13

Quantifying Slope Shifts

Changes in species presence over a longer time period can be used to assess the upslope shift of species, provided that the sampling effort covers a broad altitudinal range of habitats, as HBDB does. A cluster analysis of the sample sizes based on all available species shows three distant groups: I, sub-Mediterranean open lands and garigues; II, mesophilous forests; and III, high altitude pastures and grasslands (Figure 14.2). In only twenty-seven years, the carabid community of *Meum* pastures sampled at 1,500 a.s.l. (PMA1-77) "shifted" from group III to group I (PMA1-04), thus the new species composition and abundance reached in 2004 was more similar to the species assemblages living in thermoxeric grasslands at lower altitude than in 1977, with *Calathus fuscipes latus* populations gaining dominance upwards.

The analysis of the main ecological factors influencing species distribution (i.e., gradient analysis) is depicted in Figure 14.3, where the most evident factor is altitude, but it is likely that the most meaningful factors are correlated with altitude. The variation of sample sites was explained by

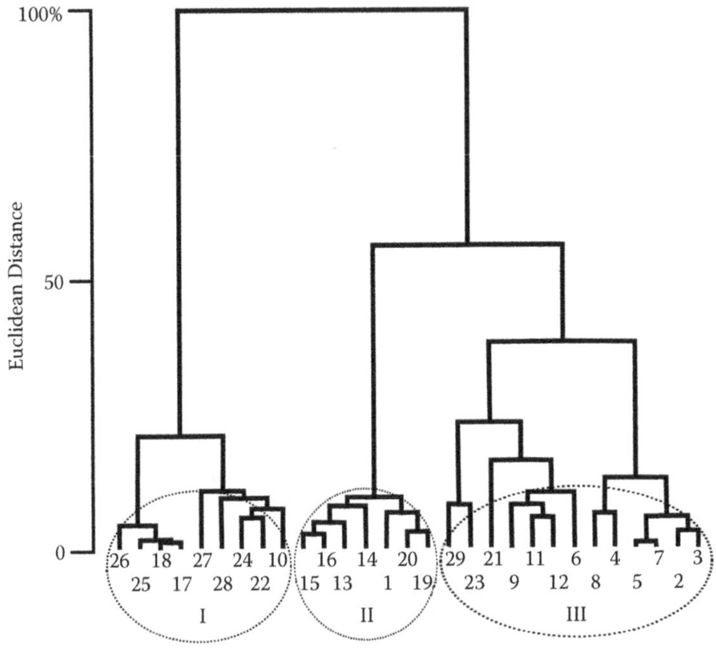

FIGURE 14.2 Classification of sample sites after Euclidean distance and Ward's agglomeration algorithm. Y axis, Euclidean distance (D) as (Dlink/Dmax)*100. Numbers follow Table 14.2.

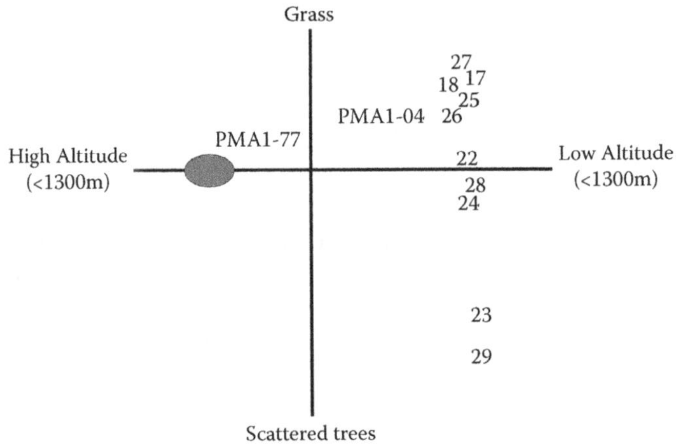

FIGURE 14.3 Ordination of sample sites after correspondence analysis, showing the critical position of the mountain pasture (PMA1). The first axis (18% of the total variance) is clearly linked to an altitudinal gradient, while the second axis (12% variance) gives the ordination of the sample sites with thermoxeric microclimate conditions, and with vegetation structure characterized by scattered trees at the bottom of the axis, and by grasslands (Mediterranean garigue) at the upper end. Numbers follow Table 14.2. Sites 1 to 8, 11 to 16, and 19 to 21 of Table 14.2 have been clumped in the area marked by the grey ellipse.

difference in altitude (first axis, 18% of the total variance). Since no site is near the origin of the first axis (i.e., no one is neutral against the first axis), it is likely that all the carabid assemblages are strongly affected by altitude, therefore by the underlying temperature gradient, and thus also by the duration of summer. Of all sites below 1,300 m, characterized by thermoxeric conditions, the species assemblages seem to be affected by the vegetation structure (second axis, 12% of the total variance), i.e., scattered trees in the sites at the bottom of the second axis versus Mediterranean garigue in the sites at the upper end of the same axis.

The variation of the species assemblage in PMA1 (highlighted in Figure 14.3) after twenty-seven years is supposedly due to a rise in temperature.

Our gradient analysis showed that in only three decades species composition of the pastures changed as if it has been shifted uphill by about 300 m. Similar results have been found in butterfly surveys in Central Europe and in the Mediterranean. Wilson et al. (2007) showed in the Sierra de Guadarrama (Central Spain) an uphill shift of lepidopteran communities of 293 metres in elevation between 1967 and 1973 and 2004 and 2005, coping with a 225 m shift of mean annual isotherms. In the Czech Republic (Konvicka et al., 2003) the analysis of 11 km square grids between 1950 and 1980 and 1995 and 2001 reveals upslope species shifts of 60 to 90 m up to 148 m. Butterfly communities on the Southern Apennine (Mt. Pollino, Calabria) show distinct changes between 1977 and 2004 (Scalercio et al., 2005), with local extinction of orophilous elements and migration of thermo-Mediterranean elements toward mountaintops. The global change impact on species assemblages is likely to be stronger in the Mediterranean countries than in other areas of the world (Sala et al., 2000; Christensen et al., 2007).

Comparing Different Mountain Ranges in Mediterranean Europe and Elsewhere

In the Mediterranean peninsulas, Spain, Italy and in the Balkans, ground beetle species show vicariant species from one mountain range to the neighboring, most steno-endemic species being terminal branches of repetitive "evolutionary pathways." This means that species of the same habitat may often have the same biological adaptations, e.g., similar life cycles, reproduction rhythms, dispersal power, size, modes of space use inside or outside the soil, etc. In other words, despite the faunae varying in a relevant manner from region to region, their life traits should be equivalent as far as adaptive responses toward changes are concerned.

The species assemblages may be characterized on the basis of the most frequent life traits, and the analysis of the life traits distribution in each sample site could help for revealing if an adaptive trend toward changes is taking place. The phenology and the range of a species have been chosen as the most meaningful traits for studying the consequences of climatic change.

Carabids living in temperate–cold climates evolved an annual life rhythm with eggs laid in spring and larvae developing in summer ("summer larvae") or even a two-year life rhythm, while an annual life rhythm with overwintering larvae could be more adapted to temperate–warm climate conditions (Paarman, 1979; Brandmayr et al., 2005).

Figure 14.4 shows a general trend in decreasing of species with summer larvae and increasing of species with overwintering larvae (sites PMA1, PAsF, and PaqF) after twenty-seven years.

Mountain pastures (e.g., PMA1), once rich in forest species able to cross pastures but often unable to

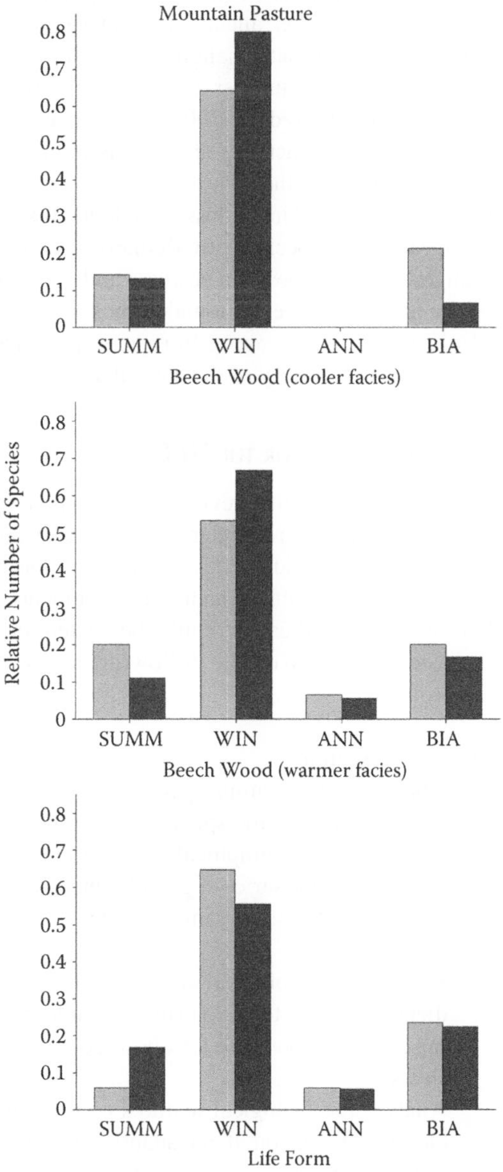

FIGURE 14.4 Variation of carabid life rhythm in three mountain ecosystems after twenty-seven years. Y axis, relative number of species; SUMM, summer larvae, i.e., eggs laid in spring and larvae developing in summer; WIN, winter larvae, i.e., eggs laid in summer–autumn and overwintering larvae; ANN, one-year life rhythm; BIA, life rhythm longer than one year. Grey bars are from 1977; black bars are from 2004.

reproduce in them (1977), were invaded by species with winter larvae (e.g., *Calathus fuscipes*) in 2004, while true forest species with biennial (slow) development of larvae, like *Pterostichus bicolor* or *Haptoderus apenninus*, were disappearing in 2004. The cooler beech forest habitat (such as PAsF in Figure 14.4) shows a decreasing number of species with summer larvae (less humidity in the soil) and an increase of Mediterranean winter-larvae taxa, as a consequence of the climate changing in the direction of higher temperatures and summer dryness. PaqF, strongly exploited by man, showed in 2004 an opposite trend (Figure 14.4), with summer larvae carabids indicating a higher water retention in the soil.

In Figure 14.5, a trend in the loss of endemic taxa (II), and the appearance of species with Mediterranean distribution range (IIIm) are shown in pastures and cool beech stands. The trend might be confounded by soil recovery in the PaqF, where the increase of European species (category III) indicates recovery of the natural soil fauna.

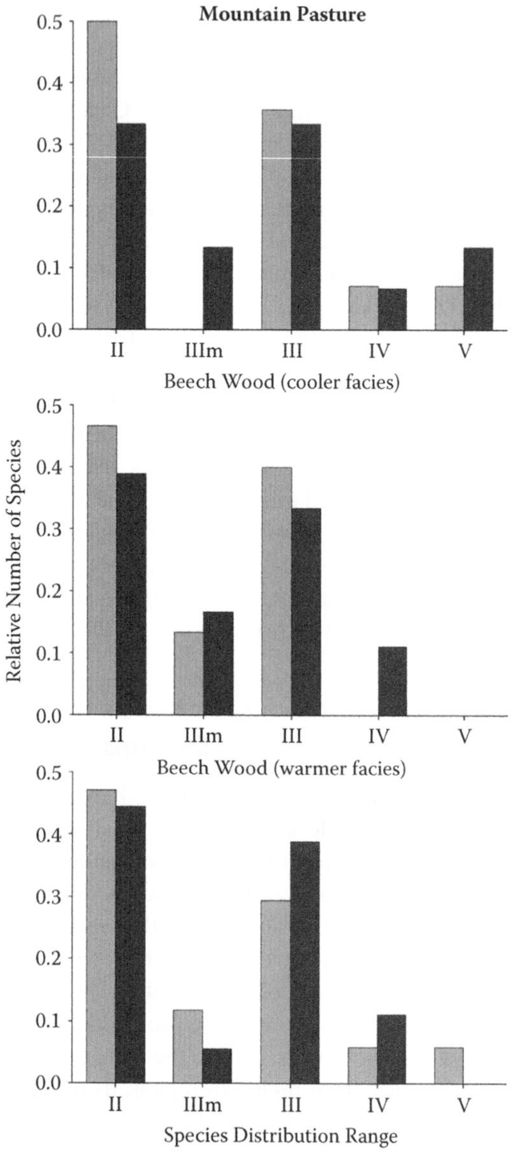

FIGURE 14.5 Chorological variation of carabid species in three mountain ecosystems after twenty-seven years. Y axis, relative number of species; II, species endemic to Italy; IIIm, species with distribution linked to the Mediterranean basin; III, European species; IV, Euroasiatic species; V, palearctic, holarctic species.

A Research Protocol for the HBDB

The procedure outlined in the previous paragraphs is based on the implementation of a georeferenced database, and it has been applied to carabid beetles for providing insights into how climate change affects natural environments.

Other biological indicators could be tested in the same way on the basis of the following a two-step research protocol.

1. "Coarse" research
   - Obtain early warning species shifts, e.g., loss of stenoendemic species, or changes of the species' geographical distribution, by comparing time series of georeferenced data collected in the same sampling site.
2. "Fine" research
   - Evaluate the ecological similarity and gradients analysis for exploring the underlying factors responsible for observed species shifts.
   - Analyze the biological features of the species for evaluation if an adaptive trend is linked to observed species shifts.

CONCLUSIONS

Italy is a hot spot of endemic species diversity in Europe. Data on vertebrate and invertebrate taxocoenoses, facilitated by databases as HBDB, are strongly needed to predict and manage the impact of global change, especially because animals react often in a faster way to environmental changes.

SUMMARY

Georeferenced databases are perhaps the most important management tool for nature conservation of the third millennium. We propose a model procedure for georeferenced database development, using the example of the Habitat Biodiversity Data Base (HBDB) in Pollino National Park in Italy. The HBDB scheme is very easy to apply, as one can directly look at species abundance structures and species occurrences on a georeferenced GIS map. Nature reserve managers will be able to check animal community structures simply by clicking on habitat polygons, and download community structures of animal and plant species assemblages of a chosen habitat. The availability of time series (annual samples) for the same habitat (site) enhances predictability and allows study of the trends in change in species abundance structures and the planning of surveys or monitoring. Moreover, the database allows the insertion of microhabitats too small to be represented in available GIS maps, such as the small *Cychrus* populations described in this paper or the today isolated *phytocoenoses* with their threatened plant species living in snow patches in the Pollino massif (Tomaselli et al., 2003). This could also improve the analysis and evaluation of the possible increase of habitat fragmentation, as a consequence of increasing thermoxeric conditions of pastures.

The HBDB is an experimental scientific tool to explore elevational gradients of diversity or species shifts within a habitat over time. Looking at species changes for several habitat types in the period 1977–2004, we observed that not all the communities react in the same way to temperature changes. Open land habitats, such as meadows and pastures, show pronounced changes in species structure, with a complete substitution of dominant carabid beetles, whereas cooler forest types as the Asyneumati-Fagetum show no substitutions in the dominant species, but thermophilic carabids appear. In other forest types (Aquifolio-Fagetum), species changes may be masked by contemporary changes in forest management (ceased logging, soil recovery) and by the consequent increase in biomass of soil invertebrates.

REFERENCES

Angelini, F. 1986. Coleotterofauna del Massiccio del Pollino (Basilicata-Calabria). *Entomologica* 21:37–125.

Ashworth, A.C. 1996. The response of arctic carabids (Coleoptera) to climate change based on the fossil record of the Quaternary Period. *Annales Zoologici Fennici* 33:125–31.

Avena, G., and F. Bruno. 1975. Lineamenti della vegetazione del Massiccio del Pollino (Appennino calabro-lucano). *Notulae Fitosociologicae* 10:131–58.

Blake, S., G.N. Foster, G.E.J. Fisher and G.L.E. Ligertwood. 1996. Effects of management practices on the caranid faunas of newly established wildflower meadows in southern Scotland. *Annales Zoologici Fennici* 33:139–147.

Brandmayr, P. 1991. The reduction of metathoracic alae and of dispersal power of carabid beetles along the ecolutionary pathway into the mountains. In *Form and function in zoology*, pp. 368–378. Lanzavecchia G., Valvassori R. (eds.). Mucchi, Modena.

Brandmayr, P., T. Mingozzi, S. Scalercio, N. Passalacqua, F. Rotondaro, and R. Pizzolotto. 2002. Stipa austroitalica garigues and mountain pastureland in the Pollino National Park (Calabria, Southern Italy). In *Pasture Landscapes and Nature Conservation*, pp. 53–66. Redecker B., P. Finck W. Härdtle U. Riecken and E. Schröder, eds. Berlin: Springer.

Brandmayr, P., T. Zetto, and R. Pizzolotto. 2005. I coleotteri carabidi per la valutazione ambientale e la conservazione della biodiversità. APAT, Manuali e linee guida, 34, Roma.

Casale, A., and A. Vigna Taglianti. 1993. I Coleotteri Carabidi delle Alpi occidentali e centro-occidentali (Coleoptera, Carabidae). *Biogeographia* 16:331–99.

Christensen, J.H., B. Hewitson, A. Busuioc A. Chen, X. Gao I. Held, R. Jones, R.K. Kolli, W.-T. Kwon, R. Laprise, V. Magaña Rueda, L. Mearns, C.G. Menéndez, J. Räisänen, A. Rinke, A. Sarr, and P. Whetton. 2007. Regional climate projections. In *Climate Change 2007: The Physical Science Basis*. Solomon, S., D. Qin, M. Manning, Z. Chen, M. Marquis, K.B. Averyt, M. Tignor, and H.L. Miller, eds. Cambridge: Cambridge University Press.

Darlington, P.J. 1943. Carabidae of Mountains and Islands: Data on the Evolution of Isolated Faunas, and on Atrophy of Wings. *Ecological Monographs*, 13:37–61.

Duelli, P., and M.K. Obrist. 1998. In search of the best correlates for local organismal biodiversity in cultivated areas. *Biodiversity and Conservation* 7:297–309.

Erwin, T.L. 1979. Thoughts on the evolutionary history of ground beetles: Hypotheses generated from comparative faunal analyses of lowland forest sites in temperate and tropical regions. In *Carabid Beetles, Their Evolution, Natural History, and Classification*, pp. 539–592. Erwin, T.L., G. E. Ball, and D.R. Whitehead, eds. W. Junk, The Hague.

Gargaglione, A.A. 2001. Geologia, Flora e vegetazione del Parco del Pollino. In *Il Parco Nazionale del Pollino*, BRN Anno XXVI-99, pp. 17–24.

Gobbi, M., F. De Bernardi, M. Pelfini, B. Rossaro, and P. Brandmayr. 2006. Epigean arthropod succession along a 154-year glacier foreland chronosequence in the Forni Valley (Central Italian Alps). *Arctic, Antarctic, and Alpine Research* 38:357–62.

Gobbi, M., B. Rossaro, A. Vater, F. De Bernardi, M. Pelfini, and P. Brandmayr. 2007. Environmental features influencing Carabid beetle (Coleoptera) assemblages along a recently deglaciated area in the Alpine region. *Ecological Entomololgy* 32:282–89.

Holdhaus, K. 1954. *Die Spuren der Eiszeit der Tierwelt Europas. Abhandlungen der Zoologisch-Botanischen Gesellschaft.* Innsbruck: Wagner.

Holdhaus, K., and C.H. Lindroth. 1939. Die europaischen Koleopteren mit boreoalpiner Verbreitung. *Annalen des Naturhistoris-chen Museums, Wien* 50:123–293.

Konvicka, M., M. Maradova, J. Benes, Z. Fric, and P. Kepka. 2003. Uphill shifts in distribution of butterflies in the Czech Republic: Effects of changing climate detected on a regional scale. *Global Ecology and Biogeography* 12:403–10.

Körner, C. 2003. *Alpine Plant Life*. Berlin: Springer-Verlag.

Kromp, B. 1999. Carabid beetles in sustainable agriculture: A review on pest control efficacy, cultivation impacts and enhancement. *Agriculture Ecosystems and Environment* 74:187–228.

Lindroth, C.H. 1949. *Die Fennoskandischen Carabidae: Eine Tiergeographische Studie.* Elanders Boktryckeri Aktiebolag, Göteborg.

Lövei, G.L., T. Magura, B. Tóthmérész, and V. Ködöböcz. 2006. The influence of matrix and edges on species richness patterns of ground beetles (Coleoptera: Carabidae) in habitat islands. *Global Ecology and Biogeography* 15:283–89.

Lövei, G.L., and K.D. Sunderland. 1996. Ecology and behavior of ground beetles (Coleoptera: Carabidae). *Annual Review of Entomology* 41:231–256.

McGeoch, M.A. 1998. The selection, testing, and application of terrestrial insects as bioindicators. *Biological Revue* 73:181–201.

Niemelä, J., D. Langor and J.R. Spence. 1993. Effects of Clear-Cut Harvesting on Boreal Ground-Beetle Assemblages (Coleoptera: Carabidae) in Western Canada. *Conservation Biology* 7:551–561.

Paarmann, W. 1979. Ideas about the evolution of the various annual reproduction rhythms in carabid beetles of the different climatic zones. In *On the Evolution of Behaviour in Carabid Beetles*, misc. papers, pp. 119–132. den Boer, P.J., H.U. Thiele, and F. Weber, eds. Landbouwhogesch, Wageningen.

Pizzolotto, R. 1994a. Ground beetles as a tool for environment management: A geographical information system based on Carabids and vegetation for the Karst near Trieste (Italy). In *Carabid Beetles, Ecology and Evolution*, pp. 343–351. Desender, K., M. Dufrene, M. Loreau, M.L. Luff, and J.-P. Maelfait, eds. Dordrecht: Kluwer Academic Publishers.

Pizzolotto, R. 1994b. Soil arthropods for faunal indices in assessing changes in natural value resulting from human disturbance. In *Biodiversity, Temperate Ecosystems and Global Change*, pp. 291–314. Boyle, T., and C.E.B. Boyle, eds. Berlin, Heidelberg, New York: Springer Verlag.

Rainio, J., J. Niemelä. 2003. Ground beetles (Coleoptera: Carabidae) as bioindicators. *Biodiversity and Conservation* 12:487–506.

Rushton, S.P., M.L. Luff and M.D. Eyre. 1989. Effects of Pasture Improvement and Management on the Ground Beetle and Spider Communities of Upland Grasslands. *Journal of Applied Ecology* 26:489–503.

Sala, O., F.S. Chapin III, J.J. Armesto, E. Berlow, J. Bloomfield, R. Dirzo, S. Huber-Sanwald, L.F. Huenneke, R.B. Jackson, A. Kinzig, R. Leemans, D.M. Lodge, H.A. Mooney, M. Oesterheld, N. LeRoy Poff, M.T. Sykes, B.H. Walker, M. Walker, and D.H. Wall. 2000. Global biodiversity scenarios for the year 2100. *Science* 287:1770–74.

Sapia, M. and F. Rotondaro. 2006. Nuova segnalazione faunistica per la Calabria: Cychrus attenuatus latialis. *Bollettino della Società Entomologica Italiana* 138:73–80.

Scalercio, S., M. Sapia and P. Brandmayr. 2006. Effetti del Global Change su ropaloceri e carabidi del Massiccio del Pollino (Lepidoptera, Coleoptera Carabidae). Proceedings of the 16th meeting of the Italian Society of Ecology, Available online at http://www.ecologia.it/congressi/XVI/articles/

Spence, J.R., D.W. Langor, J. Niemelä, H.A. Carcamo, and C.C. Currie. 1996. Northern forestry and ground beetles: The case for concern about old-growth species. *Annales Zoologici Fennici* 33:173–84.

Szyszko, J. 1983. State of Carabisae (Col.) fauna in fresh pine forest and tentative valorisation of this environment. SGGW-AG Monographs No. 28. Warsaw Agricultural University Press, Warsaw, Poland.

Thiele, H.U. 1977. *Carabid Beetles in their Environments.* Berlin, Heidelberg, New York: Springer.

Venn, S.J., D.J. Kotze, and J. Niemelä. 2003. Urbanization effects on carabid diversity in boreal forests. *European Journal of Entomology* 100:73–80.

Wilson, R.J., D. Gutiérrez, J. Gutiérrez, and V.J. Monserrat. 2007. An elevational shift in butterfly species richness and composition accompanying recent climate change. *Global Change Biology* 13:1873–87.

15 Using Georeferenced Databases to Assess the Effect of Climate Change on Alpine Plant Species and Diversity

Christophe F. Randin, Robin Engler, Peter B. Pearman, Pascal Vittoz, and Antoine Guisan

CONTENTS

Introduction 150
Climate Change in Mountain Environment (Physical Aspects) 150
Climate Change and Alpine Plant Species Distribution 150
Using Predictive Models to Assess Climate Change Impact on Plant Distribution 150
Simulating Plant Migrations in Response to Climate Change 151
Importance of Spatial Biological Databases for Predicting Future Plant Distributions 151
Aim of the Study 151
Methods 151
Study Area 151
Species Occurrence Data 152
Plant Traits Database 152
Environmental Data 152
Climate Change Scenarios 153
Statistical Modeling 153
Spatial Projections 153
Dispersal Modeling (MIGCLIM Cellular Automaton) 154
Turnover Maps—Identifying Sensitive Areas in the Landscape 155
Analyses by Species Traits 155
Results 155
Model Evaluation 155
Retained Predictors and Their Contributions 155
Turnover Maps 155
Predicted Distributions and Species Traits 155
Discussion 158
Georeferenced Biological Databases, Models, and Conservation 158
Predicting Species Distributions in a Mountain Landscape Now Requires Better GIS Environmental Predictors 159
SDMs Complementing Monitoring Databases 160
Conclusions 160
Summary 160
Acknowledgments 160
References 161

INTRODUCTION

Climate Change in Mountain Environment (Physical Aspects)

During the last two decades, evidence of global warming has accumulated, with a + 0.76°C temperature increase on average between 1850–1899 and 2001–2005 for the Earth (0.57°C to 0.95°C; IPCC, 2007). There is now very high confidence that this warming results from the parallel increase in carbon dioxide from a preindustrial value of 280 ppm to 379 ppm in 2005, at a rate that is one hundred times faster than that which dominated over the last 20,000 years (IPCC, 2007). A recent global projection of surface temperature changes in mountainous areas concluded that these areas will likely warm during the twenty-first century at a rate two or three times higher than that recorded during the twentieth century (Nogues-Bravo et al., 2007). Switzerland has already experienced warming of + 1.35°C during the twentieth century, an increase that is twice the observed temperature increase averaged over the entire northern hemisphere (Rebetez and Reinhard, 2008). During the same period, annual mean precipitation has not changed significantly. However, extreme weather events have become significantly more frequent, with long periods of either continued precipitation or drought becoming more common, while snow cover has declined at elevations < 1,300 m (Rebetez, 2002). Thus, climatic change in Switzerland is emblematic of climate change occurring in mountain regions worldwide.

Climate Change and Alpine Plant Species Distribution

Anthropogenic climatic change will certainly affect high-elevation vegetation (Beniston et al., 1996; Grabherr et al., 1995; Guisan and Theurillat, 2000a; Theurillat and Guisan, 2001; Walther, 2003). Climate warming will induce plant migration toward higher altitudes and northern latitudes. Upward shifts of the treeline ecotone since the end of the Little Ice Age (e.g., Camarero and Gutierrez 2004; Gehrig-Fasel et al., 2007; Kullman, 2001; Vittoz et al., 2008) and continuing increases in species richness in alpine and subnival vegetation in the Alps (e.g., Braun-Blanquet, 1957; Braun-Blanquet, 1975; Grabherr et al., 1994a; Hofer, 1992; Pauli et al., 2007; Vittoz et al., 2006; Walther et al., 2005a) corroborate that these shifts are already occurring. It also has been suggested that the increasing species number on high mountain summits could be the result of a natural dispersal process that was triggered by the temperature increase at the end of the Little Ice Age (Kammer et al., 2007). Nevertheless, anthropogenic climate change during the twentieth century has induced biological changes and accelerated the trend in the upward shift of alpine plants over the last twenty years (Walther et al., 2005b), and further changes are anticipated. In this chapter, we demonstrate that data stored in georeferenced biological databases can prove particularly useful in this respect.

Using Predictive Models to Assess Climate Change Impact on Plant Distribution

Species distribution models (SDM; Guisan and Thuiller, 2005; Guisan and Zimmermann, 2000) have in the last decade become important tools to estimate the impacts of climate change on plant distributions (Bakkenes et al., 2002; Thomas et al., 2004; Thuiller et al., 2005). SDMs are empirical models that statistically relate species observations with environmental variation (such as climatic and topographic variables) and can then project this relationship into geographical space yielding predictions of habitat suitability. The result of these projections can then typically be displayed within a Geographic Information System (GIS). These models can be developed to their full potential for assessing climate change impacts on biodiversity when many high-precision occurrence records of a species are available, such as are increasingly available from large biodiversity databases.

Species distribution models at a variety of scales have been used to study the potential effects of climate change. At the extent of Western Europe, Thuiller et al. (2005) model the distribution of suitable habitat for 1,350 plant species using data from the Atlas Florae Europaeae within 10 min grid cells (AFE; Lahti and Lampinen, 1999). The authors forecast that plant diversity in some European mountain ranges (e.g., at mid-elevations in the Alps) could be highly sensitive to climate change, with loss of up to 60% of species per mapping cell. Similarly, Dullinger et al. (2003) present a study of eighty-five subalpine and alpine herbacious plants of open habitat that was conducted at high resolution (20 × 20 m) in the Austrian Alps. The authors predict that up to 40 to 50% of the species could potentially disappear from certain areas owing to climate change. Finally, in a study of sixty-two alpine and nival plants, Guisan and Theurillat (2000a) predict relatively low rates (2 to 5%) of complete loss of suitable habitat, but for nearly 40% of the species they predict habitat losses of more than 90%. These results suggest that alpine and nival plants are at high risk of disappearing regionally under future climatic conditions because of habitat loss (Guisan and Theurillat, 2000a).

Simulating Plant Migrations in Response to Climate Change

The predicted warming owing to climate change will require considerable migration rates for plant species to stay under similar climatic conditions (e.g., Malcolm et al., 2002). However, so far, few studies have considered dispersal limitations and its influence on projections of species potential distribution under climate change scenarios (Broennimann et al., 2006; Carey, 1996; Carey and Brown, 1994; Dullinger et al., 2004; Iverson and Prasad, 1998; Iverson et al., 2004). Instead, most studies have used the assumption of a "universal dispersal" scenario (i.e., a species has unlimited dispersal, its future distribution becoming the entire area projected by the habitat distribution model; Bakkenes et al., 2002; Dirnböck et al., 2003; Guisan and Theurillat, 2000a; Thomas et al., 2004; Thuiller et al., 2005). This assumption likely leads to overestimation of the future suitable potential habitat distribution for many species for two main reasons. First, the landscape has become increasingly impassable owing to human land use and transportation corridors (Pitelka et al., 1997). Second, species with limited dispersal potential might not be able to follow the shift of their suitable habitats. Since universal dispersal represents an optimistic, best-case scenario, some authors (e.g., Thomas et al., 2004; Thuiller et al., 2004a; Thuiller et al., 2005) also provide for comparison a worst-case scenario, in which dispersal is considered as null. However, the differences in term of future species distributions between these two scenarios are generally great, yielding large uncertainty in the forecasts overall (e.g., Thuiller 2004). Reducing such uncertainty calls for the inclusion of seed dispersal processes when addressing the issue of climate change impact on plant distribution, which in turn requires data on species dispersal distances and rates.

Importance of Spatial Biological Databases for Predicting Future Plant Distributions

Data for calibrating predictive models of species distributions can contain information on both species presence and absence, as is derived from vegetation surveys where plots are exhaustively inventoried, or only occurrence data (i.e., presences only) arising, for instance, from georeferenced natural history collections (Graham et al., 2004; Guisan and Thuiller, 2005). Both types of data can yield models with satisfactory predictive capacity. When absence data are not available, randomly distributed "pseudo-absence" data can be used to overcome the problem of missing absences and in this way allow use of novel group-discrimination modeling techniques (Elith et al., 2006). Therefore, the availability of data on species occurrence determines whether a modeling assessment can be performed in a given study area. With increasing availability of georeferenced databases of species occurrences, data-rich assessments will be possible that should increase the reliability of predictions made with species distribution models.

Aim of the Study

In this chapter, we present a case study from the western Swiss Alps to illustrate the use of a georeferenced biodiversity database and additional species traits databases for assessing the impact of climate change on alpine plant distributions and diversity. We test the empirically based hypothesis that the future distribution of high elevation plant species is highly susceptible to climate change. We show how complementary data on plant functional traits can help to identify which areas and species are the most threatened by climate change. We also use seed dispersal data to model dynamic spatial processes by simulating plant migrations in time and space. We thus provide a more realistic estimate of extinction rates compared to projections based on the extreme assumptions of unlimited or no dispersal. We generalize our results to the study flora by running simulations for nearly 300 species. Finally, we also show how GIS land cover data can provide additional information on habitat suitability so as to help to identify corridors or barriers to seed dispersal.

METHODS

Study Area

The Diablerets study area (Figure 15.1) covers nearly all mountain massifs of the western Alps of the Canton of Vaud (Swiss state; 46°10′ to 46°30′ N; 6°50′ to 7°15′ E; > 700 km^2). The elevation ranges from 375 m in Montreux to 3,210 m on the top of the Diablerets massif. The annual mean temperature and total precipitation vary respectively from 8°C and 1,200 mm at 600 m elevation to –5°C and 2,600 mm at 3,000 m elevation (Bouët, 1985).

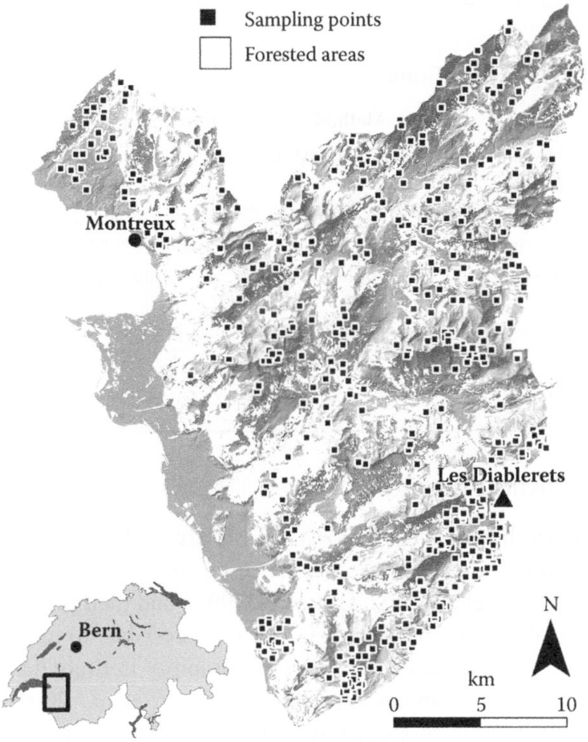

FIGURE 15.1 Diablerets study area and its location in Switzerland.

Species Occurrence Data

Species distributions data were obtained from single occurrences and vegetation plots stored in the Swiss Floristic Center (CRSF; http://www.crsf.ch) database in Geneva. For this study, we selected data from 550 vegetation plots restricted to open, nonwoody vegetation (Figure 15.1; grassland, rock, and scree vegetation; see Randin et al., 2006).

Plant Traits Database

It was not possible to obtain highly accurate data on the dispersal behavior of each species. Indeed, such data does not exist for the vast majority of the 284 species considered here. To overcome this deficit, we classified our species into the seven categories defined by Vittoz and Engler (2007) using expert knowledge and the dispersal vectors data of Müller-Schneider (1986). This procedure allowed for each species the derivation of dispersal distances with an accuracy of one order of magnitude. This is sufficient to achieve substantial improvements over models that ignore dispersal (Engler and Guisan, in revision).

TABLE 15.1
Maximum Dispersal Distance for the 50% and 99% of Seeds and Long Distance Dispersal (LDD) Events for Each Dispersal Category (1–7)

	Dispersal Distance (m)		
Type	50%	99%	LDD
1	0.1	1	2
2	1	5	20
3	2	15	200
4	40	150	1000
5	10	500	2000
6	400	1500	10000
7	500	5000	10000

Source: As defined in Vittoz and Engler (2007).

The decrease in the seed dispersal shadow and resulting colonization probability that occurs with increasing distance from a seed source was represented with negative exponential distributions that were adapted to the seed dispersal distances of species (Table 15.1). This is a common seed dispersal kernel shape (Willson, 1993), although many others exist (e.g., Clark et al., 1999; Greene et al., 2004; Nathan and Muller-Landau, 2000). Long-distance dispersal (LDD) events that allow a plant to randomly disperse at distances larger than its dispersal kernel were also assigned to species in each category (Table 15.1) and modeled as a separate process.

Environmental Data

Climate data were derived from the Swiss national meteorological station network. Monthly means for average temperature (°C) and sum of precipitation (mm) for the period 1961–1990 were used (MeteoSwiss; http://www.meteosuisse.admin.ch/web/en/climate/consultation_service.html). A 25 m digital elevation model (DEM; Swisstopo; http://www.swisstopo.admin.ch/internet/swisstopo/fr/home/products/height/dhm25.html) was used to spatially interpolate the climatic data following Zimmermann and Kienast (1999). In the current study, we employed three climatic and two topographic variables at a 25 m resolution (Table 15.2). These variables are of substantial ecophysiological significance (Körner, 2003; Pearson et al., 2002). Temperature degree-days were derived from interpolated monthly mean temperatures using a threshold value of 0° and a daily time step. Moisture index was calculated as the difference between precipitation and potential evapotranspiration, and thus

TABLE 15.2
Topographic and Climatic Variables Used to Model Species Distributions

Variables	Units	Details	Method	References
Temperature degree days	°C * day * year^{-1}	Sum of days mulitplied by daily meantemperature >0°C	ArcInfo AML	Zimmermann and Kienast (1999)
Moisture index (average of monthly values June–August)	mm * day^{-1}	Monthly average of daily water balance (precipitation-evapotranspiration)	ArcInfo AML	Zimmermann and Kienast (1999)
Global solar radiation (sum over the year)	kJ * m^{-2} * year^{-1}	Daily global solar radiation	ArcInfo AML	Zimmermann et al. (2007)
Slope	degrees	Slope inclination	DEM, ArcInfo, GRID routine	ESRI (2005)
Topographic position	Unitless	Concave vs. Convex land surface	ArcInfo AML	Zimmermann et al. (2007)

expressed the amount of soil water potentially available at a site. The sum of mean daily values over the growing season (June, July, and August) was used and represented the warmest three months of the year. The sum of the absolute potential global radiation was calculated over the entire year. This provided, in average, a better predictive power for our set of species than when it was calculated only over the growing season. Indeed, solar radiation controls other yearly processes, such as snow distribution. Topographic position and slope were derived from the 25 m resolution DEM. Positive values of topographic position express convex topographies (ridges, peaks, and exposed sites), whereas negative values indicate concave surfaces (valley bottoms or lower end of slopes).

Climate Change Scenarios

We used four different climate projections for the 2001–2100 time period developed by the U.K. Hadley Center for Climate Prediction and Research (HadCM3; Carson, 1999). These projections arose from four socioeconomic scenarios of the Intergovernmental Panel on Climate Change, A1FI, A2, B1, and B2 (IPCC; Nakicenovic and Swart, 2000).

We used the 10′ HadCM3 climatic grids to calculate monthly mean anomalies for our study area between the standard period 1961 to 1990 and twenty future time periods of five-year intervals from 2001 to 2100 for temperature and precipitation. These anomalies were then downscaled to 25 m resolution grids using a bilinear interpolation algorithm before being added to the values for the study area at present climatic conditions. With an average increase of + 7.6 K in our study area by 2100, the A1FI climate change scenario is the most extreme. B1 is mildest (+ 3.9 K); A2 (+ 6.2 K) and B2 (+ 4.4 K) are intermediate.

Statistical Modeling

Generalized Linear Models (GLM) are so far the most common modeling technique used in ecology (e.g., Bakkenes et al., 2002; Hill et al., 1999), and their predictive ability has been shown to be more resistant to transferability in space (and thus transferability in time) than other more complex algorithms (Randin et al., 2006). Models for each species were fitted using presence and absence data and GLMs with a binomial variance and a logistic link function (McCullagh and Nelder, 1989) in R version 2.6.1 (R Development Core Team, 2007). A stepwise procedure in both directions was used to select the predictors based on the Akaike Information Criterion (AIC; Akaike, 1973). Up to second-order polynomials (linear and quadratic terms) were allowed for each predictor, with the linear term being forced in the model each time the quadratic term was retained.

The predictive ability of GLMs was evaluated with a pseudoindependent internal evaluation by running a tenfold cross-validation evaluation (van Houwelingen and Le Cessie, 1990) on the training data set. The original prevalence of the species presences and absences in the data set was maintained in each fold of the cross-validation procedure. Comparisons of predicted (probabilistic scale) and observed (presence–absence) values were based on the area under the curve (AUC) of the receiver-operating characteristic plot (ROC; Fielding and Bell, 1997).

Spatial Projections

Binary presence–absence distributions were derived from the probability values of the GLM by maximizing the percentage of presences and absences correctly predicted under current climatic conditions (Pearce and Ferrier, 2000; Thuiller, 2003). A mask layer based on forests, lakes, urbanized areas, roads, and rivers also was applied to avoid spurious projections at locations that were unsuitable regardless of climate. This spatial layer is based on the VECTOR25 product of the Swiss Federal Office of Topography (Swisstopo; http://www.swisstopo.admin.ch/internet/swisstopo/en/home/products/landscape/vector25.html, last accessed on Aug. 24, 2008) transformed into a 25 m grid. In addition to this general filter, empirical observations from the sampling plots were used to associate each species with one or several of three categories of open vegetation: grassland, rock, and scree. Species were then excluded from categories where they were never observed in the calibration data set.

Dispersal Modeling (MIGCLIM Cellular Automaton)

MIGCLIM (Engler and Guisan, 2009) is a cellular automaton that allows simulating the dispersal of plant species across a landscape, while implementing a climate change scenario at the same time (see Figure 15.2 for a simplified example). Different parameters, such as seed dispersal distance and kernel shape, generation time, LDD events, or barriers to dispersal, can be specified and customized to best fit the dispersal behavior of each species. In our simulations, we specified values for dispersal distance, generation time (i.e., time required for a new established population within a pixel until it can act itself as source for seed dispersal), LDD events, and landscape fragmentation to increase realism in species dispersal behavior (Figure 15.2). Landscape fragmentation was represented by forested areas that acted as "barrier" pixels that impede dispersal through them (except LDD events) and urbanized areas, roads, rivers, and lakes that acted as "filter" pixels that represent permanently unsuitable surfaces but do not act as barriers to dispersal (Figure 15.2).

For each of our 284 species, simulations were carried out over a one hundred year period (2001–2100) and under each of four IPCC climate change scenarios (i.e., A1FI, A2, B1, B2). The initial distribution of each species

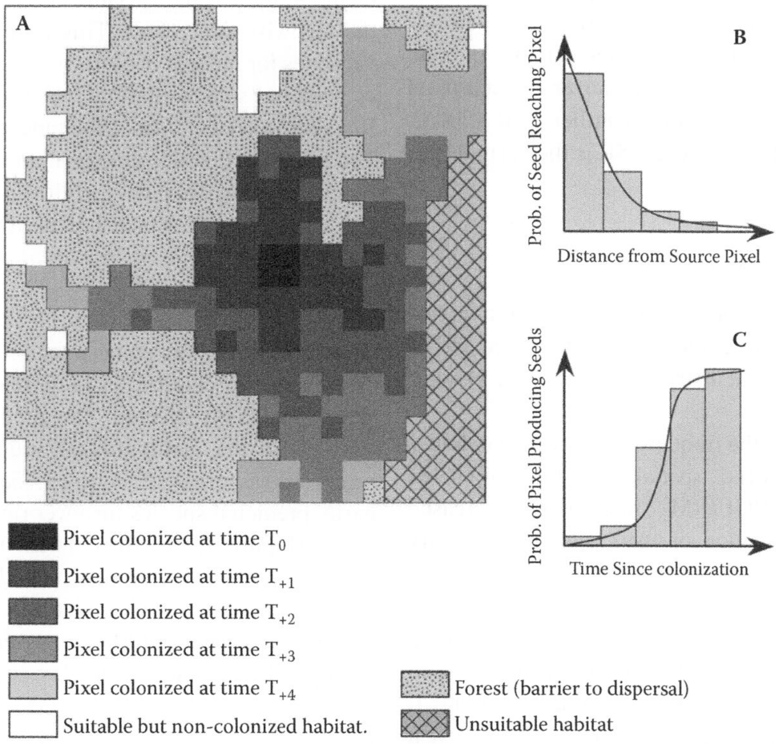

FIGURE 15.2 Example of dispersal simulation using the MIGCLIM model (A). Black pixels represent pixels occupied at T_0 of simulation, and the different shades of gray those that get colonized at times T_{+1}, T_{+2}, T_{+3}, and T_{+4}. Note that not all suitable pixels become colonized because they are either too far from the initial distribution or dispersal is impeded by forested areas that here act as barriers to dispersal. The probability of a pixel to become colonized depends upon its distance from all available source cells (B) as well as the time since those source pixels were colonized (C). Because of the discrete nature of cellular automata, these functions have to be approximated.

was defined as its suitable habitat distribution under current climatic conditions, i.e., 1961–1990 average. Change in suitable habitat distribution reflecting climate change was implemented every fifth year. Dispersal was simulated each year, except for type 1 and type 2 categories (Table 15.1), which have a too short dispersal distance to be modeled on a yearly basis, even on a 5 m resolution grid. Consequently, dispersal distances of these categories were multiplied by five, and dispersal was simulated every fifth year (this assumes a conservative "best case" approximation where spread rate = generation time * dispersal distance). Finally, because they involve random processes, each simulation was repeated five times.

Turnover Maps—Identifying Sensitive Areas in the Landscape

We summed by pixel the number of species disappearing (L) and related it to current species richness in a pixel in order to evaluate the percentage of species lost. We similarly assessed the percentage of species gained (G) per pixel when propagules of a formerly absent species successfully colonized a pixel within suitable climate space. The percentage of species turnover in each pixel then was computed as defined by Thuiller et al. (2005): $T = 100 \times (L + G)/(SR + G)$, where SR is the current species richness.

Analyses by Species Traits

We assessed the relationship between increase and decrease in a species' distribution by 2100 (expressed as a percentage of its current distribution) and three biological traits.

First, we examined the frequency with which colonizable habitat of species (i.e., spatial projections of GLMs colonized in the MIGCLIM simulations) increased or decreased as a function of their dispersal capacity (Table 15.1). Second, we tested whether persistence of colonizable habitat was related to elevation optima of species. For this we derived an index of elevational optimum for each of the 284 species, based on data from Flora Alpina (Aeschimann et al., 2005). A value (0, absent; 1, not frequent; 2, commonly present) was assigned to each elevation belt (1, collinean; 2, montane; 3, subalpine; 4, alpine; 5, nival) corresponding to the species' observed frequency in the belt. The index was then calculated as a weighted average of these values across all elevation belts between 1 for strictly collinean species to 5 for strictly nival species. Finally, we tested whether the future colonizable habitat of species is related to their growth form. The species were classified into the seven growth form categories defined in the Flora Alpina (Aeschimann et al., 2005): annuals, bulbous/rhizomatous, grass, forb, cushion, prostrate shrub, and nanophanerophytes. We then examined the frequency with which future colonizable habitat of species increased or decreased within each category of growth forms (Figure 15.3).

RESULTS

Model Evaluation

The internal cross-evaluation showed that 83% of the species GLM models attained AUC evaluation values > 0.8 and no model failed (AUC < 0.6). Thus, we obtained useful models for all species (Swets, 1988; Araújo et al., 2005).

Retained Predictors and Their Contributions

Among all predictors and across all species, temperature degree-days was the variable most frequently selected by the stepwise procedure. This procedure selected more variables for models of alpine species, with the exception of potential solar radiation, which was marginally more frequently selected for subalpine plants (Figure 15.4). The largest difference among alpine and other plants was observed in the frequency with which topographic position was selected for inclusion in SDMs.

Turnover Maps

Species turnover varied between 63 and 100% under the A1FI scenario (Figure 15.5a). Areas above the treeline were especially affected. We predicted that the summits of all the mountains we studied will experience nearly 100% turnover (Figure 15.5a). Under the B1 scenario, predicted species turnover ranged between 45 and 100% (Figure 15.5b). In contrast to predictions under the A1FI scenario, summits of mid-elevation mountain ranges showed areas with relatively low turnover values. Nevertheless, nearly 100% turnover was still predicted for high-elevation summits.

Predicted Distributions and Species Traits

A majority of the species lose suitable habitat by the end of the twenty-first century under the A1FI scenario (254 species). Moreover, most of these species (219 species, 77%) lose between 80 and 100% of their suitable habitat (Figure 15.6a) and will risk disappearing from the study

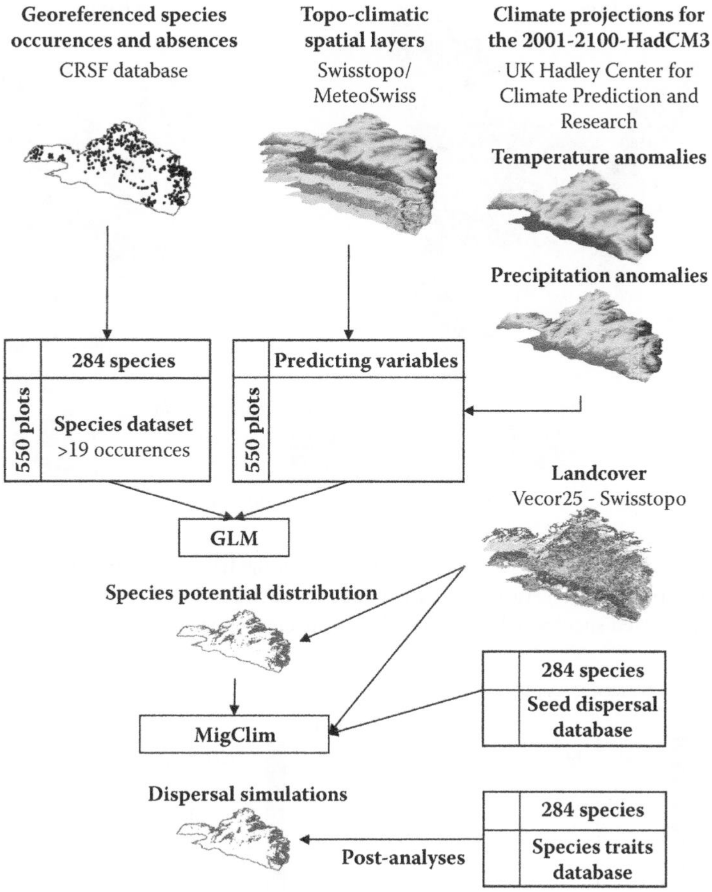

FIGURE 15.3 Conceptual framework showing the different databases used in the study.

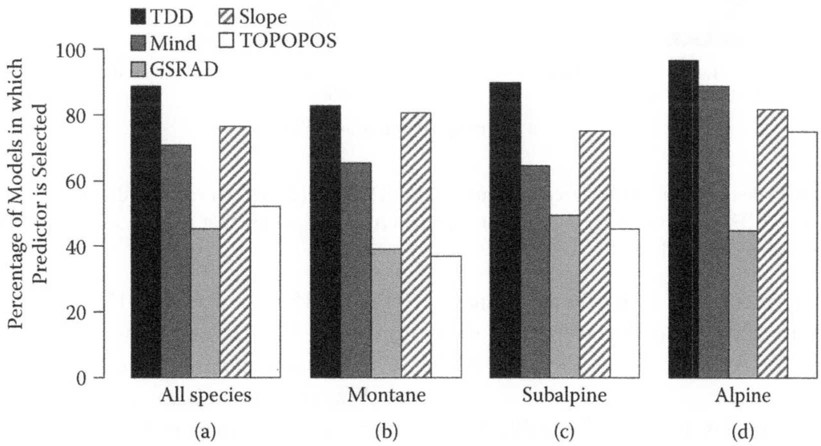

FIGURE 15.4 Percentage of times predictors were selected by stepwise for all species (a) and for species classified by their altitudinal optimum (b, c, and d). TDD, temperature degree-days; MIND, moisture index; GSRAD, potential global solar radiation; SLOPE, slope; TOPOPOS, topographic position (see "Methods" for description of the variables).

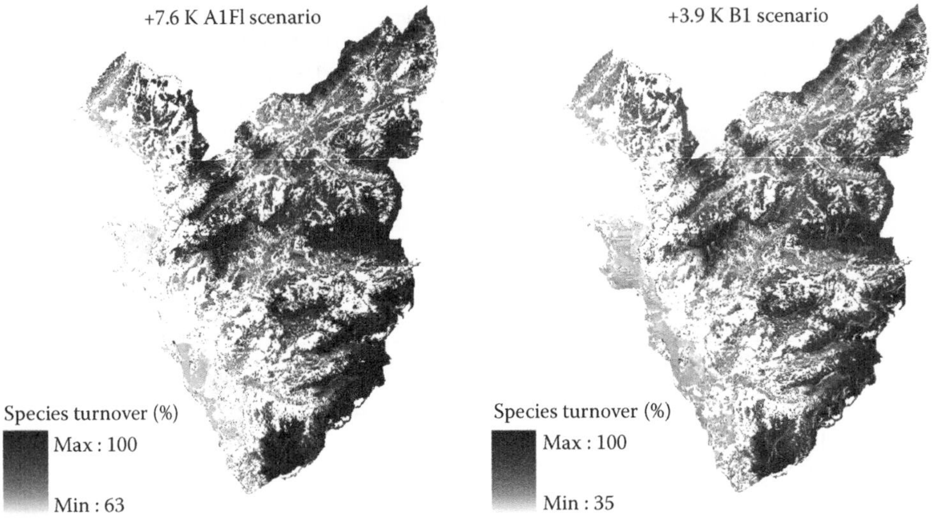

FIGURE 15.5 Species turnover under the A1F1 scenario (left) and under the B1 scenario (right). Forested areas and areas unsuitable for plant growth have been removed. The turnover is calculated for each pixel as $T = 100 \times$ (number of species lost + number of species gained) / (current predicted species richness + number of species gained).

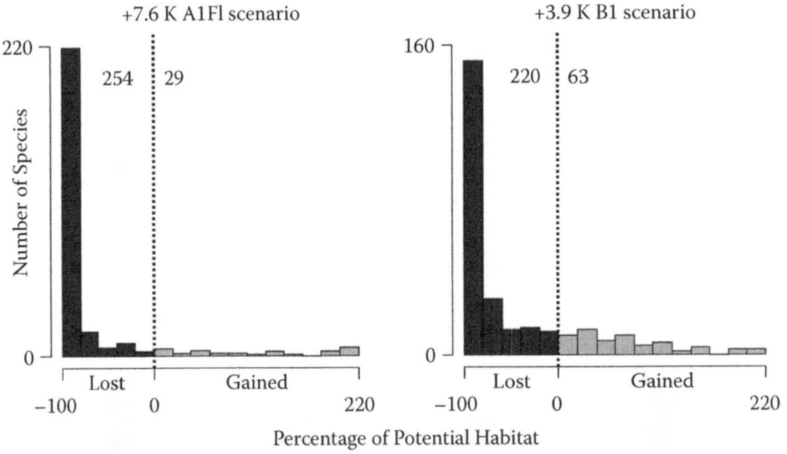

FIGURE 15.6 Frequency distribution of the percentage of potential habitat (20% intervals) predicted as lost or gained for the 284 species in 2100 under the A1FI scenario (left) and the B1 scenario (right). Species that gained > 200% of their current potential habitat are pooled in the same interval.

region. A similar pattern emerges under the B1 scenario by the year 2100, with 220 species losing part of their suitable habitat and 160 species (54%) losing > 80% (Figure 15.6b).

The proportion of species losing colonizable habitat was particularly high for short distance dispersal species (Figure 15.7a and Figure 15.7b), alpine species (Figure 15.7c and Figure 15.7d), and for species with growth forms that are common at high elevations (Figure 15.7e and Figure 15.7f), such as cushion plants, prostrate shrubs, and nanophanerophytes. This pattern was the same under both scenarios A1FI and B1.

We selected *Carex ornithopoda* to illustrate the importance of dispersal constraints and land cover in SDMs. *C. ornithopoda* is mainly found in dry meadows and has a large altitudinal distribution, from the collinean to the alpine belt. As an ant-dispersed species, (Müller-Schneider, 1986), 99% of the seeds are dispersed

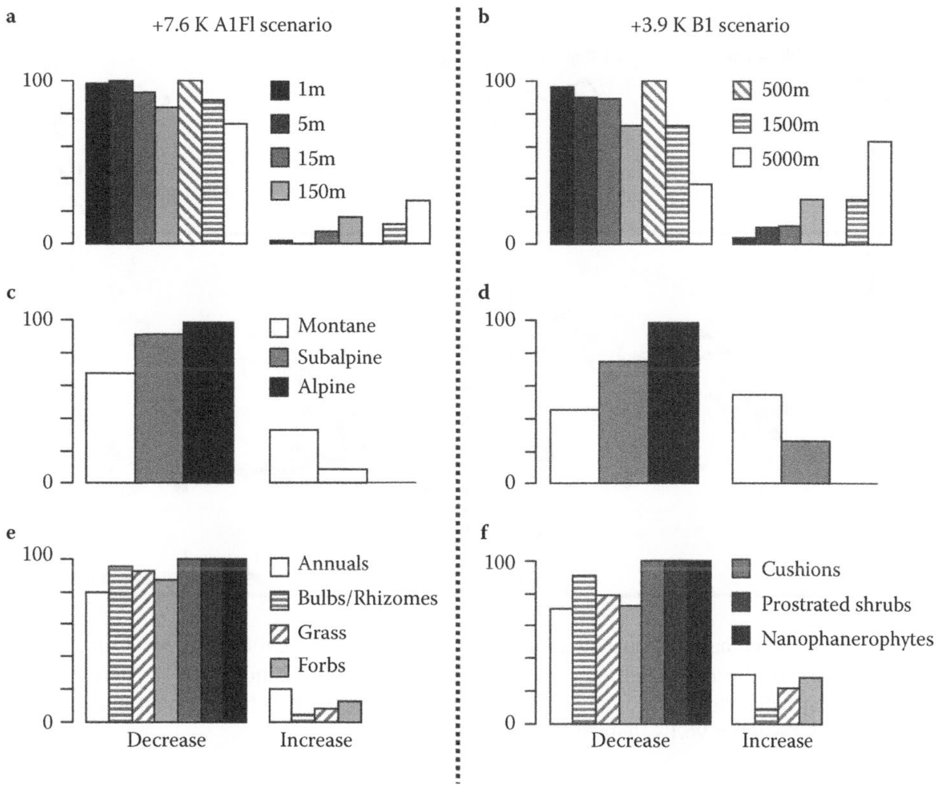

FIGURE 15.7 Percentage of species predicted to experience a decrease or increase in suitable habitat in 2100, classified by dispersal distances (a and b; values are the upper limits of the distances within which 99% of seeds are dispersed), elevation optima (c and d), and growing forms (e and f) under the A1FI scenario (a, c, e) and B1 scenario (b, d, f).

less than 15 m. Under the A1FI scenario, this species lost 93.4% of its current potential distribution. Most of the area occupied by the species in 2100 was newly colonized (Figure 15.8). Some currently suitable areas remained occupied as of 2100 at mid-elevation under the B1 scenario (Figure 15.8). In addition, the slower rate of temperature increase of the B1 scenario allowed this species to colonize higher alpine meadows (Figure 15.8). However, the highest topo-climatically suitable surfaces were primarily composed of rock and scree, thus limiting the future distribution of the species to the rare summits with appropriate edaphic conditions (Figure 15.8).

DISCUSSION

With this case study, we illustrated how observations from georeferenced biological databases can be used to derive climate change impact scenarios in a mountain landscape. Our results suggest that the flora of the western Swiss Alps is sensitive to climate change, with most alpine and nival species losing part or all of their suitable habitat by 2100. This pattern was predicted to remain the same under both the most (A1FI) and the least (B1) extreme IPCC socioeconomic scenarios. Species on high-elevation summits and alpine species in general appeared to be the most threatened.

Georeferenced Biological Databases, Models, and Conservation

SDM maps are often used as support to produce reports for conservation planners, managers, and other decision makers to anticipate biodiversity losses in alpine and other ecosystems across local and regional scales. Although models based on the four IPCC climate change scenarios consistently predict substantial impacts on plant diversity in high-elevation areas, the rates of loss of suitable habitat on mid-elevation summits vary. This is potentially important because mid-elevation summits may act as ecological traps for species (ecological trap, *sensu* Delibes et al., 2001; Remes, 2000) or as high-elevation refugia (high-elevation refugia hypothesis, *sensu* Guisan and Theurillat, 2000b). This demonstrates the

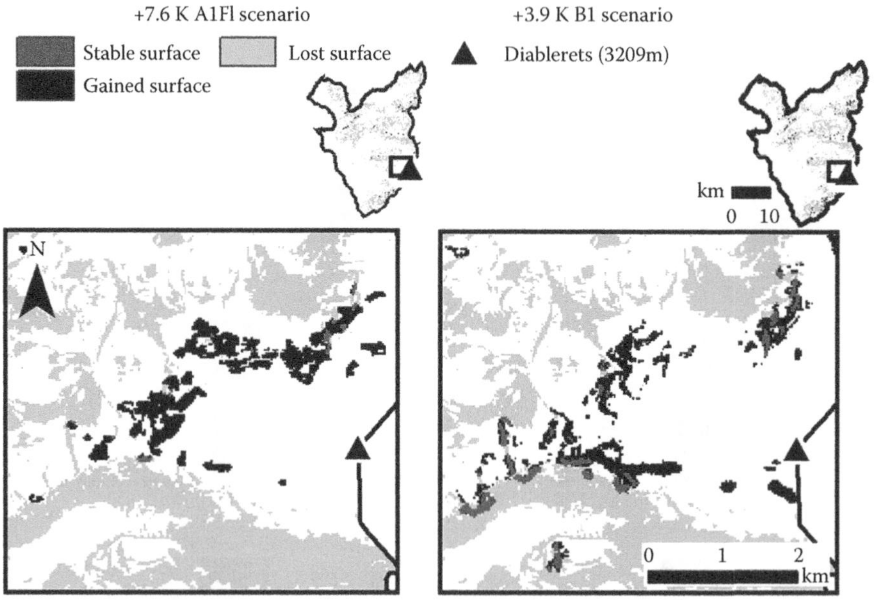

FIGURE 15.8 Change of potential habitat distribution between the year 2000 and 2100 for *Carex ornithopoda* under the A1FI scenario (left) and B1 scenario (right), and closer views of the highest summit of the study area (Diablerets, 3,209 m). A1FI: percentage of the current suitable habitat for surface stable = 0.6%, lost = 93.7%, gained = 5.6%. B1: percentage of the current suitable habitat for surface stable = 3.2%, lost = 96.1%, gained = 6.4%.

importance of sophisticated dispersal simulations based jointly on accurate databases of seed characteristics and species occurrences. The development of species occurrence databases is particularly important because the quality of SDMs depends on the abundance and quality of the original species occurrence data. Large data sets can decrease the influence bias in nonrandomly sampled records, which are the majority of the existing data in these databases, and accurate coordinates improve SDMs when linked with high resolution topoclimatic predictors. In this context, the high predictive power of our models (83% of species with AUC > 0.8) can largely be explained by the quality of the species occurrence database. The observations covered the entire elevation range of the species and thus decreased the chance of truncated response curves of species occurrence along gradients of the predictor variables (Thuiller et al., 2004b). Moreover, species occurrences have low spatial autocorrelation (mean Moran's I = 0.2), thus decreasing risk of bias during predictor selection (Dormann et al., 2007).

In our study, spatial projections were based on only one modeling technique. Although GLM is a robust technique when evaluated in a remote area (Randin et al., 2006), even more robust forecasts may be achieved with ensemble forecasts (Araujo and New, 2007).

Predicting Species Distributions in a Mountain Landscape Now Requires Better GIS Environmental Predictors

Our models allowed identifying the most important environmental drivers of plant species along an elevation gradient. Variables reflecting temperature, humidity, and curvature were more often selected for alpine species. This is not surprising, because the influence of topography on vegetation becomes more pronounced as elevation increases. This is because climatic variability is stronger and because plant life becomes more dependent on convenient microrelief to survive (Körner, 2003). In addition, a recent study by Lassueur et al. (2006) recommended use of topographic position (curvature) at high resolution. These results indicate that the use of very high resolution digital elevation models (VHR DEMs) to derivate accurate topographic predictors should be advantageous in alpine environments. Although the spatial variables we used in our models provided a good predictive power on average, we recommend the use of high resolution DEMs in combination with other DEMs at lower resolution (e.g., the 25 m resolution data set used in this study) for studies in the alpine zone. Indeed, slope and topographic position are likely to act over different measurement scales. Slope reflects large-scale geomorphic processes such as avalanches, whereas topographic position reflects water, nutrients, and snow accumulation or wind exposure that can influence species distributions at a much finer scale.

We showed that information on land cover and species dispersal constraints influenced the prediction results. For example, our simulations predicted that *Carex ornithopoda* would potentially subsist in 1.3% of its current predicted surface under A1FI (9.1% under B1). Without dispersal constraints and land cover, this species would have maintained 14.2% of its current potential distribution under A1FI (39.6% under B1). In the future, occupation by the species of larger areas than we predicted is likely owing to soil formation, but new suitable soils for *C. ornithopoda* will certainly cover only a small fraction of these high-elevation areas. Nonetheless, other land cover types resulting from human land use (Randin et al., in revision-a) and underlying geomorphologic layers (Randin et al., in revision-b) will have to be taken into account at the local scale, where they may act as barriers or corridors for plant dispersal (Butler, 2001; Dullinger et al., 2004).

SDMs Complementing Monitoring Databases

Numerous monitoring studies have assessed the impact of climate change on alpine and nival vegetation (Grabherr et al., 1994b; Pauli et al., 1996; Vittoz et al., 2006; Walther et al., 2005a). However, few studies have focused on changes within grasslands at lower altitudes (Price and Waser, 2000; Vittoz et al., 2009), although these cover a much larger area than alpine and nival belt in the Alps (Guisan and Theurillat, 2001). In order to fill this gap, new monitoring networks have already been set up to complement existing data and to produce a more precise record of climate change impacts on plant distribution and vegetation. Recent projects, such as PERMANENT.PLOT.CH (Vittoz and Guisan, 2003), GLORIA (Pauli et al., 2004), or MIREN (Dietz et al., 2006), are examples of such initiatives. Nonetheless, these monitoring studies will not allow projections of plant distributions at large spatial scales (e.g., an entire mountain range). Long-term monitoring projects, longer than a human lifetime, will be required to observe higher-level biological effects, such as those on ecosystem composition, structure, and function.

CONCLUSIONS

Large databases of geographically referenced species occurrences, traits, and environmental variation facilitate modeling of the effects of climatic change on plant distributions in mountainous landscapes. Such model–database links constitute an appropriate framework for exploring the environmental drivers of plant distribution and diversity. Our results illustrate the potential for an alpine flora to display sensitivity to climatic change. Projections of future losses of species from focal areas will assist their management and the conservation of biological diversity.

SUMMARY

Past and current climate change has already induced drastic biological changes. We need projections of how future climate change will further impact biological systems. Modeling is one approach to forecast future ecological impacts, but requires data for model parameterization. As collecting new data is costly, an alternative is to use the increasingly available georeferenced species occurrence and natural history databases. Here, we illustrate the use of such databases to assess climate change impacts on mountain flora. We show that these data can be used effectively to derive dynamic impact scenarios, suggesting upward migration of many species and possible regional disappearance when no suitable habitat is available at higher elevations. Systematically georeferencing all existing natural history collections data in mountain regions could allow a larger assessment of climate change impact on mountain ecosystems in Europe and elsewhere.

ACKNOWLEDGMENTS

We thank the Center of the Swiss Floristic Network (CRSF) in Geneva for providing access to the species occurrence database, the Centre de Conservation de la Nature (SFFN) of the Canton de Vaud for support, and the Swiss National Science Foundation (SNF Grant No 3100A0-110000) and the European Science Commission (FP6 projects ECOCHANGE and MACIS) for funding. We also thank Stefanie Maire, Dario Martinoni, Séverine Wydler, Chantal Peverelli, Valentine Hof, Grégoire Vuissoz, Lorenzo De Stefani, Roxane Milleret, Pierre Bermane Favrod-Coune, Professor Daniel Cherix, and Anny and Willy Berra for field assistance. We finally thank Dr. Miriam Duehnforth for her help on the manuscript. C.R. and A.G. benefited from interactions with colleagues within the Swiss National Centre for Competence in Research (NCCR) "Plant survival in natural and agricultural ecosystems" (http://www.unine.ch/nccr).

REFERENCES

Aeschimann, D., K. Lauber, D.M. Moser, and J.-P. Theurillat. 2005. *Flora Alpina*. Paris: Belin.

Akaike, H. 1973. Maximum likelihood identification of Gaussian autoregressive moving average models. *Biometrika* 60:255–65.

Araujo, M.B., and M. New. 2007. Ensemble forecasting of species distributions. *Trends in Ecology & Evolution* 22:42–7.

Bakkenes, M., J.R.M. Alkemade, F. Ihle, R. Leemans, and J.B. Latour. 2002. Assessing effects of forecasted climate change on the diversity and distribution of European higher plants for 2050. *Global Change Biology* 8:390–407.

Beniston, M., D.G. Fox, S. Adhikary, R. Andressen, A. Guisan, J.I. Holten, J. Innes, J. Maitima, M.F. Price, and L. Tessier. 1996. Impacts of climate change on mountain regions. In *Intergovernmental Panel on Climate Change: Second Assessment Report*. IPCC, ed. Cambridge: Cambridge University Press.

Bouët, M. 1985. *Climat et météorologie de la Suisse romande*. Lausanne: Payot.

Braun-Blanquet, J. 1957. Ein Jahrhundert Florenwandel am Piz Linard (3414 m). *Bulletin du Jardin Botanique de l'Etat á Bruxelles*. volume jubilaire Walter Robyns: 221–32.

Braun-Blanquet, J. 1975. Fragmenta Phytosociologica Raetica I: Die Schneebodengesellschaften (Klasse der Salicetea herbaceae). *Jahresbericht der Naturforschenden Gesesellschaft Graubünden* 96:42–71.

Broennimann, O., W. Thuiller, G. Hughes, G.F. Midgley, J.R.M. Alkemade, and A. Guisan. 2006. Do geographic distribution, niche properties and life forms explain plants' vulnerability to global change? *Global Change Biology* 12:1079–1093.

Butler, D.R. 2001. Geomorphic process-disturbance corridors: A variation on a principle of landscape ecology. *Progress in Physical Geography* 25:237–48.

Camarero, J.J., and E. Gutierrez. 2004. Pace and pattern of recent treeline dynamics: Response of ecotones to climatic variability in the Spanish Pyrenees. *Climatic Change* 63:181–200.

Carey, P.D. 1996. DISPERSE: A cellular automaton for predicting the distribution of species in a changed climate. *Global Ecology and Biogeography Letters* 5:217–26.

Carey, P.D., and N.J. Brown. 1994. The use of GIS to identify sites that will become suitable for a rare orchid, Himantoglossum hircinum L., in future changed climate. *Biodiversity Letters* 2:117–23.

Carson, D.J. 1999. Climate modelling: Achievements and prospects. *Quarterly Journal of the Royal Meteorological Society* 125:1–27.

Clark, J.S., M. Silman, R. Kem, E. Macklin, and J.H. Islamber. 1999. Seed dispersal near and far: Patterns across temperate and tropical forests. *Ecology* 80:1475–94.

Delibes, M., P. Ferreras, and P. Gaona. 2001. Attractive sinks, or how individual behavioural decisions determine source-sink dynamics. *Ecology Letters* 4:401–3.

Dietz, H., C. Kueffer, and C.G. Parks. 2006. MIREN: A new research network concerned with plant invasion into mountain areas. *Mountain Research and Development* 26:80–1.

Dirnböck, T., S. Dullinger, and G. Grabherr. 2003. A regional impact assessment of climate and land-use change on alpine vegetation. *Journal of Biogeography* 30:401–17.

Dormann, C.F., J.M. McPherson, M.B. Araujo, R. Bivand, J. Bolliger, G. Carl, R.G. Davies, A. Hirzel, W. Jetz, W.D. Kissling, I. Kuhn, R. Ohlemuller, P.R. Peres-Neto, B. Reineking, B. Schroder, F.M. Schurr, and R. Wilson. 2007. Methods to account for spatial autocorrelation in the analysis of species distributional data: A review. *Ecography* 30:609–28.

Dullinger, S., T. Dirnbock, and G. Grabherr. 2004. Modelling climate change-driven treeline shifts: Relative effects of temperature increase, dispersal and invasibility. *Journal of Ecology* 92:241–52.

Dullinger, S., T. Dirnböck, J. Greimler, and G. Grabherr. 2003. A resampling approach for evaluating effects of pasture abandonment on subalpine plant species diversity. *Journal of Vegetation Science* 14:243–52.

Elith, J., C.H. Graham, R.P. Anderson, M. Dudik, S. Ferrier, A. Guisan, R.J. Hijmans, F. Huettmann, J.R. Leathwick, A. Lehmann, J. Li, L.G. Lohmann, B.A. Loiselle, G. Manion, C. Moritz, M. Nakamura, Y. Nakazawa, J.M. Overton, A.T. Peterson, S.J. Phillips, K. Richardson, R. Scachetti-Pereira, R.E. Schapire, J. Soberon, S. Williams, M.S. Wisz, N.E. and Zimmermann. 2006. Novel methods improve prediction of species' distributions from occurrence data. *Ecography* 29:129–51.

Engler, R., and A. Guisan. 2009. MIGCLIM: Predicting plant distribution and dispersal in a changing climate. *Diversity and Distribution*. early view; doi 10.1111/j-1472-4642.2009.00566X.

Fielding, A.H., and J.F. Bell. 1997. A review of methods for the assessment of prediction errors in conservation presence–absence models. *Environmental Conservation* 24:38–49.

Gehrig-Fasel, J., A. Guisan, and N.E. Zimmermann. 2007. Tree line shifts in the Swiss Alps: Climate change or land abandonment? *Journal of Vegetation Science* 18:571–82.

Grabherr, G., M. Gottfried, A. Gruber, and H. Pauli. 1995. Patterns and current changes in alpine plant diversity. In Arctic and Alpine Biodiversity: Patterns, Causes and Ecosystem Consequences, pp. 167–181. Chapin A., and C. Körner, eds. Heidelberg, Germany: Springer.

Grabherr, G., M. Gottfried, and H. Pauli. 1994a. Climate effects on mountain plants. *Nature* 369:448.

Grabherr, G., M. Gottfried, and H. Pauli. 1994b. Climate effects on mountain plants. *Nature* 369:448.

Graham, C.H., S. Ferrier, F. Huettman, C. Moritz, and A.T. Peterson. 2004. New developments in museum-based informatics and applications in biodiversity analysis. *Trends in Ecology & Evolution* 19:497–503.

Greene, D.F., C.D. Canham, K.D. Coates, and P.T. Lepage. 2004. An evaluation of alternative dispersal functions for trees. *Journal of Ecology* 92:758–66.

Guisan, A., and J.-P. Theurillat. 2000a. Assessing alpine plant vulnerability to climate change: A modeling perspective. *Integrated Assessment* 1:307–20.

Guisan, A., and J.-P. Theurillat. 2000b. Equilibrium modeling of alpine plant distribution: How far can we go? *Phytocoenologia* 30:353–84.

Guisan, A., and J.-P. Theurillat. 2001. Assessing alpine plant vulnerability to climate change: A modeling perspective. *Integrated Assessment* 1:307–320.

Guisan, A., and W. Thuiller. 2005. Predicting species distribution: Offering more than simple habitat models. *Ecology Letters* 8:993–1009.

Guisan, A., and N.E. Zimmermann. 2000. Predictive habitat distribution models in ecology. *Ecological Modelling* 135:147–86.

Hill, J.K., C.D. Thomas, and B. Huntley. 1999. Climate and habitat availability determine 20th century changes in a butterfly's range margin. *Proceeding of the Royal Society of London B* 266:1197–1206.

Hofer, H.R. 1992. Veränderungen in der Vegetation von 14 Gipfeln der Berninagebietes zwischen 1905 und 1985. Berichte des Geobutanischen Institutes der ETH, Stiftung Rübel Zürich 58:39–54.

IPCC. 2007. Summary for Policymakers. In *Climate Change 2007: The Physical Science Basis: Contribution of Working Group I to the Fourth Assessment Report of the Intergovernmental Panel on Climate Change*. Solomon, S., D. Qin, M. Manning, Z. Chen, M. Marquis, K.B. Averyt, M. Tignor, and H.L. Miller, eds. Cambridge: Cambridge University Press.

Iverson, L.R., and A.M. Prasad. 1998. Predicting abundance of 80 tree species following climate change in the eastern United States. *Ecological Monographs* 68:465–85.

Iverson, L.R., M.W. Schwartz, and A.M. Prasad. 2004. How fast and far might tree species migrate in the eastern United States due to climate change? *Global Ecology and Biogeography* 13:209–19.

Kammer, P.M., C. Schob, and P. Choler. 2007. Increasing species richness on mountain summits: Upward migration due to anthropogenic climate change or re-colonisation? *Journal of Vegetation Science* 18:301–6.

Körner, C. 2003. *Alpine Plant Life*. 2nd ed. Berlin: Springer.

Kullman, L. 2001. Immigration of Picea abies into the North-Central Sweden. New evidence of regional expansion and tree-limit evolution. *Nordic Journal of Botany* 21:39–54.

Lahti, T., and R. Lampinen. 1999. From dot maps to bitmaps—Atlas Flora Europaeae goes digital. *Acta Botanica Fennica* 162:5–9.

Lassueur, T., S. Joost, and C.F. Randin. 2006. Very high resolution digital elevation models: Do they improve models of plant species distribution? *Ecological Modelling* 198:139–53.

Malcolm, J.R., A. Markham, R.P. Neilson, and M. Garaci. 2002. Estimated migration rates under scenarios of global climate change. *Journal of Biogeography* 29:835–49.

McCullagh, P., and J.A. Nelder. 1989. *Generalized Linear Models*. 2nd ed. London: Chapman & Hall.

Müller-Schneider, P. 1986. Verbreitungsbiologie des Blütenpflanzen Graubündens. Veröffentlichungen des Geobotanischen Institutes der Eidg. Techn. Hochschule, Stiftung Rübel, in Zürich.

Nakicenovic, N., and R. Swart. 2000. *Emissions Scenarios: A Special Report of Working Group III of the Intergovernmental Panel on Climate Change*, pp. 570. Cambridge: Cambridge University Press.

Nathan, R., and H.C. Muller-Landau. 2000. Spatial patterns of seed dispersal, their determinants and consequences for recruitment. *Trends in Ecology & Evolution* 15:278–85.

Nogues-Bravo, D., M.B. Araujo, M.P. Errea, and J.P. Martinez-Rica. 2007. Exposure of global mountain systems to climate warming during the 21st Century. *Global Environmental Change-Human and Policy Dimensions* 17:420–28.

Pauli, H., M. Gottfried, G. Grabbherr. 1996. Effects of climate change on mountain ecosystems: Upward shifting of alpine plants. *World Resource Review* 8:382–90.

Pauli, H., M. Gottfried, D. Hohenwallner, K. Reiter, R. Casale, and G. Grabherr. 2004. *The GLORIA field manual. Multi-summit approach. European Commission*. Luxembourg: Office for Official Publications of the European Communities.

Pauli, H., M. Gottfried, K. Reiter, C. Klettner, and G. Grabherr. 2007. Signals of range expansions and contractions of vascular plants in the high Alps: Observations (1994–2004) at the GLROIA master site Schrankogel, Tyrol, Austria. *Global Change Biology* 13:147–56.

Pearce, J., and S. Ferrier. 2000. An evaluation of alternative algorithms for fitting species distribution models using logistic regression. *Ecological Modelling* 128:127–47.

Pearson, R.G., T.P. Dawson, P.M. Berry, and P.A. Harrison. 2002. SPECIES: A spatial evaluation of climate Impact on the envelope of species. *Ecological Modelling* 154:289–300.

Pitelka, L.F., R.H. Gardner, J. Ash, S. Berry, H. Gitay, I.R. Noble, A. Saunders, R.H.W. Bradshaw, L. Brubaker, J.S. Clark, M.B. Davis, S. Sugita, J.M. Dyer, R. Hengeveld, G. Hope, B. Huntley, G.A. King, S. Lavorel, R.N. Mack, G.P. Malanson, M. McGlone, I.C. Prentice, and M. Rejmanek. 1997. Plant migration and climate change. *American Scientist* 85:464–73.

Price, M.V., and N.M. Waser. 2000. Responses of subalpine meadow vegetation to four years of experimental warming. *Ecological Applications* 10:811–23.

R Development Core Team. 2007. R 2.6.1 A language and environment.

Randin, C.F., T. Dirnbock, S. Dullinger, N.E. Zimmermann, M. Zappa, and A. Guisan. 2006. Are niche-based species distribution models transferable in space? *Journal of Biogeography* 33:1689–1703.

Randin, C.F., H. Jaccard, P. Vittoz, and A. Guisan. In revision-a. Importance of landuse in predictive habitat modeling of plant species distribution in the Western Swiss Alps. *Journal of Vegetation Science*.

Randin, C.F., G. Vuissoz, G.E. Liston, P. Vittoz, and A. Guisan. In revision-b. Introducing process-based disturbance variables in predictive models of plant distribution. *Arctic, Antarctic, and Alpine Research*.

Rebetez, M. 2002. *La Suisse se réchauffe: effet de serre et changement climatique*. Lausanne: Presse polytechniques et unversitaires romandes.

Rebetez, M., and M. Reinhard. 2008. Monthly air temperature trends in Switzerland 1901–2000 and 1975–2004. *Theoretical and Applied Climatology* 91:27–34.

Remes, V. 2000. How can maladaptive habitat choice generate source-sink population dynamics? *Oikos* 91:579–82.

Theurillat, J.P., and A. Guisan. 2001. Potential impact of climate change on vegetation in the European Alps: A review. *Climatic Change* 50:77–109.

Thomas, C.D., A. Cameron, R.E. Green, M. Bakkenes, L.J. Beaumont, Y.C. Collingham, B.F.N. Erasmus, M. Ferreira de Siqueira, A. Grainger, L. Hannah, L. Hughes, B. Huntley, A.S. Van Jaarsveld, G.F. Midgley, L. Miles, M.A. Ortega-Huerta, A.T. Peterson, O.L. Phillips, and S.E. Williams. 2004. Extinction risk from climate change. *Nature* 427:145–47.

Thuiller, W. 2003. BIOMOD—Optimizing predictions of species distributions and projecting potential future shifts under global change. *Global Change Biology* 9:1353–62.

Thuiller, W. 2004. Patterns and uncertainties of species' range shifts under climate change. *Global Change Biology* 10:2020–27.

Thuiller, W., M.B. Araujo, R.G. Pearson, R.J. Whittaker, L. Brotons, and S. Lavorel. 2004a. Uncertainty in predictions of extinction risk. *Nature* 430:34.

Thuiller, W., L. Brotons, M.B. Araujo, and S. Lavorel. 2004b. Effects of restricting environmental range of data to project current and future species distributions. *Ecography* 27:165–72.

Thuiller, W., S. Lavorel, M.B. Araujo, M.T. Sykes, and I.C. Prentice. 2005. Climate change threats to plant diversity in Europe. *Proceedings of the National Academy of Sciences of the United States of America* 102:8245–50.

van Houwelingen, J.C., and S. Le Cessie. 1990. Predictive value of statistical models. *Statistics in Medicine* 9:1303–25.

Vittoz, P., and R. Engler. 2007. Seed dispersal distances: A typological system for data analyses and models. *Botanica Helvetica* 117:109–24.

Vittoz, P., and A. Guisan. 2003. Le projet PERMANENT. PLOT.CH demande votre collaboration. Das Projekt PERMANENT.PLOT.CH bittet um Ihre Mithilfe. *Botanica Helvetica* 113:105–10.

Vittoz, P., S. Jutzeler, and A. Guisan. 2006. Flore alpine et réchauffement climatique: observation de trois sommets valaisans à travers le 20ème siècle. *Bulletin de la Murithienne* 123/2005:49–59.

Vittoz, P., C.F. Randin, A. Dutoit, F. Bonnet, and O. Hegg. 2009. Low impact of climate change on subalpine grasslands in the Swiss Northern Alps. *Global Change Biology*, 15:209–220.

Vittoz, P., B. Rulence, T. Largey, and F. Freléchoux. 2008. Effects of climate and land-use change on the establishment and growth of Cembran pine (Pinus cembra L.) over the altitudinal treeline ecotone in the Central Swiss Alps. *Arctic, Antarctic, and Alpine Research* 40:225–32.

Walther, G.-R., S. Beißner, and C.A. Burga. 2005a. Trends in the upward shift of alpine plants. *Journal of Vegetation Science* 16:541–548.

Walther, G.R. 2003. Plants in a warmer world. Perspectives *in Plant Ecology, Evolution and Systematics* 6:169–85.

Walther, G.R., S. Beissner, and C.A. Burga. 2005b. Trends in the upward shift of alpine plants. *Journal of Vegetation Science* 16:541–48.

Willson, M.F. 1993. Dispersal mode, seed shadows, and colonization patterns. *Vegetatio* 108:261–80.

Zimmermann, N.E., and F. Kienast. 1999. Predictive mapping of alpine grasslands in Switzerland: Species versus community approach. *Journal of Vegetation Science* 10:469–82.

16 The "Mountain Laboratory" of Nature—A Largely Unexplored Mine of Information
Synthesis of the Book

Eva M. Spehn and Christian Körner

CONTENTS

Introduction 165
Availability of Primary Biodiversity Data 165
Importance of Georeferenced Biodiversity Data 166
Mountains as Tools to Explore Environmental Control of Latitudinal and Elevational Diversity Gradients 166
Mining Biodiversity Data: Integrating Approaches 167
Georeferenced Databases as Management Tools for Nature Conservation 168
Niche Dynamics Modeling to Explain the Past and Predict the Future 168
GMBA Mountain Portal 168
References 168

INTRODUCTION

Steep climatic gradients in mountains over short distances offer unique test conditions for organismic adaptation and evolutionary selection. Data mining and database linkages are upcoming tools to complement, and partly replace, expensive experiments furthering ecological theory. At two successive meetings organized by DIVERSITAS' crosscutting network "Global Mountain Biodiversity Assessment" (GMBA)—one in the Central Caucasus, a hot spot of biodiversity, and one in Copenhagen at the Secretariat of the Global Biodiversity Information Facility (GBIF)—the possibilities and power of the intelligent use of existing archives of biodiversity was explored. All levels of biodiversity are examined in this volume, from genes (Chapter 11) to functional traits (Chapters 12 and 6) and species (most chapters), up to habitat and ecosystem diversity (Chapters 11 and 14). Due to the huge amount of plant data available, most chapters deal with plant diversity, but other organismic groups also receive attention, such as beetles (Chapters 10 and 14) and birds, which also provide an excellent data example due to a long tradition of assembling bird data for distribution maps (Chapters 8, 9, and 10).

AVAILABILITY OF PRIMARY BIODIVERSITY DATA

About 2 to 3 billion specimens (Krishtalka and Humphrey, 2000) are currently preserved in the world's natural history collections, a major source for primary biodiversity data (Chapter 2; Graham et al., 2004). Each of these specimens provides a single data point, documenting the occurrence of an individual of a particular species at a particular geographic locality and at a particular point in time. The Global Biodiversity Information Facility (GBIF; http://www.gbif.org) was established to encourage the provision of such data in digitized form and to enable broad access to biological databases worldwide. Biodiversity data are often local or national, with the data collected independently and maintained in unconnected electronic archives, if digitized at all, without any control for interoperability with other databases. Quite often it is only through the linking of data from different sources that scientific advance is achieved, because an individual (mostly regional) database commonly does not contain sufficient information for developing and testing general theory and furthering broad understanding. Therefore, GBIF, with all its forty-seven member countries and thirty-five international partner organizations, has committed itself to "improving the accessibility, completeness and interoperability of biodiversity data bases." GBIF has established a data portal that connects more than 150 million biodiversity records that adhere to interoperable standards for data sharing. GBIF further facilitates data mining by linking data to an electronic catalogue of names of known organisms (Catalogue of Life, http://www.catalogueoflife.org) and by developing standards in biodiversity informatics (Lane and Edwards, 2007). Many more Web functionalities exist, most of them not even imagined yet.

Chapter 3 shows how to get from primary biodiversity data to an estimate of the species number of a region, using the completeness index C. Estimating the species number of a region mainly involves debugging databases with heterogeneous origin, by e.g., consulting with data holders. High quality metadata (Chapter 4), provided along with the biodiversity data sets used, are of great help for judging the "fitness for use" for the specific research question or management decision task, as they render those who mine the data more independent from data holders.

IMPORTANCE OF GEOREFERENCED BIODIVERSITY DATA

Georeferencing is the process of converting text descriptions of locations to computer-readable geographic locations (Arzberger et al., 2004), such as Geographic Information Systems (GIS). Georeferenced species occurrence data allow for correlating species occurrence with environmental parameters and therefore offer the ability to understand effects of environmental conditions and their change. The study of large-scale patterns depends on sufficient and accurate species data, and there is an urgent need for data with precise altitude information (Chapter 17). Several authors call for systematically georeferencing of natural history collections data (Graham et al., 2004) and in mountain regions, an undertaking greatly facilitated by "Biogeomancer" (http://www.biogeomancer.org), an online tool providing the georeferences of localities, together with an estimate of precision. High precision georeferenced occurrence data and fine-scale digital climate maps are crucial for most of the studies presented here, especially for modeling the distributions of taxa that live in topographically heterogeneous environments such as mountains (high habitat diversity on a small scale) (Graham et al., 2004; Elith et al., 2007).

A large obstacle in mountain biodiversity research is the absence of adequate long-term and comparable time series of biodiversity changes, but models can complement monitoring data, and help to predict future trends in species distributions. Apart from georeferenced biodiversity data sets, the following data are required for the parametrization of models aiming at projections of future species (Chapter 15): species trait databases (such as presented in Chapter 12), seed dispersal data, preferably species dispersal distances and rates (Chapter 15). GIS data on habitat suitability will permit to account for corridors and barriers for species dispersal. Modeling niche conservation (a tool to better predict species niche shifts and migration) requires additional paleodata: fossil and modern species records, paleoclimate (only recently available in greater detail), as well as present climate (Chapter 13).

MOUNTAINS AS TOOLS TO EXPLORE ENVIRONMENTAL CONTROL OF LATITUDINAL AND ELEVATIONAL DIVERSITY GRADIENTS

It is supposed that mountains represent a globally highly replicated "experiment by nature" that offers testing a broad spectrum of hypotheses related to global biodiversity patterns (Körner, 2000). As "islands in the sky," mountains permit exploring both space and time (season length) constraints to species richness and diversification within a broad matrix of climatic conditions. Körner and Paulsen (Chapter 1) provide the tools for such analysis, a globally applicable bioclimatic reference line (the thermal tree limit), and examples for the geostatistical analysis of climatic data across latitudes, seasonality, and shrinking of land area with elevation and topographic ruggedness, all in support of large-scale comparisons. Arroyo et al. (Chapter 5) show that the alpine zone is ideally suited to study macroecological patterns, such as the effect of land area on species richness, given that alpine land area extends over large regions and exists at all latitudes. Species ranges provide a great tool to look at ecological mechanisms on a biogeographic scale. Combining altitudinal with latitudinal gradients will help to disentangle historical, evolutionary, and ecological causes. With appropriate phylogenetic information (see Chapter 11) one could even assess the effect of evolutionary history and differences in dispersal capacity on the size of latitudinal and altitudinal ranges. Fang et al. (Chapter 6) show a sampling scheme for a large-scale biodiversity monitoring program on fifty different mountain ranges all over China, also measuring climate and other environmental variables along several altitudinal gradients per mountain. Due to high replication, elevational species gradients are therefore perfectly comparable over huge latitudinal, and therefore climatic, amplitudes, reaching from tropical mountains to very dry and cold regions on the Tibetan Plateau.

Disentangling the environmental and historical factors shaping biodiversity patterns is important for understanding biodiversity. Too often such analysis is limited to elevation as a black box driver, because the true drivers (e.g., climate) are not known at sufficient precision, constraining mechanistic explanations. For example, elevational gradients of species diversity have often been reported to show a mid-elevation peak (e.g., Rahbek, 1995, 2005) with simple explanations that the elevational gradients studied covered different overlapping and unaccounted gradients of temperature, moisture, and disturbance. It is no surprise that a mechanistic explanation failed. Georeferenced databases of species, if complemented with databases on environmental factors, traits, evolutionary history, and human disturbance (e.g., Nogués-Bravo et al., 2008), allow for exploring the range and thus plasticity of adjustment of species (Chapter 5) through comparisons of altitudinal gradients with different influential covariables (water availability, season length, land cover, grazing pressure, topography [area and number of habitats, species dispersal constraints], geological substrate, and others), covarying with temperature.

Studies compiled from various mountain ranges and for various taxa need to take into account differences in scale and extent of the gradients, different regional climatic regimes, and different evolutionary histories. However, systematic comparison of diversity gradients among regions, at different scales, are necessary for a better understanding of the drivers and underlying mechanisms of biodiversity patterns (Körner, 2000). In such a comparison, Zbinden et al. (Chapter 9) found common patterns for birds in the Swiss Alps and in the Pyrenees: Above ca. 1,500 m, bird diversity declined linearly in both mountain regions, maybe due to a reduction in habitat diversity.

Sanders et al. (Chapter 10) tested different taxa along the same elevational gradient, therefore keeping regional factors and common environmental histories constant. Peaks in diversity were mostly congruent between four different groups of taxa (ants, noctuid moths, breeding birds, and beetles), but diversity patterns still differed among taxa along the elevational gradient in Great Smokey Mountains National Park. Since elevational patterns of species richness are confounded with other factors, such as climate and land area, Zhang and Sun (Chapter 7) adjusted for land area and tested how climatic variables affect the elevational pattern in the Hengduan Mountains. The climatic features at the two extreme ends of their elevational gradients, e.g., drought in the valleys and low temperature above timberline, plus the reduction of land area at both ends of the elevational gradient (most land area at medium elevation), led to hump-shaped patterns of species richness in this area, as had been found in other regions in which temperature and moisture gradients are confounded. Confounding of environmental gradients call for replications of elevational gradients differing in land area, moisture, and other nontemperature-related gradients (Körner, 2007). Fang et al. (Chapter 6) provide already exactly such a geographic and conceptual extension of the Hengduan Mountain study (Chapter 7) to more than fifty major mountains across China.

Methods to obtain elevational gradient patterns, either by local field surveys or data mining, are compared by Kessler et al. (Chapter 8). They also compare patterns at different scales, local versus regional gradients, for five groups of organisms (birds and four different plant groups) along six tropical gradients. Local and regional patterns, although often correlated, revealed different patterns depending on scale (Huettmann and Diamond, 2007). Field surveys are often spatially constrained and do not cover the full range of elevations. In the case of discontinuous or even irregular elevational trends, the selection of certain elevational segments may yield conflicting elevational patterns for scale reasons only (Nogues-Bravo et al., 2008). Archive data often permit avoiding such bias. Regional archive data did not reflect the survey data (Chapter 8), but both taken together provide insight into mechanisms driving species richness patterns.

MINING BIODIVERSITY DATA: INTEGRATING APPROACHES

Biodiversity has three components: genotypes, species, and ecosystem diversity. The project "IntraBioDiv" on plant richness of the alpine regions of the European Alps and the Carpathians used electronic databases of regional and national inventories and newly generated data on genetic, species, and habitat diversity (Chapter 11).

Aggregating these various data sources, these authors could show that species diversity is clearly correlated with habitat diversity and is higher in the southern and the western Alps than in other areas. The analysis further revealed that genetic diversity (genetic fingerprints [AFLP] of forty high-mountain species) is best correlated with present conditions, such as population density and habitat diversity, whereas endemism is more related to historical factors.

GEOREFERENCED DATABASES AS MANAGEMENT TOOLS FOR NATURE CONSERVATION

National parks often play a special role for creating species databases linked to ecological information. The Great Smoky Mountains National Park is, for instance, one of the most surveyed places in Northern America, with a long history and effort to catalogue the diversity of all life in the park (All-Taxa Biodiversity Inventory, ATBI). Sanders et al. (Chapter 10) present elevational gradient data for four different taxa, taking advantage of this species inventory. Brandmayr and Pizzolotto (Chapter 14) present a georeferenced Habitat Biodiversity Data Base (HDBD) for Pollino National Park in Italy. The HDBD is a great tool not only for scientists, but also for park reserve managers, as this database is mainly structured by habitats and provides time series (multiyear samples). Trends in species abundances over time are immediately obvious. Lavorel et al. (Chapter 12) show how their georeferenced database of plant functional trait abundance over alpine regions (ALTA) was developed for Ecrins National Park in France. A single, flexible structure can serve a variety of different purposes for scientists, naturalists, and conservation managers, provided all needs are addressed in an interdisciplinary dialogue first.

NICHE DYNAMICS MODELING TO EXPLAIN THE PAST AND PREDICT THE FUTURE

Paleodata compared with current registry data offer a great tool for predicting the response of individual species to climate change and for understanding how and why these responses vary among species. Mountains are especially well suited to look at paleodata, as fossils are well preserved. Paleodata of both biodiversity and climate revealed that mountain dwelling species show stronger niche conservation over time than lowland species (Chapter 13). Recent climate change induced species shifts within a habitat over the last thirty years in Pollino National Park in Italy (Chapter 14). Open range land (meadows and pastures) shows pronounced changes in species structure, and all dominant carabid beetles species were substituted by other, more thermophilic species, whereas no such trends could be detected in cooler forest habitats. Projections of how future climate change will further impact biological systems by upward migration of many species, or regional disappearance of species if suitable habitats are lacking, are shown in a case study from the western Swiss Alps (Chapter 15). Their simulation using a species distribution model, based on georeferenced data for ca. 300 plant species (representing the regional flora), revealed which components of the flora are sensitive to climate change. Most alpine and nival species are likely to lose part of their suitable habitats by 2100, according to this analysis.

GMBA MOUNTAIN PORTAL

GMBA is currently developing a GBIF-based mountain portal that will define the geographical and topographical coordinates and boundaries for spatial search routines in mountain regions. It will further offer a search routine by region or mountain life zones. A registry will provide for an overview of GBIF-available mountain data, as well as other mountain biodiversity databases not (yet) available at GBIF, and what their specific characteristics are. There will be links with the metadata available for each database, with a special emphasis on "fitness for use" for mountain research (such as precision of georeferences and altitudinal information). The portal is planned to be available online by the end of 2009, and will hopefully become an attractive and global venue toward the creative use of archive data for functional ecology, biogeography, evolutionary ecology, and management applications toward global sustainability.

REFERENCES

Arzberger, P., P. Schroeder, A. Beaulieu, G. Bowker, K. Casey, L. Laaksonen, D. Moorman, P. Uhlir, and P. Wouters. 2004. An international framework to promote access to data. *Science* 303:1777–78

Elith, J., C. Graham, and NCEAS working group. 2006. Comparing methodologies for modeling species' distributions from presence-only data. *Ecography* 29:129–51.

Graham, C.H., S. Ferrier, F. Huettmann, C. Moritz, and A.T. Peterson. 2004. New developments in museum-based informatics and applications in biodiversity analysis. *Trends in Ecology & Evolution* 19:497–503.

Huettmann, F., and A.W. Diamond. 2006. Large-scale effects on the spatial distribution of seabirds in the northwest Atlantic. *Landscape Ecology* 21:1089–1108.

Koerner, C. 2000 Why are there global gradients in species richness? Mountains may hold the answer. *Trends in Ecology & Evolution* 15:513.

Koerner, C. 2007. The use of "altitude" in ecological research. *Trends in Ecology & Evolution* 22:569–74

Krishtalka, L., and P.S. Humphrey. 2000 Can natural history museums capture the future? *Bioscience* 50, 611–17.

Lane, M., and J.L. Edwards. 2007. The Global Biodiversity Information Facility (GBIF). In *Biodiversity Databases: Techniques, Politics and Applications*. Curry, G.B., and C.J. Humphries, eds. Boca Raton, Fla.: CRC Press/Taylor & Francis.

Nogués-Bravo, D., M.B. Araujo, T. Romdal, and C. Rahbek. 2008. Scale effects and human impact on the elevational species richness gradients. *Nature* 453:216–20.

Rahbek, C. 1995. The elevational gradient of species richness: A uniform pattern? *Ecography* 18:200–5.

Rahbek, C. 2005. The role of spatial scale and the perception of large-scale species-richness patterns. *Ecology Letters* 8:224–39.

17 Creative Use of Mountain Biodiversity Databases

The Kazbegi Research Agenda of GMBA-DIVERSITAS

Christian Körner, Michael Donoghue, Thomas Fabbro, Christoph Häuser, David Nogués-Bravo, Mary T. Kalin Arroyo, Jorge Soberón M., Larry Speers, Eva M. Spehn, Hang Sun, Andreas Tribsch, Piotr Tykarski, and Niklaus Zbinden

CONTENTS

Aims ... 172
Data Sharing for the Mountain Research Community ... 172
The Principle of Open Access ... 172
Data Sources and Data Structure ... 172
Mountain-Specific Aspects ... 173
Additional Information (Some Useful Examples in a Mountain Context) ... 173
Visions and Suggestions for Scientific Use of Mountain Biodiversity e-Data ... 173
Mountains—A Laboratory for Understanding Basic Questions of Evolution: How Is Mountain Biodiversity Generated, Evolved, Assembled? ... 173
Are There Common Elevational Trends in Mountain Biodiversity? What Drives Them? ... 174
Are There Typical Elevational Trends in Organismic Traits across the Globe? ... 175
Are Biotic Links and Biodiversity Ratios among Organismic Groups Tighter with Elevation? ... 175
Are There Functional Implications of Mountain Biodiversity? ... 175
What Are the Socioeconomic Impacts on Mountain Biodiversity? ... 176
Effective Conservation of Mountain Biodiversity under Global Environmental Change: How Best to Assess Effects of Current Efforts and Future Trends ... 176
Mountain Terminology and GMBA Concept of Comparative Mountain Biodiversity Research ... 176
Open Access and a GBIF Portal to Shared Mountain Biodiversity Data ... 177
Summary ... 177
Acknowledgments ... 177
References ... 177

This article was first published in *Mountain Research and Development* (MRD), vol. 27 no. 3, pp. 276–281. The rights of reproduction remain with the co-copyright holders: the International Mountain Society (IMS) and the United Nations University (UNU), c/o MRD Editorial Office, Berne, Switzerland (www.mrd-journal.org).

AIMS

The Global Mountain Biodiversity Assessment (GMBA), a cross-cutting network of DIVERSITAS, aims to encourage and synthesize research on high-altitude organismic diversity, its regional and global patterns, and its causes and functions (Koerner and Spehn, 2002; Spehn et al., 2005). Existing and emerging electronic databases are among the most promising tools in this field. Gradients of altitude and associated climatic trends, topographic and soil peculiarities, and fragmentation and connectivity among biota and their varied geological and phylogenetic history are the major drivers and aspects of mountain biodiversity, and electronic archives provide avenues for testing their impact on life at high elevations. This research agenda was developed at a GMBA workshop in the Central Caucasus in July 2006. It capitalizes on expertise from different fields of biology and database experts, and was developed in cooperation with the Global Biodiversity Information Facility (GBIF). Enhancing awareness of the central role of georeferencing in database building and use is one of the central tasks of this agenda. Once achieved, this permits linkage of biological information with other geophysical information, particularly climate data. The mountains of the world exhibit different climatic trends along their slopes, with only few factors, such as the decline in atmospheric pressure, ambient temperature, and clear sky radiation changing in a common, altitude-specific way across the globe. None of the other key components of climate, such as cloudiness and, with it, actual solar radiation or precipitation and associated soil moisture, show such global trends, and hence are not altitude-specific. The separation of global from regional environmental conditions along elevational transects offers new perspectives for understanding adaptation of mountain biota. Similarly, information on bedrock chemistry and mountain topography offers test conditions for edaphic drivers of biodiversity and species radiation in an evolutionary context across geographical scales.

DATA SHARING FOR THE MOUNTAIN RESEARCH COMMUNITY

Many research projects generate biodiversity data sets that may be relevant for the wider scientific community, government and private natural resource managers, policymakers, and the public. GBIF has a mission to make the world's primary data on biodiversity freely and universally available via the Internet (www.gbif.org).

THE PRINCIPLE OF OPEN ACCESS

The U.N. Convention on Biological Diversity has called for free and open access to all past, present, and future public-good research results, assessments, maps, and databases on biodiversity (CBD Dec. VIII/11). Furthermore, all forty-seven current member countries and thirty-five international organizations in GBIF have committed themselves to "improving the accessibility, completeness and interoperability of biodiversity databases," and to "promote the sharing of biodiversity data in GBIF under a common set of standards." Added value comes from sharing data (Arzberger et al., 2004a, 2004b), but sharing requires respect of author rights and observation of certain rules as defined by GBIF standards (Stolton and Dudley, 2004). Quite often it is only through the linking of data that scientific advance is achieved. Hence, protective habits are counterproductive, given that an individual database commonly does not contain sufficient information for developing and testing theory and furthering broad understanding. Moreover, many taxonomic databases rely on the collective work of generations of scientists in a country.

DATA SOURCES AND DATA STRUCTURE

There are 1) individual-based data (primary occurrences, an individual at a place at a particular time) and 2) taxon-based data (biological taxon characteristics, such as morphology, physiology, phylogeny, ecology, genetics). These may refer to: a) vouchered primary occurrences; b) observational data; or c) literature data. The quality and use of primary species and species-occurrence data are highlighted in Chapman (2005a–c). A full, best-practice database entry should include the following types of data:

- Organismic data (conventional taxonomic information)
- Geoinformation (coordinates, altitude)
- Habitat information (edaphic, topographic, atmospheric)
- Date and time of observation, collection and recording
- Reference to a voucher or archive code
- Name of collector, observer, and recorder
- Metadata that provide information on data sets, such as content, extent, accessibility, currency, completeness, accuracy, uncertainties, fitness for purpose and suitability for use, and enable

the use of data by third parties without reference to the originator of the data (Chapman, 2005b).

MOUNTAIN-SPECIFIC ASPECTS

Given the significance of topography and elevation in mountains for local biotic conditions, reported geographical coordinates using Global Positioning System (GPS) should at least provide a resolution of seconds. Elevation should always be obtained independently of GPS. Chapman and Wieczorek (2006) provide best practices for georeferencing (assigning geographic coordinates to) a range of different location types. Should coordinates be missing, the Biogeomancer online tool (www.biogeomancer.org) may be able to reconstruct these from locality, region, or names. Elevation data can have the following structure:

- Point data (for vouchers, data loggers, climatic stations) reported as precisely as possible, with uncertainties given. In most cases, a precision of 10 m elevation is enough, although earlier GPS data will offer less precision.
- Stratified range elevation data, which offer entries for certain taxa in a step-by-step elevational catena (e.g., 100 m steps). If this is not available, at least the elevational center of the variable or taxon should be provided.
- Full range or amplitude data (maximum and minimum elevation) with uncertainties. Range data are critical for making up lists of species for different elevational bands. The midpoint is insufficient.

Note that such information becomes almost useless if uncertainties in the observation are not identified. One way of getting around this is to quote the data within range width (100 m, 200 m, 1,000 m). Uncertainty associated with georeferenced localities along elevational gradients can be measured with post-hoc, three-dimensional georeferencing (Rowe, 2005).

ADDITIONAL INFORMATION (SOME USEFUL EXAMPLES IN A MOUNTAIN CONTEXT)

- Plants: Biological attributes such as size (height), life form, flower features, current phenology, seed size, growth form, and other special attributes. These data can sometimes be obtained from taxonomic sources and stored in relational databases.
- Animals: Biological attributes, such as size (width, length, etc.), trophic habit, interactions (prey, mutualistic species, host, phenology, life stage).
- Abundance or frequency measures (e.g., random sample of quadrats) and information on rareness, conservation status, dominant associates, and population structure, if available.

VISIONS AND SUGGESTIONS FOR SCIENTIFIC USE OF MOUNTAIN BIODIVERSITY E-DATA

The power of openly accessible, interconnected electronic databases for scientific biodiversity research by far exceeds the original intent of archiving for mainly taxonomic purposes, as will be illustrated by the following examples. Each example starts with a scientifically important question or hypothesis (what?) and continues by providing a motive (why would we want to know this?), and suggestions about how to approach this task by data mining and data linking. The application of a common mountain terminology (a convention) is an essential prerequisite for communication (Figure 17.1).

MOUNTAINS—A LABORATORY FOR UNDERSTANDING BASIC QUESTIONS OF EVOLUTION: HOW IS MOUNTAIN BIODIVERSITY GENERATED, EVOLVED, ASSEMBLED?

What? The origin and assembly of mountain biota have to be understood in a historical context. For a given mountain area: Where did its taxa arise, and how were taxa assembled over time? How many of the extant species resulted from the radiation of lineages that evolved within the area, as opposed to the radiation of lineages that were introduced from other areas or even continents or other ecosystems? How important has long-distance dispersal been for the assembly of mountain biota, and how and when did evolutionary lineages migrate from one mountain area to others? What are the main sources of long-distance dispersal events? Has the capacity of long-distance dispersal itself been a factor in the rapid radiation of alpine lineages?

Why? Mountains are islands of varying size, and thus present a good opportunity to ask questions about genesis of mountain biota, the impact of competition from other biota on speciation rates, and adaptive evolution. Where arid climates have developed at lower elevations, alpine areas can act as "conservation areas for phylogenetic

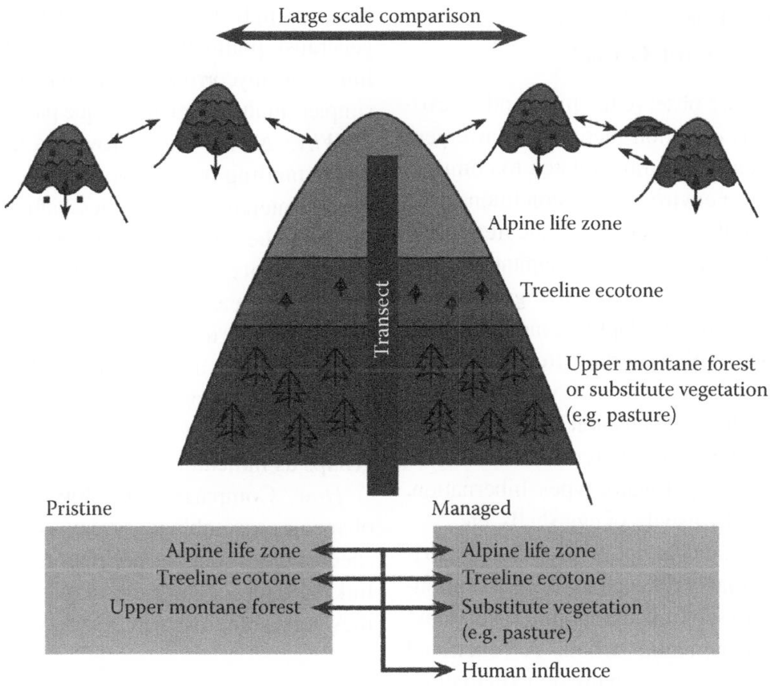

FIGURE 17.1 The GMBA concept of vertical and horizontal comparison of mountain biota.

lineage" for lowland lineages (see Hershkovitz et al., 2006). Mountains have acted (and will act) as refugia for species survival during extreme climatic events, including for ancient phylogenetic lineages. Rapid rates of speciation have been documented in recent phylogenetic studies for genera in high-elevation areas (e.g., Hughes and Eastwood, 2006). Rapid evolution also is a factor for predictions related to climate change.

How? Combine data from phylogenetic and phylogeographic databases, regional species lists, classification by elevation (e.g., selection of alpine species), geographic distribution, and species range limits. Information on resilience of a species to change (life form, life cycle characteristics, reproduction, and phenological data).

Are There Common Elevational Trends in Mountain Biodiversity? What Drives Them?

What? The overarching issue is to challenge the common notion that species richness in alpine areas is necessarily low. Life conditions change with elevation in global, but also in very regional, ways, and equal steps in elevational climatic change are associated with decreasing available land area per step (belt; Koerner, 2000). Furthermore, land surface roughness (habitat diversity) commonly increases with elevation. Finally, mountains represent archipelagos of contrasting connectivity and island size. How is biodiversity influenced by these four aspects of elevation (climate, area, fragmentation, and roughness)?

Why? The wide amplitude of climatic conditions and topographies across the world's mountains offers an unparalleled opportunity for developing and testing biodiversity theory. How does species richness in mountains change with latitude or elevation; do reductions in species richness on opposite-facing slopes parallel altitudinal gradients and similar temperature gradients (Figure 17.1)? Ratios of trends in various taxonomic groups make it possible to distill biotic interdependencies or at least correlative associations. Such biodiversity ratios can serve as predictive tools. The climatic relatedness of emerging trends can assist in projections of climate change impacts.

How? The major tool is selective comparison of stratified biodiversity data for various organismic groups across elevational transects of major mountain systems. Key problems to be solved are the confounding between altitude-specific (global) and region-specific (local) climatic trends, and the geological age and spatial extent of mountain systems. Links with fine-resolution GIS and world climate databases are essential.

Are There Typical Elevational Trends in Organismic Traits across the Globe?

What? Across the globe, we observe the independent evolution of certain traits as elevation increases (convergent evolution). Are these trends and traits related to common elevational gradients under environmental conditions (e.g., temperature) or do they reflect specific climatic trends that are not common to all mountains (e.g., precipitation), and would they thus also be found at respective gradients at low elevations? Would common edaphic conditions (e.g., presence of scree) alone explain certain trends? Typical traits to be explored are size and mass of organisms, special functional types such as the cushion plant life form, giant rosettes or woolly plants, certain reproductive strategies, plant breeding systems, pollinator types, hibernation, dispersal characteristics, diffusivity of egg shells, etc.

Why? Most of these traits cannot be modified experimentally, and thus presumably reflect long-term evolutionary selection. We need to separate taxonomic relatedness from independent environmental action. A functional interpretation would require a mechanistic explanation: are cushion plants abundant at high elevations because of loose, poorly developed substrate, insufficient moisture, strong wind, too low temperature, short seasons, or certain combinations of these? Is high pubescence truly and generally more abundant at high elevation, and if so, under which high-elevation environmental conditions does this trend become enhanced?

How? The compressed width of climatic belts in mountains offers "experiments by nature" to test such hypotheses over short geographical distances (Koerner, 2003) by comparing trends in traits across a suite of mountain transects in areas of contrasting geological and evolutionary history and different climates. A test across different phylogenetic groups would reveal taxonomic relatedness. A comparison across different latitudes could separate seasonality and absolute altitude (pressure), effects because low temperatures such as those at treeline are found at 4,000 m near the equator and 500 m above sea level at the polar circle.

ARE BIOTIC LINKS AND BIODIVERSITY RATIOS AMONG ORGANISMIC GROUPS TIGHTER WITH ELEVATION?

What? Functional interactions between organisms (trophic, mechanical, physiological, and pathogenic) drive coexistence and competition among taxa. Do these ties become looser or tighter as elevation increases? For instance, does generalist pollination increase with elevation? Are such links (e.g., mycorrhization, predation, facilitation) becoming simpler (multiple versus unique partner organisms)?

Why? Alpine areas provide a unique opportunity for understanding how coevolution developed. Functionally, the maintenance of species richness and mutualism is known to be critical for maintaining plant fitness in harsh environments. As biodiversity of montane environments usually decreases with elevation, it may be more and more difficult to find a host for any specialized organism, and having a wider range of hosts could be favorable. Biodiversity ratios are a promising (to be explored) tool for rapid inventory works (the diversity of key taxonomic groups as indicators).

How? Comparisons of altitudinal patterns of diversity of species assemblages, use of known data on mutualistic species (e.g., specific pollinators, prey, mycorrhiza), and linking data for different taxonomic groups (e.g., butterflies versus angiosperm diversity).

ARE THERE FUNCTIONAL IMPLICATIONS OF MOUNTAIN BIODIVERSITY?

What? What is the contribution of mountain biodiversity to ecosystem integrity, i.e., slope stability? What is the functional redundancy in traits among organisms in a given area, what is their sensitivity to stress and disturbance (insect outbreaks, avalanches)?

Why? Ecosystem integrity on steep mountain slopes and in high-elevation landscapes is mainly a question of soil stability, which in turn depends on plant cover. The insurance hypothesis of biodiversity suggests that the more diversity (e.g., genetic diversity, morpho-types) there is, the less likely it is that extreme events or natural diseases will lead to a decline in ecosystem functioning or a failure of vegetation to prevent soil erosion. In steep terrain, more than anywhere else, catchment quality is intimately linked to ecosystem integrity. The provision of sustainable and clean supplies of water is the most important and increasingly limiting mountain resource.

How? Old versus new inventory data, recent loss or gain of certain plant functional types (e.g., trees). Recent land cover change (remote sensing evidence, Normalized Differenced Vegetation Index [NDVI]). Apart from information on composition of vegetation and functional traits of taxa (e.g., rooting depth, root architecture, growth form), geographical information is needed (geomorphology: slope, relief, soil depth;

climate, precipitation, evapotranspiration, extreme rain events, snow cover duration). Comparison of different mountain regions (e.g., presence or absence of woody or nonwoody vegetation). Spatial land cover information can be used to develop scenarios at landscape scale.

What Are the Socioeconomic Impacts on Mountain Biodiversity?

What? Humans shape mountain vegetation by clearing land, grazing, abandoning, collecting, etc., which may increase or decrease mountain biodiversity (Spehn et al., 2005) and, through this, affect slope processes, erosion, water yield, and inhabitability. Are areas with traditional burning regimes, in combination with grazing, poorer in species of flowering plants, butterflies, and wild ungulates than grazed areas in which burning is not a tradition? Do these trends interact with precipitation? Is high human population density at high elevations related to the specific loss of woody taxa? Is the biological richness of inaccessible microhabitats (topography-caused "wilderness") a measure or good reference of potential biodiversity of adjacent, transformed land?

Why? Of all global change effects, land use is the predominant driver of changes in mountain biodiversity. By comparing areas of historically contrasting land use regimes, we can learn how these human activities shape biota. Ratios of wilderness biodiversity to adjacent managed biodiversity indicate the actual impact of land use. The abundance of red list taxa or medicinal plants can be related to human population pressure and land use intensity.

How? Linking thematic databases for land cover type, population density, and climate with regional biodiversity inventories. Global comparisons across different climates and land use histories should permit distilling certain overarching trends. Comparison of intensively used high-elevation rangeland in regions of contrasting natural biodiversity should illustrate the significance of regional species pools for biodiversity in transformed landscapes (e.g., Caucasus versus Alps). A comparison of rangeland biodiversity in geologically young (steep) mountain regions with that in geologically old (smooth) mountain landscapes could reveal interactive influences of landscape roughness and land use on biodiversity.

Effective Conservation of Mountain Biodiversity under Global Environmental Change: How Best to Assess Effects of Current Efforts and Future Trends

What? Which is the minimum altitudinal range required for protected areas in mountain regions? What are the minimum habitat size and requirements for long-term viable (metapopulations) under high mountain conditions and under future climate change? Which are the best diversity–area relationships in high mountain environments for conservation purposes? What is the relevance of connectivity through gene flow for geographically isolated populations on high mountains? Which are suitable indicators and the most likely drivers of biodiversity change in protected areas in mountains?

Why? With many global mountain biodiversity hotspots increasingly threatened, efforts are under way to preserve these unique biota, largely by establishing a system of protected areas on mountains (Koerner and Ohsawa, 2005). Relevant variables for conservation biology, such as minimum range, viable population size, and connectivity, become especially critical in high mountain environments, where range sizes are generally small and where populations are often geographically isolated. In combination with population, genetic, ecological, and phylogeographic data for species of high conservation concern, analysis of such comparative data from different mountain ranges should provide guidelines for critical habitat sizes and minimum coverage of elevational ranges, with the overall task of maximizing the evolutionary potential through phylogenetic diversity and of capturing unique elements of mountain biota (see box).

Mountain Terminology and GMBA Concept of Comparative Mountain Biodiversity Research

GMBA distinguishes between three elevational belts and a transition zone:

- The montane belt extends from the lower mountain limit to the upper thermal limit of forest (irrespective of whether forest is currently present).
- The alpine belt is the temperature-driven treeless region between the natural climatic forest limit and the snowline that occurs worldwide. Synonyms for alpine are "'Andean" or "Afro-alpine."
- The nival belt is the terrain above the snowline, which is defined as the lowest elevation where snow is commonly present all year round (though not necessarily with full cover).
- The treeline ecotone is the transition zone between the montane and alpine belts.

How? For conservation planning, it will be important to integrate occurrence data across multiple organismic groups from different mountain areas, which need to be analyzed in combination with other biotic and abiotic data using information such as in the Global Database of Protected Areas of International Union for Conservation of Nature (IUCN) and World Conservation Monitoring Centre (WCMC).

OPEN ACCESS AND A GBIF PORTAL TO SHARED MOUNTAIN BIODIVERSITY DATA

The GBIF has already established biodiversity information networks, data exchange standards, and an information architecture that enables interoperability and facilitates mining of biodiversity data. GBIF's technical expertise is an essential prerequisite for this project, and we welcome the idea of creating a specific GBIF data portal on mountain biodiversity. GMBA, in turn, can help to encourage mountain biodiversity researchers to share their data within GBIF in order to increase the amount and quality of georeferenced data on mountain biodiversity provided online. These tasks also are in line with the implementation of the program of work (PoW) for the Global Taxonomy Initiative (GTI) and for mountain biological diversity of the Convention on Biological Diversity (CBD).

SUMMARY

Georeferenced archive databases on mountain organisms are very promising tools for achieving a better understanding of mountain biodiversity and predicting its changes. The Global Mountain Biodiversity Assessment (GMBA) of DIVERSITAS, in cooperation with the Global Biodiversity Information Facility, encourages a global effort to mine biodiversity databases on mountain organisms. The wide range of climatic conditions and topographies across the world's mountains offers an unparalleled opportunity for developing and testing biodiversity theory. The power of openly accessible, interconnected electronic databases for scientific biodiversity research, which by far exceeds the original intent of archiving for mainly taxonomic purposes, has been illustrated. There is an urgent need to increase the amount and quality of georeferenced data on mountain biodiversity provided online in order to meet the challenges of global change in mountains.

ACKNOWLEDGMENTS

We gratefully acknowledge the hospitality of the Institute of Botany, Georgian Academy of Science (George Nakhutsrishvili, Otar Abdalaze). Funds were provided by the Swiss National Science Foundation (SNSF), DIVERSITAS (Paris), and individual travel grants to participants from their home institutions.

REFERENCES

Arzberger, P., P. Schroeder, A. Beaulieu, G. Bowker, K. Casey, L. Laaksonen, D. Moorman, P. Uhlir, and P. Wouters. 2004a. Promoting access to public research data for scientific, economic, and social development. *Data Science Journal* 3(29):135–152. www.jstage.jst.go.jp/article/dsj/3/0/135/_pdf. Accessed on May 18, 2009.

Arzberger, P., P. Schroeder, A. Beaulieu, G. Bowker, K. Casey, L. Laaksonen, D. Moorman, P. Uhlir, and P. Wouters. 2004b. An international framework to promote access to data. *Science* 303:1777–78.

Chapman, A.D. 2005a. Principles and Methods of Data Cleaning, version 1.0. Report for the Global Biodiversity Information Facility, Copenhagen. Copenhagen, Denmark: Global Biodiversity Information Facility. www.gbif.org/prog/digit/data_quality. Accessed on Jan. 5, 2007.

Chapman, A.D. 2005b. Principles of Data Quality, version 1.0. Copenhagen, Denmark: Global Biodiversity Information Facility. www.gbif.org/prog/digit/data_quality. Accessed on Jan. 5, 2007.

Chapman, A.D. 2005c. Uses of Primary Species—Occurrence Data, version 1.0. Copenhagen, Denmark: Global Biodiversity Information Facility. www.gbif.org/prog/digit/data_quality. Accessed on Jan. 5, 2007.

Chapman, A.D., and Wieczorek J., eds. 2006. *Guide to Best Practices for Georeferencing*. Copenhagen, Denmark: Global Biodiversity Information Facility. www.gbif.org/prog/digit/Georeferencing. Accessed on Jan. 5, 2007.

Hershkovitz, M.A., M.T.K. Arroyo, C. Bell, and F. Hinojosa. 2006. Phylogeny of Chaetanthera (Asteraceae: Mutisieae) reveals both ancient and recent origins of the high elevation lineages. *Molecular Phylogenetics and Evolution* 41:594–605. doi:10.1016/j.ympev.2006.05.003.

Hughes, C., and R. Eastwood. 2006. Island radiation on a continental scale: Exceptional rates of plant diversification after uplift of the Andes. *Proceedings of the National Academy of Sciences* 103:10334–39. doi:10.1073/pnas.0601928103.

Koerner, C. 2000. Why are there global gradients in species richness? Mountains might hold the answer. *Trends in Ecology & Evolution* 15:513–14.

Koerner, C. 2003. *Alpine Plant Life*. 2nd ed.. Berlin, Germany: Springer.

Koerner, C., M. Ohsawa, et al. 2005. Mountain systems (Chapter 24). In *Ecosystems and Human Wellbeing. Current State and Trends: Findings of the Condition and Trends*

Working Group. Millennium Ecosystem Assessment, Vol 1, pp. 681–716. Hassan, R., R. Scholes, and N. Ash, eds. Washington, D.C.: Island Press.

Koerner, C., and E.M. Spehn, eds. 2002. *Mountain Biodiversity: A Global Assessment*. London: Parthenon Publishing Group.

Rowe, R.J. 2005. Elevational gradient analyses and the use of historical museum specimens: A cautionary tale. *Journal of Biogeography* 32:1883–97

Spehn, E.M., M. Liberman, and C. Koerner, eds. 2005. *Land Use Change and Mountain Biodiversity*. Boca Raton, FL: CRC Press.

Stolton, S., and N. Dudley. 2004. Sharing Information with Confidence—"The Biodiversity Commons": Past Experience, Current Trends and Potential Future Directions. IUCN (The World Conservation Union). http://www.conservationcommons.org/media/document/docu-h0xjc6.doc. Accessed on Jan. 5, 2007.

Index

A

Akaike Information Criterion (AIC), 51, 53, 73, 153
All-Taxa Biodiversity Inventory (ATBI), 76, 77, 85, 168
Alpine Functional Traits Database, 108–110
Alpine life zone
  in South American Andes, 30–31
  as template for testing biogeographic theory, 29–30
Alpine plants, species and diversity, 150–161
  analyses by species traits, 155
  climate change and species distribution, 150
  climate change in mountain environment, 150
  climate change scenarios, 153
  dispersal modeling, 154
  environmental data, 152
  georeferenced biodiversity databases, 158
  importance of spatial biological databases, 151
  methods of study, 151–155
  model evaluation, 155
  plant traits database, 152
  predicted distributions and species traits, 155–158
  predicting species distributions, 159
  results of study, 155–158
  retained predictors and contributions, 155
  SDMs complementing monitoring databases, 160
  simulating plant migrations, 151
  spatial projections, 153
  species occurrence data, 152
  statistical modeling, 153
  study area, 151
  turnover maps, 155
  using predictive models to assess climate change impact, 150
Alps, plant functional traits database for, 107–119
  Alpine Functional Traits Database, 108–110
  discussion, 115–118
  ecosystem measurements, 111
  effects on ecosystem properties, 114
  functional community responses to management of subalpine grasslands, 113
  methods, 108–111
  overview, 107–108
  results, 112–114
  species-level variation in species trait values, 112
  statistical analyses, 111
  study sites, 110–111
  vegetation and plant functional trait measurements, 111
Altitudinal gradient, species richness of breeding birds along, 65–72
  Catalonia, 67
  differences in environment, 68
  methods, 66–68
  observed species richness, 69
  overview, 65–66
  relationship between species richness and environmental variables, 69–71
  results, 68–71
  species set, 67
  statistical analysis, 67
  Switzerland, 66
Antarctica, 1, 5
Anthropogenic influence, 2

C

Carpathian Mountains, 89–104, 167
Catalonia, species richness in, 67
China, plant diversity in mountains, 39–46
  aims of PKU-PSD project, 40–41
  altitudinal patterns of species richness, 44
  data analysis, 42
  data collection, 42
  database, 42
  latitudinal pattern of species richness, 44
  methods of study, 42–43
  nested sampling design, 42
  overview, 39–40
  relationship between species richness and climate, 45
  results of PKU-PSD project, 44–45
  results of study, 44
  study sites, 42
Climate change, 30, 65, 116, 125–127, 132
  Alpine plants and, 149–161
  in mountain environments, 150
  Southern Italy mountains and, 137–146
  species distribution and, 150
Community weighted means (CWM) traits, 111, 111–116
Costa Rica, 17, 58, 61, 83

D

Data quality and limitations, 13–14
Data–Information–Knowledge–Wisdom (DIKW) hierarchy, 11–12, 14
Digital elevation model (DEM), 50, 99, 152–153, 159
Directory Interchange Format (DIF), 26
Dispersal modeling, 154
DIVERSITAS, research agenda, 171–177
  aims, 172
  common elevational trends in organismic traits, 175
  data sources and structure, 172
  datasharing, 172
  functional implications of mountain biodiversity, 175–176
  mountain-specific aspects, 173
  open access and GBIF portal, 177
  principle of open access, 172
  scientific uses of mountain biodiversity e-data, 173
  typical elevational trends in organismic traits, 175

E

Ecological Metadata Language (EML), 26–27
Environmental diversity, assessing, 99–102

F

Federal Geographic Data Committee (FGDC), 25–28
Fossil species occurrence data, 127

G

Generalized additive models (GAMs), 67–68
Genetic diversity, assessing, 92–97
Global Biodiversity Information Facility (GBIF), 1, 14, 25, 27, 28, 89, 134, 165–166, 168, 177
Global Biodiversity Information Network (GBIN), 22
Global Mountain Biodiversity Assessment (GMBA), 2, 8
Global mountain geostatistics for testing, 4–8
  global area × seasonality patterns, 5
  global ruggedness patterns, 5–6
  latitudinal and altitudinal patterns of "warmth", 6–8
GMBA mountain portal, 168
Great Smoky Mountains National Park, diverse elevational diversity gradients in, 75–85
  discussion, 83

methods, 76–78
overview, 75
patterns of diversity, 76–78, 78–79, 83
results, 78–83
study area, 76
underlying causes of diversity gradients, 78, 79–83, 84
Greenland, 2

H

Habitat Biodiversity Data Base (HDBD), 139, 168
Hengduan Mountains, elevational pattern of seed plant species richness in, 49–54
analysis of data, 51–52
area data, 50
climate data, 50–51
discussion, 52–54
elevational belts, 50
results, 52
source of data, 50
study area, 50

I

ICE algorithm, 19–20
IntraBioDiv, 89–103
assessing environmental diversity, 99–102
assessing genetic diversity, 92–97
assessing species diversity, 97–99
biodiversity research and integration of data, 89–90
initiative for cross-nation database for mountain biodiversity in Europe, 90–92
potential impact and dissemination of data and results, 102–103
relationships of levels of biodiversity with glacial refugia, 102

K

Kazbegi Research Agenda, 13, 14, 28, 172–177

L

Last Glacial Maximum (LGM), 126–133
Leaf nitrogen content (LNC), 112–114, 118
Local field surveys vs. "mining" of archive data, 57–61
discussion, 59–61
methods, 58
overview, 57–58
results, 58–59

M

Mean annual biotemperature (MAT), 50–54
Mean annual rainfall (MAR), 50–54
Metadata, 25–37
dealing with problems, 34–35
explained, 25–28
need for, 25
problems associated with in fine-tuned macroecological work, 32–33
Senecio database to show ecological trends across, 31–32
trends in altitude vs. latitude in *Senecio*, 33–34
underlying causes of putative pattern, 35–36
Mexican National Commission on Biodiversity (CONABIO), 22
MIGCLIM cellular automation, 154
Mountain biodiversity, 1–10
anthropogenic influence, 2
climactic reference line, 3
evolution and, 1–2
global area × seasonality patterns, 5
global mountain geostatistics for testing, 4–8
global ruggedness patterns, 5–6
growing season, 3
latitudinal and altitudinal patterns of "warmth", 6–8
montane belt, 3
superiority of mountains as test system, 2
treeline ecotone, 3–4
Mountain species numbers assessments, 17–22
debugging heterogeneous-origin databases, 18–20
methods, 18
results and discussion, 20–22
Mountains, as mine of information, 165–168
availability of primary biodiversity data, 165
georeferenced biodiversity databases, 168
GMBA mountain portal, 168
importance of georeferenced biodiversity data, 166
mining biodiversity data, 167
niche dynamic modeling, 168
overview, 165

N

National Biological Information Infrastructure (NBII), 26–28
New York Botanical Garden (NYBG), 31, 35
Niche dynamics, determining with species occurrence databases, 125–134
data layers for present and past climates/ landscapes, 127
explanations for differential niche conservatism, 131
fossil species and occurrence data, 127
material and methods, 127–129
medeling approach, 128
next steps, 132
overview, 125–127
results, 129
study species and occurrence data, 127
use of occurrence databases, 130
North American Biodiversity Information Network (NABIN), 17, 22

O

Ordinary least squares (OLS) models, 51–53
Organisation for Economic Co-operation and Development (OECD), 14, 26

P

Plant diversity, mountains of China, 39–46
aims of PKU-PSD project, 40–41
altitudinal patterns of species richness, 44
data analysis, 42
data collection, 42
database, 42
latitudinal pattern of species richness, 44
methods of study, 42–43
nested sampling design, 42
overview, 39–40
relationship between species richness and climate, 45
results of PKU-PSD project, 44–45
results of study, 44
study sites, 42
"Point-radius" method, 13
Potential evapotranspiration (PET), 50–54
Primary biodiversity data, 11–14
advantage of sharing, 12–13
data quality and limitations, 13–14
Pyrenees Mountains, 66, 77, 167

R

Red Mexicana de Biodiversidad (REMIB), 17, 22

S

Senecio species
database to show ecological trends across, 31–32
trends in altitude vs. latitude in, 33–34
Sharing data, advantages of, 12–13
Slope shifts, quantifying, 142–144
South American Andes
alpine life zone in, 30–31
Senecio database to show ecological trends across, 31–32
Southern Italy Mountains, climate change, 137–146
comparing different mountain ranges, 144
data collection, 139
database, 142
database construction, 139
overview, 137
quantifying slope shifts, 142–144
research protocol for HBDB, 145

study area, 138
study results, 139–145
Species diversity, assessing, 97–99
Species occurrence databases, 125–134
Specific leaf area (SLA), 112–114
Switzerland, species richness in, 66

T

TROPICOS database, 31, 35, 37